Sustainable Civil Infrastructures

About this Series

Sustainable Infrastructure impacts our well-being and day-to-day lives. The infrastructures we are building today will shape our lives tomorrow. The complex and diverse nature of the impacts due to weather extremes on transportation and civil infrastructures can be seen in our roadways, bridges, and buildings. Extreme summer temperatures, droughts, flash floods, and rising numbers of freeze-thaw cycles pose challenges for civil infrastructure and can endanger public safety. We constantly hear how civil infrastructures need constant attention, preservation, and upgrading. Such improvements and developments would obviously benefit from our desired book series that provide sustainable engineering materials and designs. The economic impact is huge and much research has been conducted worldwide. The future holds many opportunities, not only for researchers in a given country, but also for the worldwide field engineers who apply and implement these technologies. We believe that no approach can succeed if it does not unite the efforts of various engineering disciplines from all over the world under one umbrella to offer a beacon of modern solutions to the global infrastructure. Experts from the various engineering disciplines around the globe will participate in this series, including: Geotechnical, Geological, Geoscience, Petroleum, Structural, Transportation, Bridge, Infrastructure, Energy, Architectural, Chemical and Materials, and other related Engineering disciplines.

More information about this series at http://www.springer.com/series/15140

Louay Mohammad
Editor

Advancement in the Design and Performance of Sustainable Asphalt Pavements

Proceedings of the 1st GeoMEast International Congress and Exhibition, Egypt 2017 on Sustainable Civil Infrastructures

Springer

Editor
Louay Mohammad
Louisiana State University
Baton Rouge, LA
USA

ISSN 2366-3405 ISSN 2366-3413 (electronic)
Sustainable Civil Infrastructures
ISBN 978-3-319-61907-1 ISBN 978-3-319-61908-8 (eBook)
DOI 10.1007/978-3-319-61908-8

Library of Congress Control Number: 2017946469

Printed on acid-free paper

This Springer imprint is published by Springer Nature
The registered company is Springer International Publishing AG
The registered company address is: Gewerbestrasse 11, 6330 Cham, Switzerland

Preface

Toward building sustainable and longer civil infrastructures, the engineering community around the globe continues undertaking research and development to improve existing design, modeling, and analytical capability. Such initiatives are also the core mission of the Soil-Structure Interaction Group in Egypt (SSIGE) to contribute to the ongoing research toward sustainable infrastructure. This conference series "GeoMEast International Congress and Exhibition" is one of these initiatives.

Ancient peoples built their structures to withstand the test of time. If we think in the same way, our current projects will be a heritage for future generations. In this context, an urgent need has quickly motivated the SSIGE and its friends around the globe to start a new congress series that can bring together researchers and practitioners to pursue "Sustainable Civil Infrastructures." The GeoMEast 2017 is a unique forum in the Middle East and Africa that transfers from the innovation in research into the practical wisdom to serve directly the practitioners of the industry.

More than eight hundred abstracts were received for the first edition of this conference series "GeoMEast 2017" in response to the Call for Papers. The abstracts were reviewed by the Organizing and Scientific Committees. All papers were reviewed following the same procedure and at the same technical standards of practice of the TRB, ASCE, ICE, ISSMGE, IGS, IAEG, DFI, ISAP, ISCP, ITA, ISHMII, PDCA, IUGS, ICC, and other professional organizations who have supported the technical program of the GeoMEast 2017. All papers received a minimum of two full reviews coordinated by various tracks' chairs and supervised by the volumes editors through the Editorial Manager of the SUCI "Sustainable Civil Infrastructure" book series. As a result, 15 volumes have been formed of the final +320 accepted papers. The authors of the accepted papers have addressed all the comments of the reviewers to the satisfaction of the tracks' chairs, the volumes editors, and the proceedings editor. It is hoped that readers of this proceedings of the GeoMEast 2017 will be stimulated and inspired by the wide range of papers written by a distinguished group of national and international authors.

Publication of this quality of technical papers would not have been possible without the dedication and professionalism of the anonymous papers reviewers. The names of these reviewers appear in the acknowledgment that follows. For any additional reviewers whose names were inadvertently missed, we offer our sincere apologies.

We are thankful to Dr. Hany Farouk Shehata, Dr. Nabil Khelifi, Dr. Khalid M. ElZahaby, Dr. Mohamed F. Shehata, and to all the distinguished volumes' editors of the proceedings of the GeoMEast 2017. Appreciation is extended to the authors and tracks' chairs for their significant contributions. Thanks are also extended to Springer for their coordination and enthusiastic support to this conference. The conference staff acknowledges the assistance of Ms. Janet Sterritt-Brunner, Mr. Arulmurugan Venkatasalam in the final production of the 15 edited volumes "Proceedings of GeoMEast 2017".

Contents

Development and Finite Element Analysis of Piezoelectric-Based Prototypes for Harvesting Energy from Roadway Pavement

Hossein Roshani[1], Samer Dessouky[2(✉)], Arturo Montoya[2], A.T. Papagiannakis[2], and Ala Abbas[3]

[1] Center for Transportation Research, University of Texas at Austin, Austin, TX, USA

[2] Department of Civil and Environmental Engineering, University of Texas at San Antonio, San Antonio, TX, USA

samer.dessouky@utsa.edu

[3] Department of Civil Engineering, University of Akron, Akron, OH, USA

Abstract. Energy harvesting is a process that captures unused ambient energy such as heat, vibration, stress or movement that would otherwise be lost. Highway pavements infrastructure exposed to energy-potential resources from vehicle vibrations and traffic loading strains that could be harvested. Piezoelectric transducers (PZT) are potential materials for harvesting energy from pavements as they convert mechanical loading strains into electric voltage. For this purpose, this study is aimed to evaluate two types of piezoelectric prototypes integrated within asphalt mix; cylindrical disks and thin film sheet, suitable for compression and bending state of stresses, respectively. Finite element (FE) analysis was conducted to simulate prototype response in different geometry under dynamic loading. The evaluation of the prototype involves laboratory testing of their power output as a function of stress, and FE simulation of their mechanical behavior. Results suggested that power output is highly dependent on loading frequency and magnitude and PZT geometry but could be promising in powering low-watt LED traffic lights and wireless sensors embedded in pavement particularly in remote areas.

1 Introduction

Energy harvesting, or energy scavenging technique is the attempt to gather, store, and distribute ambient energy for human and civil purposes (Sodano et al. 2004). These techniques are becoming more and more necessary as the energy demands continue to increase. To minimize the dependency of non-renewable energy, thus there is a need to seek other sources of energy within the environment that has the potential for harvest and storage.

In recent years, there has been increasing interest in energy harvesting through transduction. Three technologies have been used for this purpose, namely electromagnetic, electrostatic and piezoelectric. Piezoelectric transduction appears to be the

L. Mohammad (ed.), *Advancement in the Design and Performance of Sustainable Asphalt Pavements*, Sustainable Civil Infrastructures, DOI 10.1007/978-3-319-61908-8_1

most promising, given its widest power density (Cook-Chennault et al. 2008). A number of recent studies explored the use of piezoelectric transduction for harvesting energy from roadways (Xiong 2014; Kim et al. 2015; Zhao et al. 2014). Work by Xiong et al. at Virginia Tech University in Virginia, USA produced a piezoelectric harvesting system consisting of multiple cylindrical piezoelectric elements that are compressed by the action of traffic tires. Under a traffic volume of 4,000 vehicles per day (167 vehicles/hour), this system generated voltage ranging from 400 to 700 V and electric currents ranging from 0.2 to 0.35 mA. The corresponding power output was obtained by multiplying voltage by current, yielding a power range between 0.08 and 2.1 W per system (Xiong 2014). Work by Kim et al. at Georgia Southern University involved laboratory testing using an asphalt pavement analyzer (APA) machine. Two piezoelectric materials were tested; one manufactured by *Noliac* and the other by *Kinetic*. The APA wheel loads at three levels were applied; 222, 445 and 889 N. The maximum resulting voltages for the Noliac at those loads were 5, 10, and 15 V, while for the *Kinetic* were 5, 10 and 20 V, respectively. Assuming a traffic level of 600 vehicles/hour at a speed of 45 mph, such a system could produce up to 2.67 mW of power hourly (Kim et al. 2015).

Zhao et al. at Tongji University in China studied power generation from several types of piezoelectric sensor configurations. These included multiple lead zirconate titanate (i.e., PZT) prismatic elements referred to as "piles" with circular, square or hexagonal cross sections, as well as commercially available cymbal-shaped and bridge-shaped elements. They performed finite element (FE) analysis to study the effect of the shape of the PZT piles in producing electric power output, concluding that the circular cross section piles were preferable. Power generators involving multiple piles were analyzed. Stress analysis combined with theoretical calculations established that generators with 8–16 piles each can be used to harvest significant amounts of electrical power under heavy traffic (Zhao et al. 2014).

A report presented by the California Energy Commission (CEC) to independently evaluate the feasibility of piezoelectric technology in harvesting energy from roadways (Hill et al. 2013). It evaluated some of the pilot systems developed by Universities, as well as commercially available harvesting systems. Three commercially available systems were evaluated in terms of their vendor output claims, namely, *Treevolt* (Treevolt 2016), *Genziko* (Genziko 2016) and *Innowattech* (Edel-Ajulay 2010). In evaluating the cost effectiveness of these systems, the CEC report recommended using the "levelized" cost of energy (LCOE) produced by the various harvesting systems. The LCOE is defined as the average total life-cycle cost to construct, operate and maintain a power-generating system divided by its total energy output over its service life. This report recommended further field testing of this technology to validate its feasibility.

The piezoelectric effect has been observed in a number of ceramic materials such as lead-zirconate ($PbZrO_3$), lead-titanate ($PbTiO_2$), barium-titanate ($BaTiO_3$), and lead-zirconate-titanate (PZT). These materials show a polarized electrostrictive effect (Haertling 1999). There are two forms in which piezoelectric functions; the first is the direct piezoelectric effect in which the materials are able to transform mechanical strain into electrical charge. The current applications of direct piezoelectric effect are in the form of sensors manufacturing and energy harvesting. The second form is the reverse effect, in which the piezoelectric materials convert electrical charge into mechanical strain. Nowadays, different technologies take advantage of the reverse piezoelectric effect in the production of actuators (Sodano et al. 2004).

This study contributes to fill in the gap of the lack of research studies on the effectiveness of using piezoelectric to harvest energy from roadways. In this study, two piezoelectric prototypes were fabricated for the purpose of harvesting energy from roadway pavement. The prototypes are designed to meet roadway loading and material conditions.

2 Fabrication of Prototypes and Experimental Plan

2.1 Prototype I

The prototype I included two copper plates mounted with electric wires, two polystyrene sheets, and eight piezoelectric disks housed in asphalt specimen in the UTM (Fig. 1a). The copper plates are 6.35 mm in thickness and 152 mm in diameter and they act as electrodes to collect the generated voltage from compressing the piezoelectric disks (Fig. 1d). Two holes were built on two sides of the plates to attach the polystyrene sheet. The polystyrene sheets were used to keep the piezoelectric disks

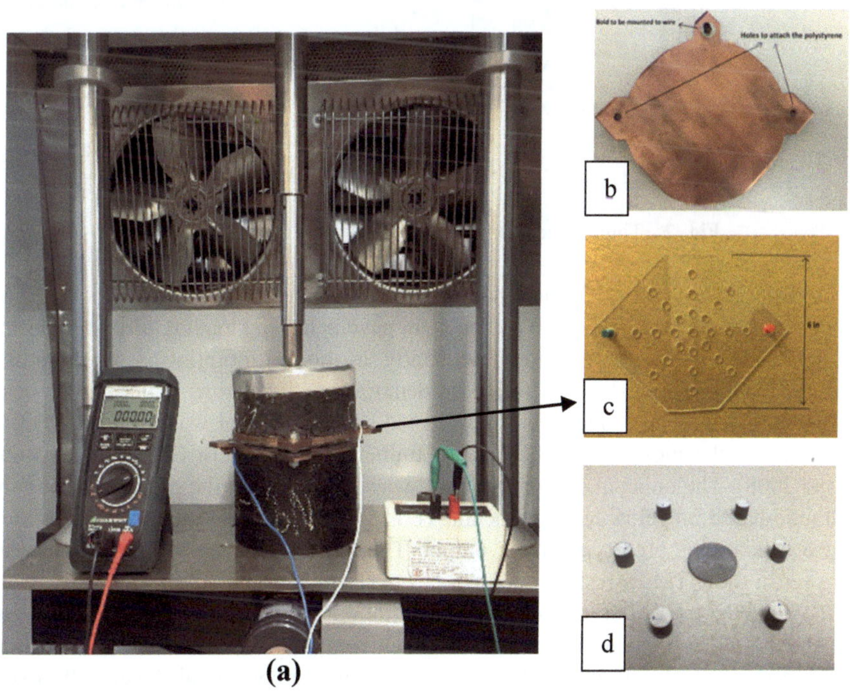

Fig. 1. (a) The experiment setup in UTM with prototype I components: (b) copper plate with 15.24 cm diameter, (c) the polystyrene sheet with pre-holes, (d) the piezoelectric disks (diameter of 8 mm and thickness of 8 mm)

fixed from slipping and lateral movements. The polystyrene materials are lightweight, non-conductive and mechanically resilient to resist horizontal stresses and lateral movements. The polystyrene sheets are 2 mm in thickness and cut in same diameter of copper plates. As can be seen in Fig. 1c, three circular patterns of holes have been chosen to hold the piezoelectric disks. The holes' diameters were 0.2 mm greater than the piezoelectric diameters. The sheets were fastened to the plates through mounting screws. The piezoelectric disks were 8 mm in height and diameter as shown in Fig. 1d. Figure 2 also shows a three-dimensional view of the prototype's design for assembly. The electrical charge gathered on copper plates are transferred to a multi-meter device using wires mounted to the copper plates.

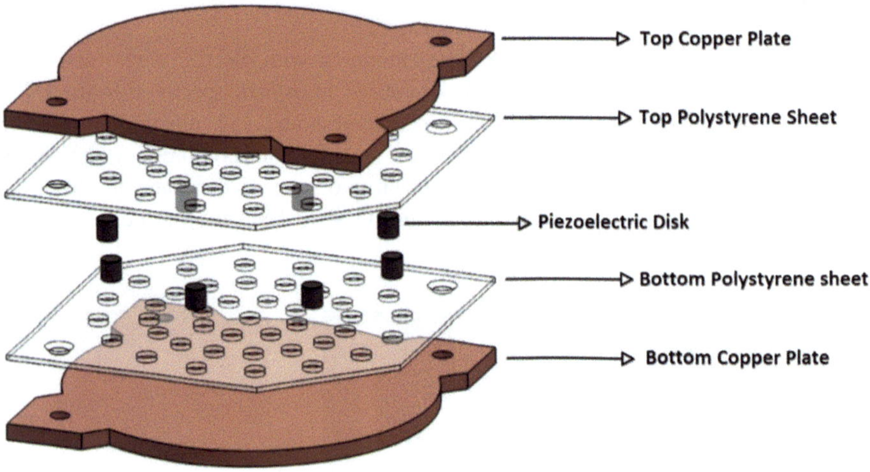

Fig. 2. Three-dimensional view of prototype assembly design

As can be seen in Fig. 3, the loading imposed on the prototype changes based on the position of tire with respect to prototype in pavement. The loading time starts as the tire approaches the prototype and subsequently, as the tire footprint passes the prototype, the loading diminishes. Therefore, to demonstrate the loading pattern in the field, the laboratory loading modes were represented by a sinusoidal wave shape, which is proven to be the most accurate shape for traffic loading pattern (Papagiannakis and Masad 2008). The uniaxial compression test was performed on the prototype using the hydraulic-based universal testing machine (UTM) to simulate and the testing variables and to validate the prototype power output. The loading frequency and peak loading were set in the UTM as testing variables.

To simulate the prototype to near field conditions, asphalt mix cores were used on top and bottom of the prototype. The top and bottom thicknesses are 38 mm and 127 mm respectively. These thickness represent an overlay asphalt layer with a supporting layer underneath in a typical pavement structure system. It is suggested that the thickness of these cores has negligible effect in the response of the prototype due to the uniaxial state of stress imposed on the prototype.

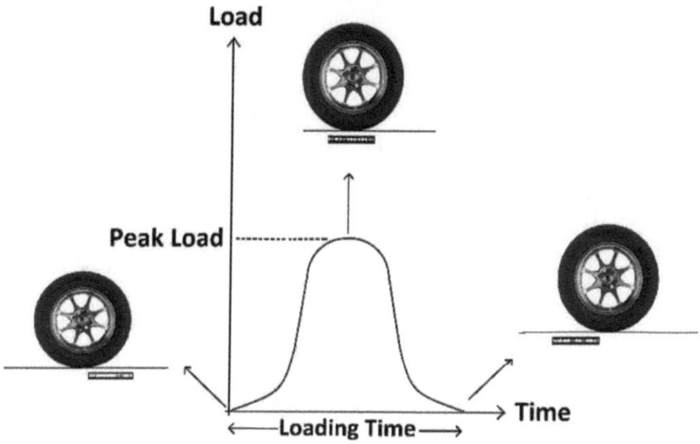

Fig. 3. Load shape and correspondent different positions of tire with respect to energy harvester

For prototype I, the effect of speed and tire loading were studied. Traffic speed is represented in the experiment through varying the load frequency. Increasing the frequency results in decreasing of loading time that is associated with increasing speed. Three frequencies of 5, 10, and 20 Hz were considered in the experiment that correspond to 200, 100, and 50 ms of loading time, respectively. Tire loading is determined as another variable that controls the output power. Five different peak loads of 1.0, 1.5, 2.0, 2.5, and 3 kN were selected to actuate the piezoelectric disks. These loads are sufficient to generate a relationship between load magnitudes and their corresponding output power. Most importantly, they are within the capability of the UTM hydraulic system to obtain accurate responses.

2.2 Prototype II

At the bottom of thin pavement layers the asphalt mix is exposed to cycles of tension and compression in the form of flexural and bending stresses under traffic loading. These loading cycles can be used to activate piezoelectric thin films placed properly at the bottom of the mix to generate power. For this purpose, prototype II design was created, and the flexural fatigue test was performed according to AASHTO T 321. Using the beam fatigue device (Fig. 4). Prototype II consists of a metalized piezoelectric film sheet 90 (length) × 63 (width) × 0.110 (thickness) mm attached to the bottom center of a compacted asphalt mix beam (Fig. 4). This area of beam was selected at where the maximum bending moment and tensile stress occurs during the experiment. The piezoelectric film sheets were utilized because of their flexibility under bending moment. Table 1 shows the specifications of the piezoelectric film sheet. The beam length is 380 mm with cross section of 50 mm width and 63 mm height. The beam is made of a dense-graded asphalt mix using PG 70-22 asphalt binder and local limestone aggregates from San Antonio, Texas. A metal-concrete epoxy was used to attach the metalized piezo film to the beam. The epoxy was non-conductive, therefore the piezoelectric film was not connected electrically to the beam and draining the

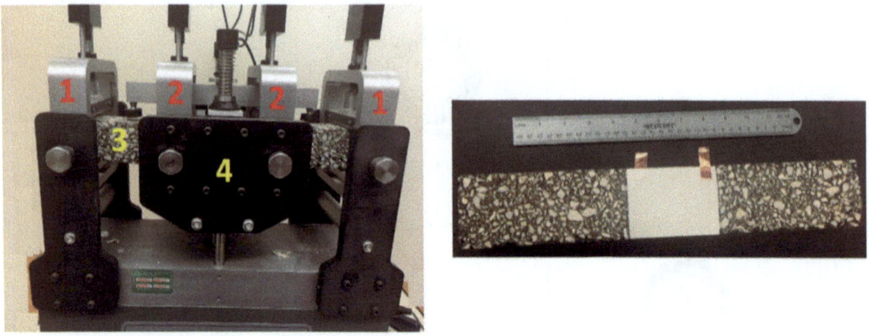

Fig. 4. Beam fatigue device showing prototype II of a metalized piezoelectric film sheet attached to the asphalt beam

Table 1. Properties of the piezoelectric material used in prototype II (Measurement Specialties 2016; PIEZOTECH 2016)

Young's modulus	Piezoelectric charge constant, d_{33}	Piezoelectric charge constant, d_{31}	Tensile stress at break	Piezoelectric voltage constant, g_{33}
1×10^9 N/m^2	15×10^{-12} C/N	-6×10^{-12} C/N	30 MPa	180×10^{-3} Vm/N

electric charge to other testing instruments. Two small pieces of copper foil tape was attached to each face of the film as electrode to gather the electric charge built up on the surfaces of piezoelectric sheet. Electrical wires were clipped to the copper foils for measuring the electric charge through the multi-meter.

Haversine load with different frequencies of 5, 10, and 20 Hz was applied on the beam. The amplitude of the applied load was based on the strain value in the middle center of the beam. In this study, eight different strains were applied; 160, 180, 200, 220, 240, 260, 280, and 300 micro-strain. The test was performed in room temperature (22 °C). Amperage reading was performed usually around the 20th second of the test (200 cycles) in which the output current approximately became stable.

3 Results and Discussions

In both prototypes the output current/voltage was measured and monitored by a multi-meter. To maximize the electric power output, the external resistance (R_L) in the circuit should be equal to the internal resistance (R_S) of the prototype, defined as impedance matching (Thompson 2009). The internal resistance of the prototype is calculated by the following equation:

$$R_s = \frac{1}{2\pi cf} \tag{1}$$

where c is the internal capacitance of the prototype, and f is the load pulses per second (frequency). The internal capacitance of the prototype depends on the number of piezoelectric disks. The theoretical calculated resistance is approximate; therefore, another attempt has to be made to measure the optimum resistance R_S. Several electrical load values close to the calculated R_s were applied to the circuit using an external resistance substitution box. The current was measured for different resistance values, and the power (P) was calculated based on the following equation:

$$P = R_L I^2 \quad (2)$$

where I is the current measured by the multi-meter (Croft and Summers 1987). Figure 5 shows the output power generated by prototype I including 8 piezoelectric disks at five different tire loads and three frequencies. It should be mentioned that the energy harvester and its top layer of asphalt pavement are not under confinement as compared to pavement materials in real field conditions, therefore the loss in the vertical stress along the depth is negligible and does not have significant influence on the output power. As suggested in these figures, increasing the loading magnitude is associated with increasing the output power. Also, increasing the frequency minimizes the loading time on the piezoelectric disks which results into increasing the output power. These results are consistent with the findings of the research conducted by Roshani et al. (2016) that shows the inverse relationship between power and time. The results help to determine the potential locations for installation of these types of energy harvesting modules in the roadways. It can be suggested that, the road sections such as interstate highways with high traffic speed and volume could be a good candidate. On the other hand, choosing the locations such as parking lots and intersections controlled by traffic signals might decrease the efficiency of piezoelectric energy harvesting modules because of longer loading time.

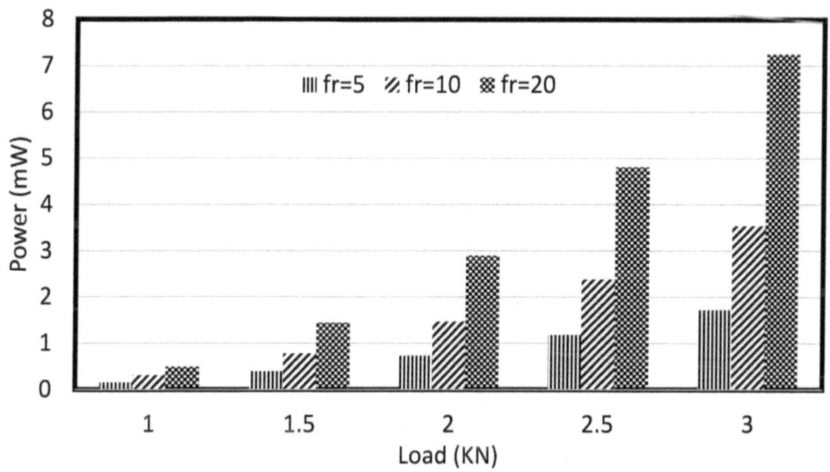

Fig. 5. Maximum output power generated by prototype I at different loads and frequencies

As mentioned previously, a thin film of piezoelectric materials was glued to the bottom of asphalt beam to capture the tensile stresses results from the three-point loading. In this case, the maximum bending moment and tension occurs in the bottom of the beam. According to the AASHTO T 321, the maximum tensile stress can be obtained by the following equation:

$$\sigma_t = \frac{(0.375P)}{bh^2} \tag{3}$$

where P is the load (N), b is the beam width (m), and h is the beam height (m). The tensile stresses in the piezoelectric film sheet versus the power were plotted in Fig. 6. As can be seen, increasing the tensile stress and loading frequency is associated with increasing the output power. It is worth to mention that the tensile stress caused by an eighteen-wheeler truck at 5-cm depth in the pavement is about 630 kPa that generates a power of around 0.001 mW which is very low comparing to prototype I. It also can be concluded that the relationship between the tensile stress and output power in this case has nearly a non-linear form.

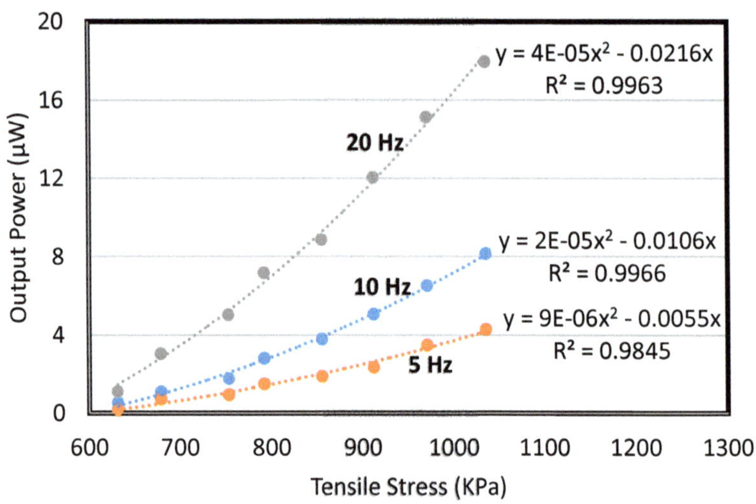

Fig. 6. Generated power in the piezoelectric film versus different tensile stress levels at different frequencies

3.1 Finite Element Analysis

The testing on prototype I was supplemented by a three-dimensional FE simulation using ABAQUS in order to provide a detailed insight of the vertical displacement and stress distribution induced on the piezoelectric. The geometric and material properties of the assembled prototype previously described were used to generate the sandwich-like model shown in Fig. 7. All parts of the assembly were modeled with

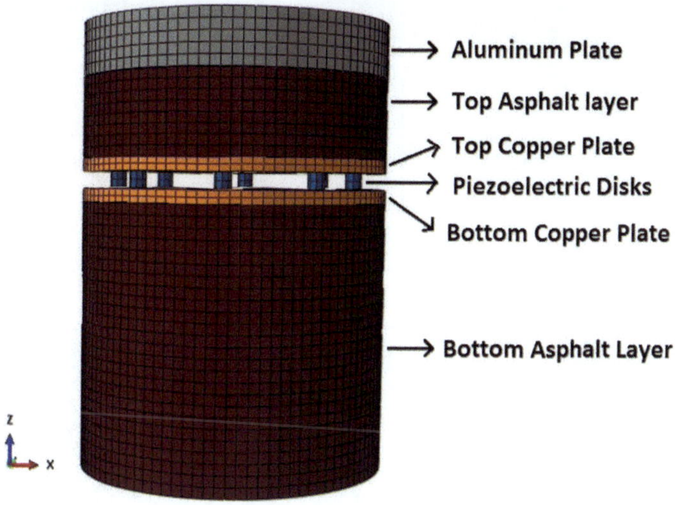

Fig. 7. 3D finite element model of the assembled prototype

linear elastic isotropic material properties. The assemblage was discretized into a total of 36,480 eight-node (brick) elements. A 1 kN concentrated force was applied at the center of the top aluminum disk. The assembled prototype was free to deform radially, while roller boundary conditions (in the z-direction) were imposed to the bottom surface to replicate experimental setup conditions.

The analysis results show that the load path mechanism is analogous to that of a slab-column structure. As the load is transferred from the piezoelectric disks to the bottom copper plate, the vertical displacement decreases in the radial direction. Figure 8 shows that the vertical-induced stress of the piezoelectric disks is approximately even on the top surface (σ_z = 219 MPa to 264 MPa). Stress variations are observed along the length of the piezoelectric devices due to the slight bulging effects as shown in Fig. 8. Nonetheless, it can be concluded that the mechanical behavior of the eight disks is equivalent and will produce comparable voltage.

The prototype II containing a compacted asphalt beam and a piezoelectric film sheet was simulated to evaluate the output voltage and tensile stress on the film. An 8-node brick element (C3D8R) for the beam and 8-node brick element (C3D8E) for the film with a total of 47,859 elements and 55,934 nodes were used in the analysis, respectively. The corresponding applied loads to cause tensile strains ranged from 160 to 300 micro-strains, were demonstrated using the AASHTO T 321 and the UTM software. The beam was supported at the two-end and assumed to have no movement along z-direction as can be seen in Fig. 9. The middle section of the beam was allowed to move downward in vertical direction and return to the original horizontal shape following a sinusoidal pattern that induces lateral strain at the bottom of the beam. The top face of the piezoelectric film was considered to be electrically connected to the ground, which is considered as boundary condition with zero voltage. For simulation, eight different static loading steps were defined to deform the beam as shown in Fig. 9.

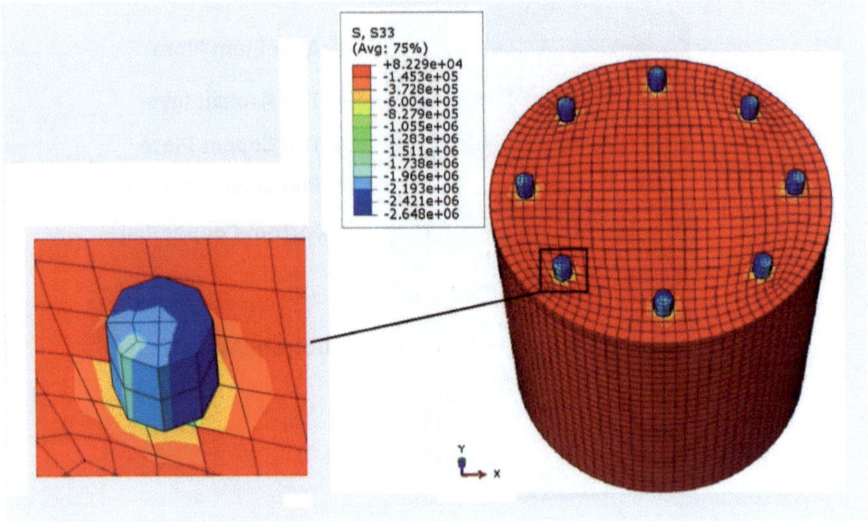

Fig. 8. Vertical stresses on deformed piezoelectric disk

(a)

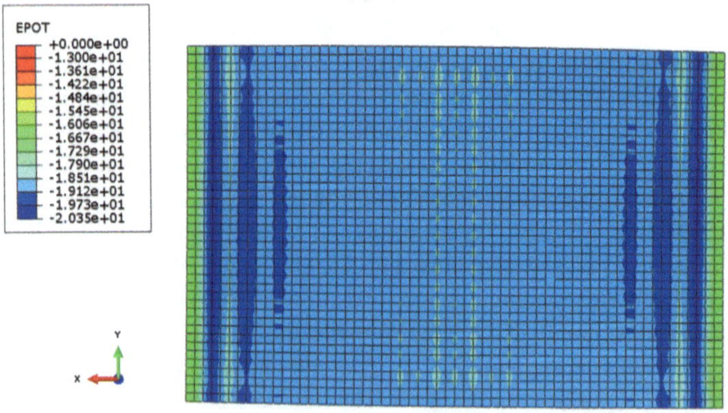

(b)

Fig. 9. (a) Deformed shape of prototype II under applied loading and (b) electric potential (voltage) on the bottom surface of piezoelectric film

The FE simulation showed that the voltage across the bottom face of the film sheet, as shown in Fig. 9b, is determined to be almost constant. It is also suggested that the voltage at this face is equal to the total output voltage.

The voltage generated in each micro-strain level were plotted in Fig. 10. The figure suggests that increasing output voltage is associated with increasing applied strain. The actual output voltage measured by the multi-meter was lower (about 30 percent) compared to the voltage obtained from FE modeling because of the voltage loss in the wiring and devices resistance.

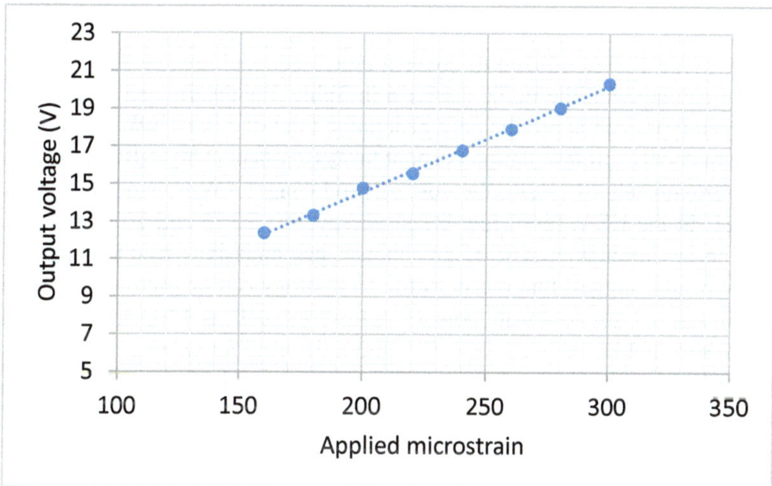

Fig. 10. Output voltage versus the applied micro-strain

3.2 Economic Analysis

The capability of electric power production from the prototypes was studied using thc laboratory obtained power versus stress curves highlighted previously. A passenger car and commercial truck with tire pressure of 206 and 690 kPa, respectively were considered. Considering 2 axles for a passenger car and 5 axles for a truck, the frequency of tire load will be 2 and 5 Hz, respectively. The traffic volume and composition was considered as an input to the energy analysis using a busy interstate highway in the US as shown in Table 2. In order to take advantage of the entire tire print of a truck wheel,

Table 2. Traffic assumptions

Traffic property	Assumed value
Average Annual Daily Traffic (AADT)	320,000
Number of lanes	3
Percent cars/middle lane	0.40
Percent trucks/middle lane	0.70
Number of truck axles/day	111,200
Number of car axles/day	102,400

Table 3. Calculated power, and energy output for prototype I including four PZT disks under the traffic described in Table 2

Parameters	Value
Stress from car tire on each PZT disk	16.87 MPa
Stress from truck tire on each PZT disk	56.3 MPa
Watts/car tires/day	318
Watts/truck tires/day	1,725
Watt-hrs from truck tires/year	104,94
Total kWatt-hrs/year	114.61

four of prototype I were utilized to cover a cross sectional area of 30.5 × 30.5 cm^2. Energy estimation using the traffic scenario in Table 2 is determined in Table 3.

Four prototypes can approximately generate a total of 114.61 kW-h in a year. The initial cost materials and installation is about 960$ (Roshani et al. 2016). With a service time of approximately 20 years, the cost of producing 1 kW-h using piezoelectric is about 0.42$. In the USA, fossil fuel-based energy cost $0.15 and $0.30 per kWatt-hr ignoring the environmental cost of greenhouse gas emissions resulting from power plants. The results suggest that piezoelectric technology can be used to harvest energy from roadways but for low-watt power application such as LED traffic lighting or wireless sensors embedded in pavements without the need for external grid power. This is particularly attractive in rural areas, where electric grid power is not available at roadside.

4 Conclusions

This paper presented an effort to develop energy harvesting prototypes based on piezoelectric elements embedded into pavements. The major conclusions of the study are:

- Increasing traffic speed and loading increases power output generated from piezoelectric transducers.
- For maximum power, energy harvesting should target vertical compressive stresses in pavement instead of tensile stress.
- Finite element suggested that loading on symmetrically arranged piezoelectric disks in prototype I will produce comparable voltage for each disk.
- The power output of piezoelectric elements was shown to be a non-linear (i.e., quadratic) function of stress suggesting that increasing stresses at piezoelectric through changing prototype design or number of disks will lead to improved power output.
- The results available to date suggest that this technology shows promising in powering LED traffic lights and wireless sensors embedded into pavement structure.

Acknowledgements. Funding for this study was provided by the Texas Department of Transportation.

References

Cook-Chennault, K.A., Thambi, N., Sastry, A.M.: Powering MEMS portable devices—a review of non-regenerative and regenerative power supply systems with emphasis on piezoelectric harvesting systems. J. Smart Mater. Struct. IOPsci. (2008). doi:10.1088/0964-1726/17/4/043001

Croft, T., Summers, W.: American Electricians' Handbook, 11th edn. McGraw Hill, New York (1987)

Edel-Ajulay, J.: Innowattech; harvesting energy and data. In: 1st International Symposium. The Highway to Innovation, Tel Aviv (2010)

Haertling, G.H.: Ferroelectric ceramics: history and technology. J. Am. Ceram. Soc. (1999). doi:10.1111/j.1151-2916.1999.tb01840.x

Hill, D., Agrawal, A., Tong, N.: Assessment of Piezoelectric Materials for Roadway Energy Harvesting. Report prepared by DNV KEMA Energy and Sustainability Inc. CSC-500-2013-007, California Energy Commission (2013)

Kim, S., Shen, J., Ahad, M.: Piezoelectric-based energy harvesting technology for roadway sustainability. Int. J. Appl. Sci. Technol. **5**(1), 20–25 (2015)

Measurement Specialties, Metallized Piezo Film Sheets. http://meas-spec.com/WorkArea/DownloadAsset.aspx?id=5323 Accessed 21 Jan 2016

Papagiannakis, A.T., Masad, E.: Pavement Design and Materials. Wiley, Newark (2008)

PIEZOTECH, Piezoelectric Films Technical Information. http://www.piezotech.fr/image/documents/22-31-32-33-piezotech-piezoelectric-films-leaflet.pdf. Accessed May 14 2016

Roshani, H., Dessouky, S., Montoya, A., Papagiannakis, A.T.: Energy harvesting from asphalt pavement roadways vehicle-induced stresses: a feasibility study. J. Appl. Energy (2016). doi:10.1016/j.apenergy.2016.08.116

Sodano, H.A., Inman, D.J., Park, G.: A review of power harvesting from vibration using piezoelectric materials. Shock Vib. Dig. (2004). doi:10.1177/0583102404043275

Thompson, S.P.: Dynamo-electric Machinery; a Manual for Students of Electrotechnics. BiblioBazaar (2009)

Treevolt. www.treevolt.com. Accessed 5 Jan 2016

www.genziko.com/technology/road-power-generation/. Accessed 5 Mar 2015

Xiong, H.: Piezoelectric Energy Harvesting for Public Roadways. Ph.D. Dissertation, Department of Civil and Environmental Engineering, Virginia Polytechnic Institute and State University (2014)

Zhao, H., Tao, Y., Niu, Y., Ling, J.: Harvesting energy from asphalt pavement by piezoelectric generator. J. Wuhan Univ. Technol. Mater. Sci. Ed. (2014). doi:10.1007/s11595-014-1023-3

A Bitumen-Based Prototype to Predict the Workability of Asphalt Concrete Mixtures

Aboelkasim Diab[1(✉)] and Zhanping You[2]

[1] Department of Civil Engineering, Aswan University, Aswan 81542, Egypt
daali@mtu.edu
[2] Department of Civil and Environmental Engineering,
Michigan Technological University, Houghton, MI 49931, USA

Abstract. Having a reliable and repeatable method for measuring the workability of asphalt mixtures for a wide range of asphalt blends is important. The major objective of this paper is to develop a new bitumen-based prototype for the measurement of the workability of different asphalt blends. Correlation and prediction of the mixing temperature with the aid of the developed prototype have been established as a secondary objective of this study. At first, a principle was presented as a basis to develop the workability prototype. The device developed in this study utilized a commercially available motorized Vane Shear Test (VST) apparatus which is mainly used to measure the shear strength of soil. To resemble the aggregate effect while mixing, a specially designed spindle was developed. The VST has been modified to suit the present purpose of the workability test. An accurate torque meter was installed to measure the torque required to rotate the spindle in the blend at a constant revolution. The device was tested with asphalt blends of different ranges of workability. The workabilities of polymer-modified and water-foamed asphalts were evaluated at temperatures of 100, 140 and 180 °C, respectively. It was found that the bitumen-based prototype was able to differentiate the workability in light of the constituents of the studied blends. In addition, the prototype helped in roughly predicting the mixing temperatures of the studied blends based on the concept of Workability Index (WI).

Keywords: Workability · Viscosity · Workability index · Mixing temperature · Modified asphalt · Warm mix asphalt

1 Introduction

The workability of asphalt concrete mixtures affects the construction activities as well as the performance of the pavement in the long-term. Workability is a term used to describe the ease with which the asphalt mixture can be mixed, placed and compacted on the roadway. In other words, the workability can be defined as the movement of an asphalt mixture through construction equipment, handwork, and compactibility on the roadway. Workability is an essential component in acquiring the required density and air voids within a compacted mixture (Airey et al. 2008; Celik and Atis 2008). For asphalt blends that are extremely harsh, and in this way are less workable, it can be more difficult to construct smooth asphalts. Poorly compacted asphalt pavements might

L. Mohammad (ed.), *Advancement in the Design and Performance of Sustainable Asphalt Pavements*, Sustainable Civil Infrastructures, DOI 10.1007/978-3-319-61908-8_2

encounter issues, for example, stripping, raveling and fatigue, which basically result from the increase in the air void content as well as the increase in the rate of oxidative aging of bitumen, and consequently, decreasing the service life of the asphalt pavement (Horan et al. 2012). The higher the temperature, the better the mixture workability, the reason being that the viscosity of the binder decreases as the temperature increases. However, increasing the mixture temperature to get proper workability is not always the best strategy. Unfortunately, excessively increasing the production temperatures of asphalt mixture may cause serious problems, such as compaction problems and an increase in oxidation of the asphalt cement. On the other hand, a better compaction of asphalt pavement results in longer pavement life, lower pavement maintenance, and better pavement performance (McLeod 1967).

Because of the numerous points of interest they have, the utilization of additives is expanding in the asphalt industry, and the workability of their asphalt blends has turned out to be a more imperative issue. Historically, the viscosity of bitumen has been used to decide the mixing and compaction temperatures of the asphalt-aggregate combinations (Yildrim et al. 2000; Bennert et al. 2010). However, there may be a few challenges associated with this; for example, shear rate dependencies (non-Newtonian behavior), non-linearity, or the chemical and/or physical mechanisms between bitumen and different mineral types or additives can alter the rheological properties at different temperatures (Yildrim et al. 2000). In any case, with the utilization of additives in asphalts, the workability changes considerably since the modifiers tend to change the viscosity of the blend. Consequently, compacting modified asphalt mixtures to achieve the desired density may be different from mixtures that utilize unmodified binders. Different testing methods widely used to benchmark the mixing and compaction temperatures, including workability testing, have been discussed in detail elsewhere (Yildrim et al. 2000; West et al. 2010). It was demonstrated that no reliable method for determining the production temperatures has been adopted by all pavement practitioners. The equiviscous principle that is commonly used to determine asphalt production temperatures is no longer feasible for many polymer-modified asphalts as it concludes exaggerated mixing and compaction temperatures (Shuler et al. 1992; Shenoy 2001). Also, in spite the capability of increasing the workability at lower temperatures compared to Hot Mix Asphalt (HMA), the determination of the production temperatures of Warm Mix Asphalt (WMA) is still questionable. Generally, overestimated mixing temperatures can stiffen the bitumen in the form of volatilization and oxidation (Clark 1958). An increase in temperature by 10 °C approximately doubles most oxidation mechanisms (Fink 1958). In particular, polymer additives can break down in bitumen at excessively high temperatures; however, the temperature ranges that cause these detrimental effects are not clearly established (Linde and Johansson 1992; Airey and Brown 1998; Shenoy 2001). Also, numerous studies (e.g. Azari et al. 2003) have demonstrated the significant effect of compaction temperatures on the mechanical performance of compacted asphalt specimens. The workability testing could be considered a simulation of the real process in the mixing plant; therefore, an extension of the current study was performed as an initiative to predict the mixing temperatures of different asphalt mixtures from the developed workability prototype.

Measuring the workability of asphalt mixtures is an important but also a challenging task. To have a standardized protocol for asphalt mixtures, it is important that the entire protocol is extremely well-defined since any variation could yield significantly different results leading to inaccurate conclusions as to the impact of other parameters. The repeatability of the end test results is, therefore, a critical factor that received much attention in this study. The workability of asphalt mixtures is influenced by many distinctive parameters, for example, production temperatures, aggregate properties (i.e. mineralogy, gradation, the size and shapes) and the type of bitumen used (Tayebal et al. 1998; Delgadillo and Bahia 2008). The influence of these parameters may also vary from case to case, depending on specific mixture characteristics or circumstances. The requirements that assist pavement engineers in the material design and production levels ought to be appropriate for a wide range of materials and boundary conditions. Therefore, sensitive experimental strategies and analytical or numerical fashions, primarily based on fundamental understanding, may be needed to convey sufficient control and predictability in bitumen and asphalt mixture design and production. There had been some attempts to measure the workability of asphalt mixtures directly from the loose mixture (i.e. aggregates coated with bitumen); however, different technical issues such as device geometry challenges or lack of repeatability led these to be not fully successful (Gudimettla et al. 2004). Marvillet and Bougault (1979) introduced an instrument to measure the workability of HMA. The primary principle of this device was a paddle immersed in a container of HMA sample and the torque to maintain the paddle rotating inside the blend at a constant speed was measured. The relationship between temperature and the corresponding measured torque was taken as the workability curve. Generally, they found that as the viscosity decreases, the workability and hence compactability increases for the same mixture. Except for the binder content, the angularity of the aggregate particles and dust content affected the workability results. Also, they warned that if different mixtures have the same workability, it does not guarantee that the mixtures undergo the same compactability. The poor repeatability and scattered data for several of the regressions indicate that the previous trial tests may not be fully dependable in establishing specific relationships between temperature and workability for different asphalt binders (West et al. 2010). There are still some fundamental and obvious problems to overcome.

2 Research Motivation and Objectives

The workability of asphalt mixtures is a basic component in getting the design air void content and, hence, the desired density of the mixture, which leads to greater durability and better performance of asphalt pavements. Researchers have found that compared to the equiviscous principle for determining mixing and compacting temperatures of modified asphalt mixtures, the workability tests predict rationale production temperatures. Previous attempts to measure the workability of asphalt mixtures have found that although the constituents of each mixture are held constant, the workability results cannot be considered repeatable as a range of workability resulted at same temperatures. Therefore, on account of the significance of workability in asphalt development, the current paper intends to present a bitumen-based prototype to quantify the workability of different asphalt mixtures in a standard way. A simple mathematical principle was presented first to obtain a fundamental basis of the current prototype to compare the conditions in the target workability prototype versus a real plant mixing. The current procedure is based only on the bitumen phase and involves the aggregate effect. As a secondary objective, this prototype was used to roughly determine the mixing temperatures of polymer-modified and foamed WMA blends.

3 Principle of Workability

The analysis of workability while preparing asphalt mixtures is very complicated. Presenting a principle for the workability is important before going through with the prototype, to give a basis for its development. The process of mixing and the required torque can be described mathematically in a simple way to enable the prediction of torque for different conditions. The general principle of workability illustrated in this section is based on a simplified derivation. Consider different-sized aggregate particles in contact with each other with the aid of interposed bitumen films. A conceptual diagram of bitumen coated aggregate particles impacting each other during mixing and the internal forces associated with it is shown in Fig. 1. The engineering properties of aggregates, including shape, size, and gradation, in addition to bitumen content, are very important in having the desired workability of asphalt concrete mixtures (Tayebal

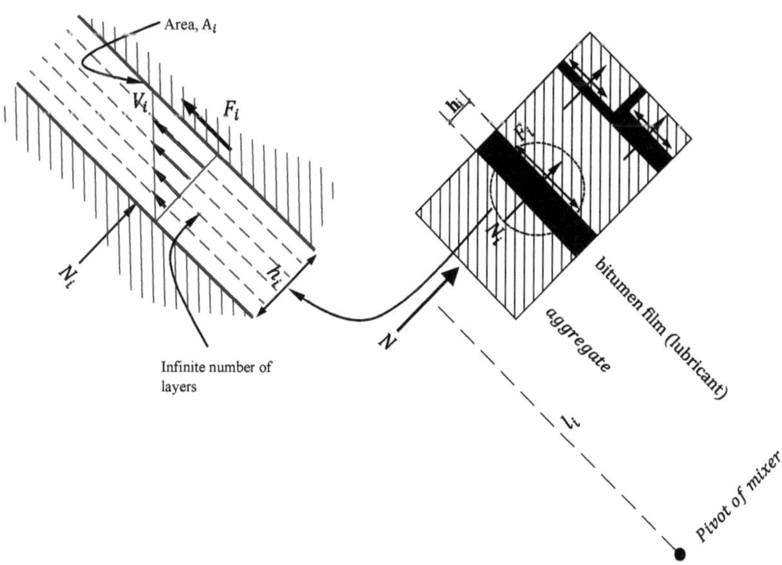

Fig. 1. A Conceptual diagram of aggregate particles and bitumen while mixing and internal forces

et al. 1998; Delgadillo and Bahia 2008). Aggregate occupies approximately 90–95% by weight and 75–85% by volume of an asphalt mixture; therefore, the frictional regime between aggregate particles in the loose stage and during mixing is highly dominant. With the interpose of bitumen layer between aggregate particles, a lower friction coefficient results due to lubrication (Hanz et al. 2011). During mixing, the mobility and manipulation of aggregate particles around one another undergo different lubrication regimes, depending on the bitumen film thickness (or bitumen content), which changes the coefficient of friction between the aggregate particles. The friction force due to the collision of the aggregate particles while mixing is controlled by shearing forces in the lubricant layer. The exerted shear forces depend on the viscous characteristics of the lubricant, as well as the shear velocity distribution in the bitumen film (Stachowiak and Batchelor 2013). Therefore, the friction function, $f(\mu_i)$ has been suggested instead of a constant friction coefficient, μ_i, to consider the variation in the coefficient of friction while mixing. The behavior of the friction coefficient due to the change in the properties or thickness of the lubricant layer can be explained by the Stribeck curve (Stachowiak and Batchelor 2013).

Due to the application of a normal force, N_i, the magnitude of the friction force, F_i, that can be exerted between any two plane surfaces (i) of particles in contact and in motion relative to each other can be calculated using the Coulomb equation.

$$F_i = f(\mu_i)N_i \tag{1}$$

From the basic principle of the relationship between shear rate, $\dot{\gamma}_i$, and shear stress, τ_i, for a Newtonian fluid, the tangential reaction force, F_i, is a product of contact area (A_i), local viscosity (η_i) and relative velocity between the surfaces (V_i) divided by film thickness, h_i i.e.:

$$F_i = \frac{\eta_i V_i A_i}{h_i} \tag{2}$$

From the above equation, it can be concluded that, physically, the lateral reaction force between two aggregate particle surfaces, F_i, is the result of many different mechanisms, which depend on contact geometry and topology, properties of the bulk and surface of the aggregate particles, displacement and relative velocity of the particles and viscosity of the present lubricant or bitumen. Therefore, the normal force, N_i, between two adjacent particle surfaces (i) while mixing can then be calculated as:

$$N_i = \frac{\eta_i V_i A_i}{h_i f(\mu_i)} \tag{3}$$

The total torque, q, required to keep mixing aggregate particles displaced at different radial distances l_i from the pivot of mixing (Fig. 1) can be calculated as follows:

$$q = \sum N_i l_i \tag{4}$$

Considering the torque is temperature (T) and time (t) dependent, substitution yields:

$$q(T,t) = \sum \frac{\eta_i V_i A_i}{h_i f(\mu_i)} l_i \tag{5}$$

Different properties of asphalt mixtures (such as WMA) change during the mixing time. The workability index, $WI = q(T,t)^{-1}$, can be used for a relative comparison of the change in workability during mixing of different asphalt mixtures. The WI shows how easy or how difficult a bitumen-aggregate blend is to handle during the mixing process. Unlike the modified mixtures, Eq. (5) infers that the WI of the unmodified mixtures is close to the linearity with respect to temperature.

From Eq. (5), it can be concluded that assuming the position of each aggregate particle or l_i is constant, the torque or the workability depends on different factors and, therefore, should be studied while developing the workability device. For the unmodified mixtures, the torque depends on the amount and properties of aggregate. Also, the properties of the asphalt binder used are functions of temperature. It is also quite clear that the viscosity of the lubricant is an important parameter for calculating the workability of asphalt mixtures. The asphalt binder content affects the torque needed to keep the aggregates slipping while mixing. The asphalt binder thickness is mainly dependent on its content, considering two bitumen films made up of equispaced layers; for same surface velocity, different thicknesses of bitumen film will undergo different velocity gradients, V_i/h_i. Thicker films need less force to shear as they contain more single layers compared to the thinner films (Stachowiak and Batchelor 2013). Therefore, the viscous resistance and hence the torque are inversely proportional to the film thickness of bitumen. To compare the results between different asphalt mixtures, the properties of aggregates should be kept constant. Although it can be recognized as a limitation, including the interactions between bitumen and aggregate would complicate the process. According to the above discussion, the aggregate properties enormously affect the workability of asphalt mixture. This conclusion is consistent with the experimental work by Gudimettla et al. (2004). Therefore, to reduce errors in measuring the workability of different blends as well as maintaining the possibility of repeatability, the sample preparation should be standardized. Also, to accurately predict the mixing temperatures of an asphalt mixture, a unique practical temperature-workability relationship should be gained, which is not possible based on the previous discussion.

4 Configurations of Developed Workability Prototype

4.1 Description of the Proposed Prototype

A pictorial view of the whole workability prototype developed herein can be seen in Fig. 2. To apply torque, the device developed in this study made use of a commercially available motorized VST apparatus that is mainly used to measure the shear strength of soil. The VST apparatus has been modified to suit the workability test purpose. The torsional spring has been replaced with a rigid steel rod to directly transfer torque measured with the aid of a sensitive torque meter. The VST apparatus was placed on the top surface of an oven. A 60 cm length and 0.5 cm diameter steel rod was attached to the VST apparatus, and the rod was impeded through the top vent of the oven. A specially designed spindle was immersed in a specially designed cylindrical bowl (70 mm diameter and150 mm height) containing the asphalt blend and was attached to the bottom of the rod. The oven was used to maintain the blend at a specific temperature in degrees Celsius. All the data of the torque meter were obtained and recorded by an automated sensitive torque meter attached to a data reader. The values of the torque were measured at different times excluding the initial readings of the first two seconds close to the stationary position as the torque values are very high and, therefore, not representative. The temperatures used in this study were chosen as 100, 140 and 180 °C, which is the temperature range over which asphalt mixtures are normally subjected to through different construction stages. Once the equipment was developed, the spindle configuration was fixed for any type of blend to be tested, which maintains the results comparable for different asphalt blends. Moreover, research was employed to evaluate the impact of various blend constituents on the workability of asphalt.

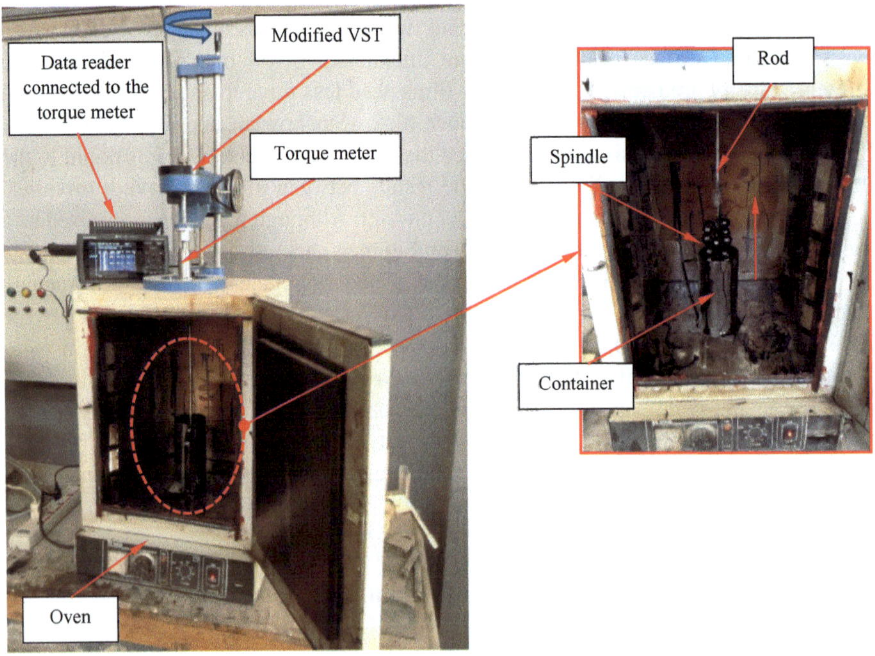

Fig. 2. Pictorial view of the proposed method

4.2 Design and Fabrication of the Spindle

The design of the spindle was built on the findings drawn from Eq. (5). Keeping the bitumen volume constant, the significant differences in the workability were hypothesized to be caused by different shapes and sizes of aggregates and the aggregate–bitumen interactions. To make sure all samples consist of the same sized aggregates and distribution, an appropriate amount of metallic balls in asphalt blends were chosen to ensure a representative aggregate distribution. The particle size is another parameter which has an important influence on the workability of asphalt blends. It can be even more important when the modified asphalt blends are tested at different concentrations for comparing the effect of additive content on the workability of asphalt. Therefore, care should be taken to take a representative sample of the asphalt blend. Each aggregate particle influences its surrounding bitumen layer and its characteristics, such as thickness. Stiffness can be altered by changing the type of aggregate. Mixtures with the same aggregate and asphalt binder would have different workabilities at multiple testing trials. An asphalt mixture can be considered a granular paste in which the asphalt binder plays the media role. The same proportions of one aggregate can have different sizes, and this changes the mobility (i.e. workability) of the particles within the bitumen as the interfacial properties between the particles and bitumen change. This hypothesis may be dominant in modified asphalts, especially for adhesion promoter additives which act on the surface of the aggregate. Various shapes of aggregate might be present during crushing in the quarries starting from rounded to flaky and elongated aggregates. Also, the arrangement or distance between aggregate particles while mixing is one of the most important parameters affecting the workability of its granular paste. While mixing, the position of different aggregate particles varies from time to time. Changes in the internal structure of the asphalt mixture can lead to changes in the workability of the blend. Aggregate characteristics and gradation may overwhelm binder effects. The fabricated spindle (Fig. 3a) is a unit composed of five rows spaced at a 15 mm vertical distance; each row is formed by joining the center of four-connected 20 mm diameter aluminum balls ninety degree angles. Odd rows (3 rows) were rotated 45 degrees in the horizontal plane with respect to the even ones (2 rows). The geometry of the fabricated spindle is also shown schematically in Fig. 3b. The height of the bitumen sample was kept at 85 mm, just covering the upper balls and leaving a 5 mm gap between the lowest point of the lower balls and the bottom of the container. The geometry and the gap between the aluminum balls and the inner wall of the cylindrical container were chosen as 5 mm to avoid the influence of the boundaries on the measurements. The sample container is heavy enough to prevent rotation along with the spindle while testing. In the current modified VST apparatus, the rotational speed increased to a possible speed of 1/18 rpm, which corresponds to a maximum peripheral or tangential speed of 0.175 mm/s, and for a 5 mm gap between the ball and inner wall of the container, a maximum instantaneous shear rate of 0.035 s^{-1} will be yielded. Because of the importance of the volumetric characteristics of the aggregate and bitumen, the amount of aggregate content was calculated by volume and weight. Although it was difficult adding more balls, the aluminum balls represent approximately 35% by volume and 85% by weight of the bitumen blend used, which were hypothesized to be close enough to resemble realistic aggregate properties (e.g. percentage).

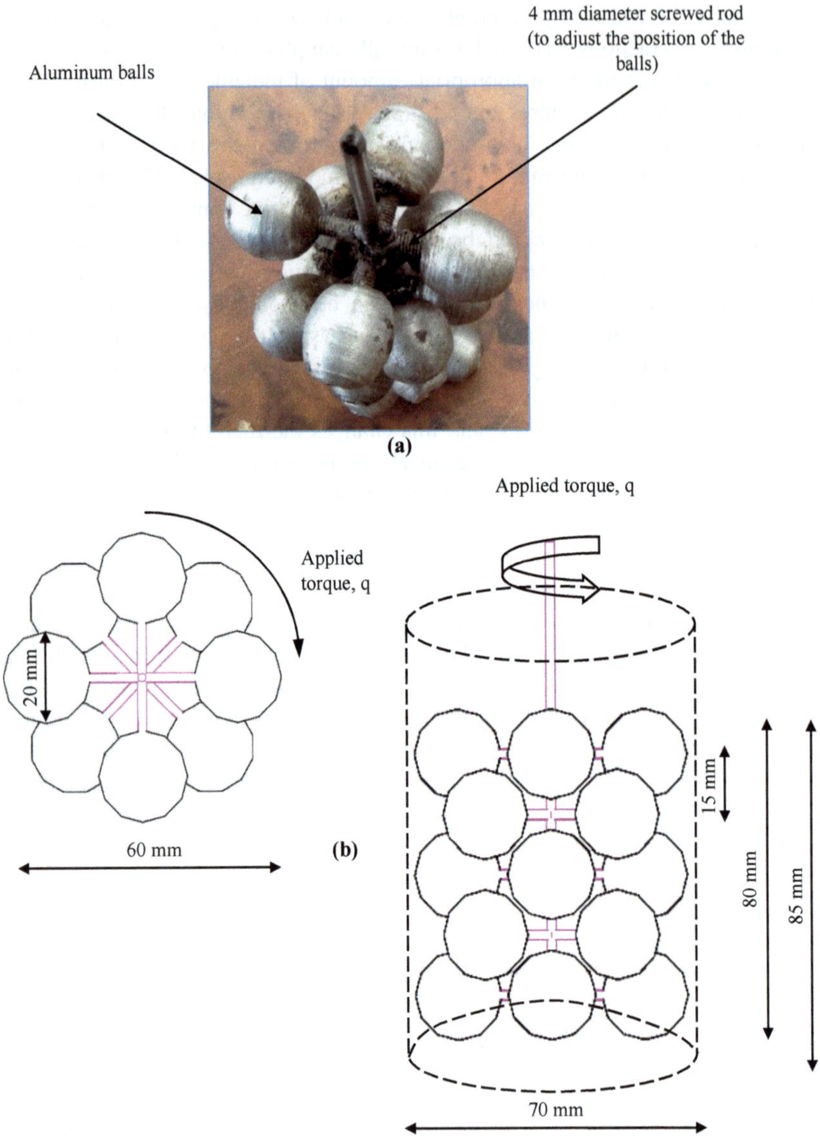

Fig. 3. Configuration of the designed spindle: (a) an oblique view of the fabricated spindle and (b) a schematic showing the geometry of the designed spindle (plan view (left side) and elevation view)

5 Workability of Polymer-Modified and Warm Mix Asphalt

Due to the wide utilization of polymer additives and water foaming in the asphalt industry, the workability measurement of their blends turned out to be an imperative issue and was determined with the aid of the developed prototype. For simplicity, it is usually assumed that the modified asphalt binders are Newtonian, i.e. their viscosity is constantly proportional to the shear rate, however, this leads to an inaccurate estimation of the mixing and compaction temperatures resulting from the equiviscous principle. Recognizing the potential problems with using this method, the linear temperature-viscosity relationship assumed for unmodified binders may not be valid for modified asphalt binders. According to Yildrim et al. (2000), the viscosity of polymer-modified asphalt binders is significantly affected by the shear rate; this effect was very small in the case of unmodified asphalt binders and can be ignored during the viscosity measurements. Regarding the WMA and foaming technology in particular, the inclusion of pressurized air and quantity of cold water into a hot bitumen phase produces volumetric effects that make asphalt testing a challenge. The air bubbles formed after evaporation of cold water into hot bitumen result in foamed asphalt. After the injection, the foamed bitumen expands rapidly, and after some time (about a few minutes in the case of water-based foaming) it reaches its maximum volume followed by a rapid collapse process and a slow, asymptotic return to its original volume. This makes measuring the viscosity of the foamed binder difficult, or in other words, not feasible. No bubbles should be present in the asphalt binder when tested using the current viscosity protocols. The foamed asphalt can be considered a two-phase system where the bitumen phase encloses another phase of air bubbles formed by the evaporation of cold water into the hot bitumen. The bitumen phase in this system should have the same viscosity as that of the tank bitumen at the same temperature. However, the overall viscosity or workability of the foamed bitumen (if we consider the asphalt-air two-phase system as a whole) is lower than liquid tank bitumen at the same temperature.

6 Materials and Preparations

In the current study, the developed prototype was verified with the aid of three asphalt blends: two different polymer-modified bitumen blends (PMB1 and PMB2) and water foamed bitumen. Paving bitumen (penetration 60/70) commonly used in Egypt from Suez refinery was used to produce the blends. The polymer-modified blends were prepared in an appropriate container equipped with a high shear mixer. A known quantity of the asphalt binder was added to the container and heated to 180 °C, followed by the addition of 3% (by the weight of bitumen) of the polymer additive. Each bitumen-polymer blend was continuously stirred for 2 h at 2000 rpm. In the case of foamed prepared bitumen, there are several lab-scale foaming machines, however, to prepare the foamed bitumen, the procedure discussed in (Goh and You 2011) has been adopted in this study. The syringe-based system for generating foam is less expensive and capable of producing good quality foam, provided that a field foaming effect can be ensured. The most important is ensuring that the bitumen is heated to the desired temperature before injecting the cold water. In this study, the water was injected at rates of 1.5 and 3% (by the weight of bitumen) at the bottom of the container, which results in an acceptable foaming. However, in order to produce acceptable foam, the water must be pressurized suddenly in a short period of time (usually within a second). Afterward, a spatula was used to disperse the bubbles generated in the entire volume of bitumen. The prepared blends were then immediately poured in the spindle container for testing. For the polymer modified bitumen, the torque data was recorded after the blend reached target temperatures (after 1 h heating), while for the foam prepared bitumen, a single point torque was recorded every 20 s at the target temperature.

7 Discussion

7.1 Workability Index (WI)

After the prototype configuration was finalized, further testing was done for verification. The philosophy behind utilizing the polymer-modified and WMA, in particular, is that their workabilities are still questionable. This information was then examined to decide the impact of individual constituents on the workability of the studied blends. The torque required to maintain a constant rotational rate of a specially designed spindle while submerged in the samples of asphalt blends at a specific temperature was measured. Plots of the WI data at the temperature points are shown in Fig. 4. The WI at different temperatures shows that the blend gets more workable as the temperature increases and vice versa. As for the polymer-modified bitumen, a one point workability value was measured as shown. The polymer-modified bitumen is less workable than the tank bitumen. The workability of WMA is particularly time-dependent. The workability data for the WMA were recorded at different times (except for the first 2 s) and plotted as shown in Fig. 5. One of the promising benefits marketed about using WMA is the improved workability at decreased production temperatures. There is a great interest in discerning how this increased volume affects the workability of asphalt. According to the derived Eq. (5) and the experimental results presented herein, it can be seen that the WI, which is the reciprocal of the torque, q, is affected by different factors. For example, the foaming effect increases the volume of bitumen, and therefore, the thickness of bitumen film changes with the time of mixing; and the film thickness decreases with the increase of mixing time, which increases the applied

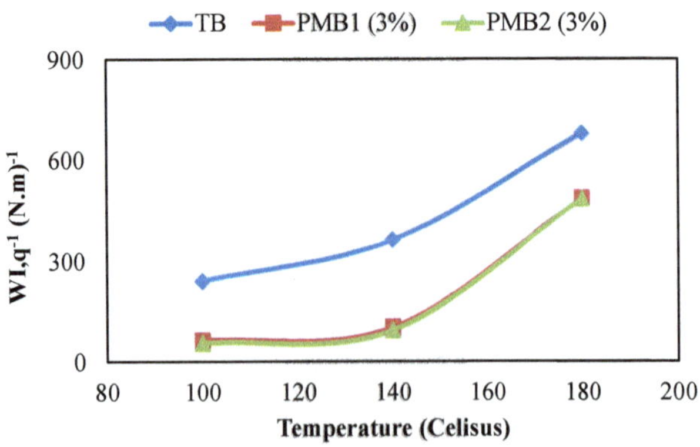

Fig. 4. Workability index for 3% polymer-modified bitumen (PMB1 and PMB2) and tank bitumen (TB)

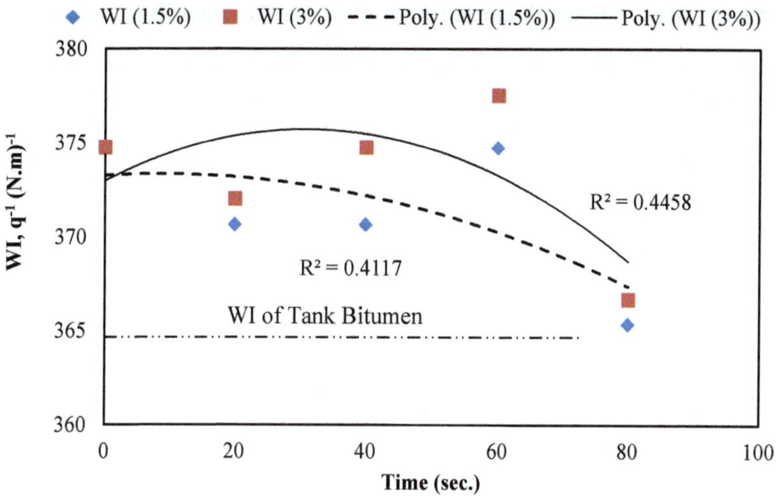

Fig. 5. Workability index vs. testing time for water foamed bitumen (at 140 °C)

torque with time increase. It is also quite obvious from Fig. 5 that mixtures foamed with 3% of water get more workable compared to 1.5% foamed mixtures. Water and air mixed with the bitumen resulted in foam multiple times larger than the original volume. In the real mixing, the contact area between colliding aggregates also changes. As partial lubrication increases due to the inclusion of bubbles in bitumen, aggregate particle contact decreases and wear is decreased, thus increasing the acceleration of the moving part. This shows that mixtures prepared through foamed WMA are more workable in the early stage of mixing compared to mixtures prepared through HMA mixing. In the case of foamed bitumen, the figure shows WI values at 140 °C only. The WI gradually decreases with respect to time until the WI attains the lowest value. Depending on the foaming temperature, foaming is effective over a short period of time. This conclusion has been stated in many research papers (e.g., Jenkins et al. 2000; Diab et al. 2014). The scatter of WI values could be due to the movement of the spindle in a non-homogenous medium, as the bitumen phase is randomly disconnected by air bubbles. Despite the scattered results, a reasonable trend of the WI can be registered. Higher mixing speeds would most likely be better for proper analysis of this scenario. The current mixing process can be different than that in the real mixing plants; for in reality, a mass mixing undergoes turbulent mixing actions in the plant mixers, corresponding to an extremely wide range of applied shear rates (West et al. 2010). The developed prototype is limited to workability results at a low shear rate; however, the results at higher shear rates are not predictable and need further investigation. Finally, the presented prototype was able to differentiate the workability of polymer-modified and foamed WMA.

7.2 Shear Stress Versus Shear Rate for Polymer-Modified Bitumen

First, the relationship between shear rate and shear stress for polymer-modified bitumen was discussed based on the experimental results. The behavior of shear stress versus shear rate was verified experimentally for different levels of shear rates and temperatures. The rotational viscometer (Brookfield DV2T) that measures the viscosity based on the principle that the fluid is sheared between two surfaces following the AASHTO T 316-11 (2011) was used for this testing. The velocity of the #27 spindle can be varied so that the relationship between shear rate and shear stress, from which the ratio of both measures of viscosity, can be recorded. The shear stresses corresponding to shear rates at rotational speeds ranging from 5 to 40 rpm in 5 rpm increments were recorded from the rotational viscometer at different temperatures (100, 140, and 180 °C) and are plotted in Fig. 6a–c, respectively. Much of the possible shear rates with the Brookfield DV2T have been used according to each individual binder and temperature combination allowed.

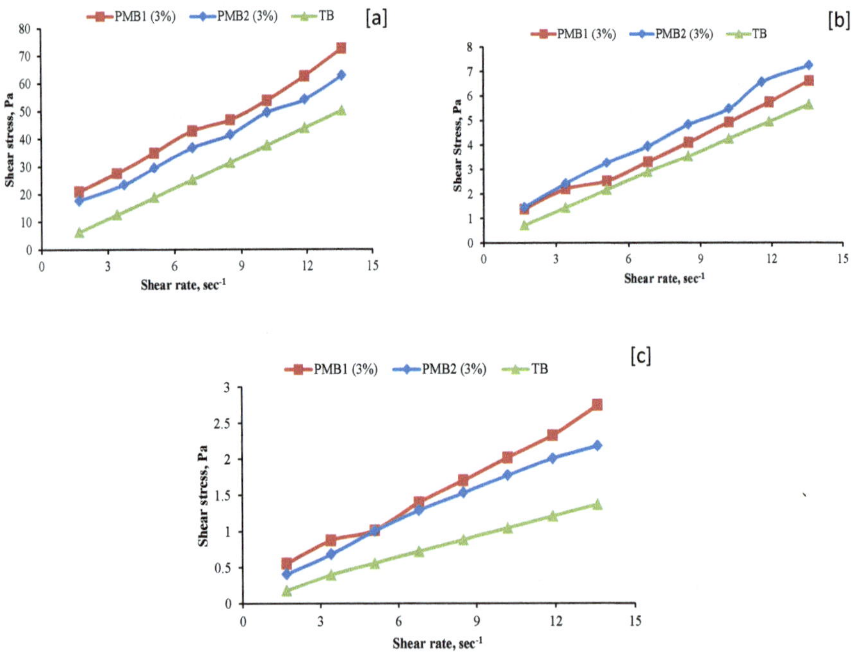

Fig. 6. Shear stress -shear rate characteristic for 3% polymer-modified bitumen (PMB1 and PMB2) and tank bitumen (TB): (a) 100 °C, (b) 140 °C, and (c) 180 °C

Generally, fluids like the tank or neat bitumen compared to the modified bitumen have a loose and homogenous molecular structure which is not significantly affected by shearing. For fluids like the tank bitumen, theoretically, the shear stress is constantly proportional to the shear rate. On the other hand, when the fluids in which a portion of the substance is composed of suspended polymer molecules, the non-Newtonian behavior prevails. During the process of shearing in polymer-modified bitumen, separated long molecules, which are randomly orientated, tend to align, changing the apparent viscosity of the whole fluid (Yildrim et al. 2000; Stachowiak and Batchelor 2013). The relationship between shear stress and shear rate is no longer linear, and deviates from the Newtonian behavior with the increase in temperature, making the estimation of mixing and compaction temperatures for polymer-modified asphalt a challenge. Care is needed with non-Newtonian fluids, as this concept is mainly calibrated for Newtonian fluids as the dominance of non-Newtonian characteristics increases the complexity in determining the mixing and compaction temperatures from the viscosity-temperature curve (equiviscous principle). Although approximate analysis predicts a straight proportionality between shear stress and shear rate, for polymer-modified bitumen this proportionality is lost and a different pattern is observed at different temperatures. Therefore, the workability, as a genuine characteristic, seems to be a distinctive alternative for predicting the production temperatures of different asphalt mixtures.

7.3 Predicting Mixing Temperature of Modified Asphalt Mixtures

The WI of tank bitumen hypothesized as a desired level of workability was used as a reference for the polymer-modified blends (Fig. 4). Table 1 shows the mixing temperatures of tank bitumen and polymer-modified bitumen blends based on the equiviscous principle and WI concept suggested herein. For the PMB1, when compared with the equiviscous principle, the WI data resulted in a predicted mixing temperature about the same. However, in the case of PMB2, the suggested method resulted in less predicted mixing temperature; by about 5 °C as compared to that estimated from the equiviscous principle. In the case of foamed WMA, the WI is higher than that for the tank bitumen. Although there is no proven data, since the WI data is limited to 140 °C in this study, at lower temperatures it is likely that the WMA blends undergo higher WIs, too, resulting in lower mixing temperatures. Therefore, as long as the WI of foamed WMA is higher than that of the tank bitumen, the temperature can be considered a suitable mixing temperature. It can also be seen that every studied blend has a unique relationship between temperature and WI based on the bitumen type. In the case of WMA, the testing temperature seems suitable as a mixing temperature. Mixture tests should be performed in conjunction with the aforementioned WI results to analyze the effects of mixing temperature on the aggregate coating, and the desired mixture and

Table 1. Predicted mixing temperatures of polymer-modified mixtures

	Range of mixing temperatures, °C	
	Equiviscous principle	WI method
Tank bitumen	161.63–165.13	–
3% polymer-modified bitumen		
PMB1	178.3–182.68	180
PMB2	184.47–187.47	180

mechanical properties of the corresponding mixtures; however, this is out of the scope of this research. In another study, Diab (2016) proved the influence of production temperatures on the mechanical properties of the mixtures and cautioned a care should be taken to select the appropriate temperatures to suit different factors (e.g. performance and workability). Determining the mixing and compaction temperatures of asphalt mixtures is a complex issue and needs separate additional studies. In an attempt by Gudimettla et al. (2004), they developed a different device and determined the compaction temperature at which the mixing torque begins to change dramatically as the lowest desirable compaction temperature. However, the initiative to determine the mixing temperature in this paper is just a preliminary attempt that is limited by the use of some data analysis of the developed workability prototype, and more effort is needed to make it more conclusive.

8 Summary and Conclusions

The desired outcome of this study was to develop a fundamental-based prototype to simply measure the workability of asphalt mixtures based on bitumen testing. Also, there was a need to present a method for determining the mixing temperature of different asphalt mixtures. A device that effectively distinguishes between the workabilities of different asphalt mixtures is important. Unfortunately, acceptable laboratory and test conditions have yet to be developed to quantify the workability of asphalt mixtures. The current paper presents a bitumen-based prototype in an attempt to differentiate the workability of different asphalt mixtures. There has been a limited research effort to quantify the workability of asphalt mixtures in the literature. Traditionally, the temperature-viscosity curve is used to determine the mixing and compaction temperatures of asphalt mixtures based on the equiviscous principle. However, it was found that the viscosity of modified binders significantly moves away from (linear) Newtonian toward the non-linear behavior compared to the tank bitumen, especially at higher temperatures. From this issue, researchers paid a lot of attention to the workability of asphalt mixtures in lieu of the equiviscous principle. Before designing the prototype, a simple fundamental derivation was performed to study the factors that could affect the workability of asphalt mixtures. The workability of asphalt mixtures depends on different factors, including binder type, aggregate properties, modifiers, and temperature of the mixture. Therefore, a special spindle was designed and fabricated to obtain repeatable and comparable results by diminishing the sensitivity of the workability results to aggregate properties (the size or gradation, distribution, particle shape, and particle mineralogy). The torque required to maintain the spindle rotation at a specific speed in a sample of asphalt binder was measured. A research effort to evaluate the workability of polymer-modified asphalt binder, as well as water-based foamed WMA, is summarized. The workability prototype presented was able to rank the workabilities of the studied blends in a rational order and compared favorably with one another. In this instance, the foaming results in an improved workability of the mixture which can subsequently allow a decrease in the production temperatures. The prototype aided a rough estimation of the mixing temperature of polymer-modified and foamed bitumen.

9 Recommendations

This prototype worked best in scenarios where the mixing speeds are low; however, due to the limited speeds of the current prototype, the results at higher speeds should be investigated.

Acknowledgments. The authors would like to thank the department of civil engineering laboratories at Aswan University, Egypt for providing the facilities that helped in making this prototype possible.

References

AASHTO T 316-11: Viscosity determination of asphalt binder using rotational viscometer. In: AASHTO Standard Specifications for Transportation Materials and Methods of Sampling and Testing. American Association of State Highway and Transportation Officials, Washington, DC (2011)

Airey, G.D., Brown, S.F.: Rheological performance of aged polymer modified bitumens. J. Assoc. Asph. Paving Technol. **67** (1998)

Airey, G.D., Hunter, A.E., Collop, A.C.: The effect of asphalt mixture gradation and compaction energy on aggregate degradation. Constr. Build. Mater. **22**, 972–980 (2008)

Azari, H., McCuen, R.H., Stuart, K.D.: Optimum compaction temperature for modified binders. J. Transp. Eng. **129**(5) (2003)

Bennert, T., Reinke, G., Mogawer, W., Mooney, K.: Assessment of workability and compactability of warm-mix asphalt. Transp. Res. Rec.: J. Transp. Res. Board **2180**, 36–47 (2010)

Celik, O.N., Atis, C.D.: Compactibility of hot bituminous mixtures made with crumb rubber-modified binders. Constr. Build. Mater. **22**, 1143–1147 (2008)

Clark, R.G.: Practical results of asphalt hardening on pavement life. In: AAPT, vol. 27 (1958)

Delgadillo, R., Bahia, H.U.: Effects of temperature and pressure on hot mixed asphalt compaction: field and laboratory study. J. Mater. Civ. Eng. **20**(6), 440–448 (2008)

Diab, A.: Studying viscosity of asphalt binders and effect of varied production temperatures on engineering properties of HMA mixtures. Can. J. Civ. Eng. (2016)

Diab, A., You, Z., Ghabchi, R., Zaman, M.: Effects of regular-sized and nanosized hydrated lime on binder rheology and surface free energy of adhesion of foamed warm mix asphalt. J. Am. Soc. Civil Eng. **27**(9), 04014254 (2014)

Fink, D.F.: Research studies and procedures. AAPT **27** (1958)

Goh, S.W., You, Z.: Moisture damage and fatigue cracking of foamed warm mix asphalt using a simple laboratory setup. In: ASCE T&DI Congress 2011: Integrated Transportation and Development for a Better Tomorrow, Proceedings of the First T&DI Congress (2011)

Gudimettla, J.M., Cooley, L.A., Brown, E.R.: Workability of hot-mix asphalt. J. Transp. Res. Board **1891**, 229–237 (2004)

Hanz, A., Mahmoud, E., Bahia, H.U.: Asphalt lubricity test evaluation and relationship to mixture workability. In: 90th Annual Meeting of the Transportation Research Board, pp. 23–27, Washington, DC (2011)

Horan, R.D., Chang, G.K., Xu, Q., Gallivan, V.L.: Improving quality control of hot-mix asphalt paving with intelligent compaction technology. Transp. Res. Rec. J Transp. Res. Board **2268**(1), 82–91 (2012)

Jenkins, K.J., Molenaar, A.A.A., de Groot, J.L.A., van de Ven, M.F.C.: Developments in the uses of foamed bitumen in road pavements. Heron **45**(3), 167–175 (2000)

Linde, S., Johansson, U.: Thermo-oxidative degradation of polymer modified bitumen. In: Polymer Modified Asphalt Binders. ASTM STP 1108 (1992)

Marvillet, J., Bougault, P.: Workability of bituminous mixes: development of a workability meter. Proc. Assoc. Asph. Paving Technol. **48**, 91–110 (1979)

McLeod, N.W.: Influence of viscosity of asphalt-cement on compaction of paving mixes in the field. Bituminous and Concrete Mixes. Highway Research Record, No 158 (1967)

Shenoy, A.V.: Determination of the temperature for mixing aggregates with polymer-modified asphalt. Int. J. Pavement Eng. **2**(1) (2001)

Shuler, T., Hanson, D., McKeen, R.: Design and construction of asphalt concrete using polymer modified asphalt binders. In: Polymer Modified Asphalt Binders, ASTM STP 1108 (1992)

Stachowiak, G., Batchelor, A.W.: Engineering Tribology. Butterworth-Heinemann (2013), ISBN 978-0-12-397047-3

Tayebal, A., Malpass, G., Khosla, N.: Effect of mineral filler type and amount on design and performance of asphalt concrete mixtures. Transp. Res. Rec. **1609**, 36–43 (1998)

West, R.C., Watson, D.E., Turner, P.A., Casola, J.R.: Mixing and Compaction Temperatures of Asphalt Binders in Hot-Mix Asphalt. NCHRP Report 648 (2010)

Yildrim, Y., Solaimanian, M., Kennedy, T.W.: Mixing and Compaction Temperatures for Hot Mix Asphalt Concrete. Texas Department of Transportation, Research Report Number 1250-5 (2000)

The Use of Wastewater in Construction of Base Course Layers in Pavement Structures

Farid H. Abed[1], Munir D. Nazzal[2(✉)], Mousa F. Attom[1], Magdi E. El-Emam[1], Nouran ElMessalami[1], and Saif Al-Dabagh[1]

[1] Civil Engineering Department, American University of Sharjah, 26666 Sharjah, United Arab Emirates
[2] Civil Engineering Department, Ohio University, Athens, OH 45701, USA
nazzal@ohio.edu

Abstract. Large amounts of wastewater are produced annually and treated for use in agriculture, irrigation and groundwater replenishment. The produced treated wastewater (TWW) can be used as an alternative to the millions of gallons of fresh water that are typically used in pavement construction. This study was conducted to investigate the potential of using treated wastewater in base course layer in pavements. Chemical analysis tests were conducted both on the treated wastewater samples and on tap water to determine their properties. Samples of the base course material were prepared and soaked using the two different TWW samples as well as the tap water. Proctor compaction and California Bearing Ratio (CBR) tests were conducted on the prepared samples, and the results have shown that TWW can be a suitable replacement for fresh water in base course layers in pavement construction.

1 Introduction

The UAE is listed by the United Nations as a high-rank country when it comes to water stress, which is a situation that occurs when the availability of water is not in balance with the demand. The UAE has one of the highest water consumption rates in the world, while resources are limited. According to the Abu Dhabi Environment Agency, 72% of water used in the UAE comes from groundwater, 21% comes from desalination and only 7% is coming from treated wastewater. Treated wastewater is typically used for irrigation, yet it can be also used to replace the millions of liters of fresh water typically consumed during the construction of pavements and bridges. Treated wastewater is mostly used for agricultural and landscape irrigation or replenishing groundwater. However, during the past years few studies have been conducted to explore its use in construction (e.g. Al-Jabri et al. 2011).

Several researches have been carried out in the past investigating the use of recycled aggregate and other recycled materials in concrete and pavements, but very limited research was conducted on the use of recycled water in concrete production and pavement construction. A study was conducted by Ismail and Al-Hashmi (2010) to investigate the possibility of totally replacing the fresh water used in concrete composites with industrial wastewater released from polyvinyl acetate resin manufacturing plant. Concrete composites were tested to investigate the effect of using various

L. Mohammad (ed.), *Advancement in the Design and Performance of Sustainable Asphalt Pavements*, Sustainable Civil Infrastructures, DOI 10.1007/978-3-319-61908-8_3

Polyvinyl acetate wastewater PVAW/C ratios of 0.30, 0.35, 0.40, and 0.45 on the slump, compressive strength, flexural strength, and dry density of the concrete mixes using control mixes and PVAW-concrete mixes. The results have shown that the slump values of the PVAW-concrete mixes were less than the slump of the control mixes, but were slightly increased with increasing PVAW/C ratios. The strength values for all PVAW/C ratios were close to or slightly higher than that for the control mix, but decreased slightly with increasing the PVAW/C ratios. The hard density values were increased with increasing the PVAW/C ratios. The waste materials leaching test showed no leaching of the PVAW toxic constituents, mainly represented by COD measurement, implying that using PVAW is non-hazardous for concrete mixes and could meet the requirements of clean construction products.

Other studies examined the use of recycled or waste water from concrete plants in the manufacture of ready-mixed concrete. In a study by Tsimas and Zervaki (2011), recycled wastewater from a ready-mixed concrete plant was chemically analyzed and compared with the concrete mixing water specifications for use in concrete. The results showed that the recycled wastewater did not affect the concrete workability, setting time or water demand of cement paste. Also, the compressive strength of the concrete specimens were at 7 and 28 days than the control specimens. It was concluded that the recycled wastewater meets the ASTM and EN standards and can be used in concrete since using 100% of the recycled wastewater in concrete did not alter the concrete performance.

In summary, previous research studies have investigated the suitability of using wastewater from washing car stations and concrete plant in the production of cement concrete mixtures only (Chatveera et al. 2006; Chatveera and Lertwattanaruk 2009; Low et al. 2007; Sandrolini and Franzoni 2000). However, to the knowledge of the authors, very limited research have been conducted on the use of treated wastewater or recycled water in the construction of base course layers in pavement structures. This study investigates the possible use of treated wastewater (TWW) obtained from two treatment plants in the UAE in the compaction of base course materials.

2 Objectives

The objective of this study was to examine and compare the effects of using treated wastewater, obtained from two different sources, as well as tap water on the compaction characteristics and mechanical properties of a base course material used in the construction of pavements.

3 Laboratory Testing Program

3.1 Properties of Used Water Samples

Chemical analysis tests were conducted on the tap water as well as on water samples obtained from different wastewater treatment plants in Sharjah and Dubai Emirates. The chemical analysis includes tests for sulphate content, chloride content, pH value,

COD, 5-days BOD, total dissolved solids, total suspended solids, ammonia nitrogen, total hardness and calcium carbonate. The results of the tests are as shown in Table 1 and they show that the obtained water samples met all the specifications of ASTM C94 as well as the United States Environmental Protection Agency (USEPA) guidelines.

Table 1. Water chemical analysis tests results

Test	Test method	Test result			USEPA guidelines	ASTM spec.
		Tap water	Sharjah rec.	Dubai rec.		
Sulfate content as SO_4 (g/L)	BS1377:1990 Part 3 CI. 5.5	0.03	0.1	0.23	–	<3
Chloride content (%)	BS1377:1990 Part 3 CI. 7.2	0.01	0.01	0.07	–	<0.05
pH value at 22.3 °C	BS1377:1990 Part 3 CI. 9	7.9	8	7.0	6–9	–
Chemical oxygen demand (mg/L)	APHA 5220 B	<10	<10	10	–	–
BOD 5 days (mg/L)	APHA 5220 B	6	7	7	≤ 30	–
Total dissolved solids (mg/L)	APHA 2540 C	170	670	1760	–	<50000
Total suspended solids (mg/L)	APHA 2540 D	12	14	60	≤ 30	
Ammonia nitrogen as NH_3–N (mg/L)	Salicylate method	0.01	0.01	7.9	–	–
Total hardness (mg/L)	APHA 2340 B/IHTP 17	64	172	470	–	–
Calcium carbonate as $CaCO_3$ (mg/L)	APHA 3120 B	48	69	250.5	–	–

3.2 Properties of Base Course Material

The base course material used in this study is unbound granular limestone material commonly used in the construction of base course layers in the UAE. Figure 1 presents the grain size distribution of the considered base course material. Based on the American Association of State Highway and Transportation (AASHTO) classification system and the Unified Soil Classification System (USCS), the considered base course material is classified as A-1-b and GW-GM, respectively.

Other tests performed on the base course material include specific gravity (G_s), absorption, and abrasion tests in accordance with ASTM standards D 792, D 698, D 570, and D 1883 respectively. Table 2 shows the results of these tests. It is noted that the selected base course material has a relatively low absorption value and a Los Angeles abrasion value of 30% which is less than the maximum value of 40% for aggregates used in the construction of pavements.

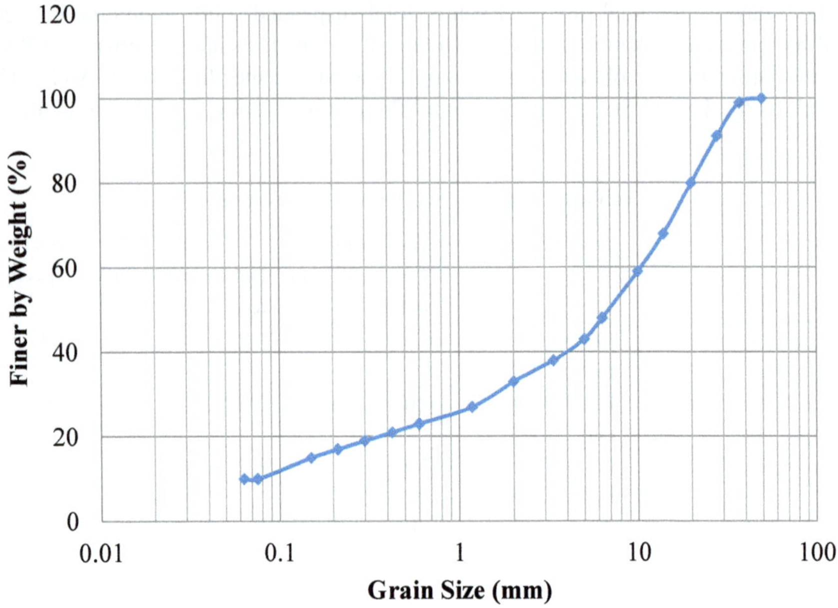

Fig. 1. Grain size distribution of the used base course material

Table 2. Properties of the used base course material

Property	Limestone
G_s	2.71
Absorption (%)	1.0
LA abrasion (%)	30
AASHTO classification	A-1-b
USCS classification	GW-GM
Coarse aggregate angularity (%)	100

4 Results Discussions

4.1 Effect of Wastewater on Compaction Characteristics of Base Course Material

Modified Proctor Compaction tests were conducted on the unbound granular limestone base course material using the two treated wastewater samples as well as the tap water. The obtained compaction curves for the tested samples are shown in Fig. 2. It is noticed that Dubai wastewater resulted in a slightly higher maximum dry unit weight as compared to the other water types. However, the values obtained were close for all the tested samples and the use of treated wastewater did not significantly change the compaction characteristics of the considered base course material. The maximum dry unit weight was about 22 kN/m^3 and optimum moisture content was about 5.5%.

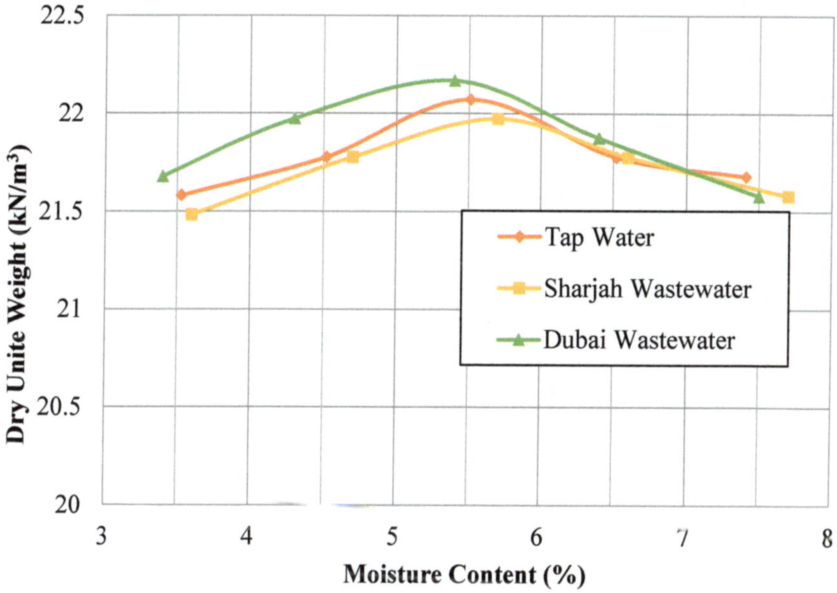

Fig. 2. Modified proctor compaction test results

4.2 Effect of Wastewater on CBR of Base Course Material

California Bearing Ratio (CBR) tests were conducted on the base course material prepared using the tap water and the treated wastewater samples obtained from Sharjah and Dubai. The CBR test was conducted as it helps determine the strength properties of base course materials and the CBR value is the main design input parameter in many pavement design procedures. CBR tests were conducted both on unsoaked samples and on samples soaked for 96 h, and the results are shown in Fig. 3. It could be noted that the CBR values of unsoaked samples were relatively high and that the use of wastewater did not significantly affect the measured CBR values of unsoaked specimens. It could also be noted that 96 h of soaking resulted in a reduction in the CBR values of all samples; however, this reduction was slightly more pronounced in the samples prepared and soaked using the treated wastewater. The largest reduction in CBR value due to soaking was observed with the sample prepared with Sharjah treated wastewater. The results are shown in Fig. 4. Despite this, the CBR values of all tested samples were still above 100% after soaking.

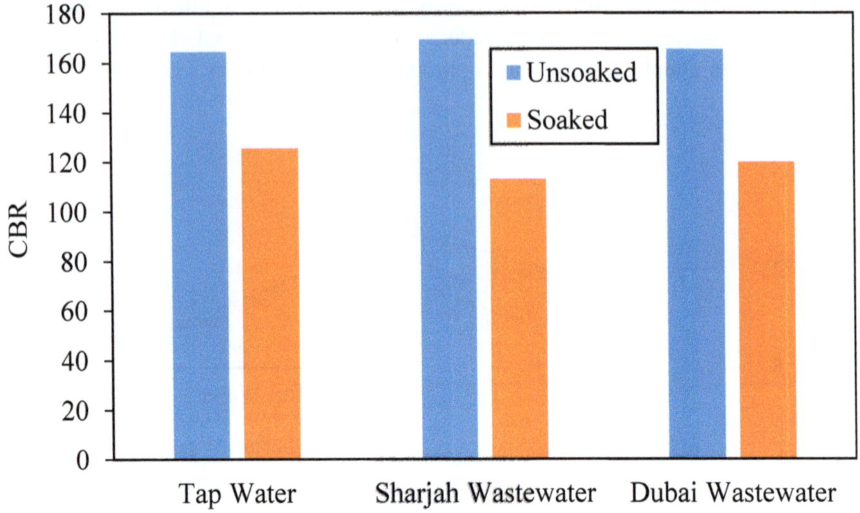

Fig. 3. CBR values of tested materials

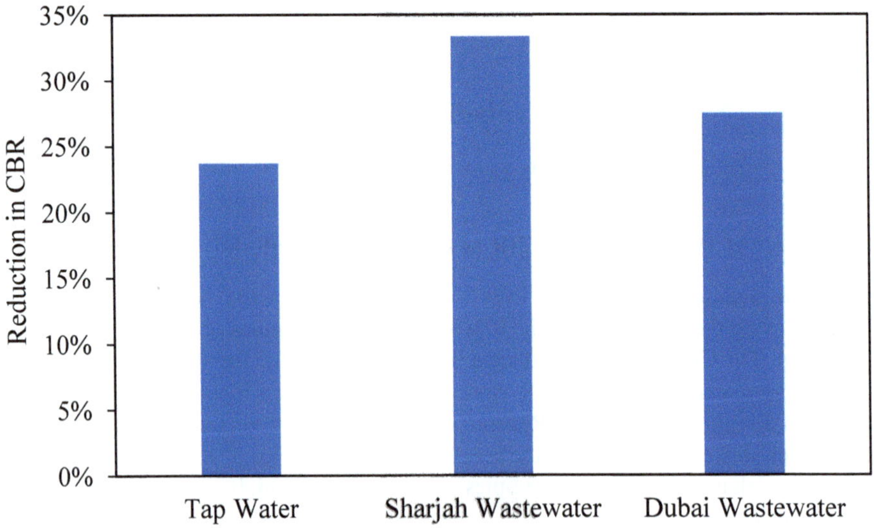

Fig. 4. Reduction in CBR values due to 96 h soaking

5 Conclusions

This study investigated the potential of using treated wastewater as an alternative to tap water in the construction of base course layers in pavements. Unbound granular limestone materials were mixed and compacted with tap water and with treated wastewater obtained from two treatment plants within UAE. Laboratory tests were conducted on the prepared samples and the following conclusions can be made from the results:

- The treated wastewater obtained from Dubai treatment plant contained more sulfate and chloride contents as well as higher total dissolved and suspended solids as compared to both the tap water and the treated wastewater obtained from Sharjah treatment plant. However, both Sharjah and Dubai treated wastewater satisfied all ASTM C 94 specifications and USEPA guidelines.
- The proctor compaction and CBR results of the samples prepared using the two types of treated wastewater were comparable to those prepared using tap water.
- All samples showed a reduction in the CBR values after 96 h of soaking and the reduction was slightly more pronounced in the base course materials compacted and soaked using the Sharjah treated wastewater.
- Treated wastewater obtained from treatment plants can be a suitable alternative for tap water for use in compaction and construction of base course layers in pavement structures.

Acknowledgements. The project is fully funded by Gulf Ecosystems Research Center (Project No. GER14-8) at American University of Sharjah in collaboration with Bee'ah. The authors are highly acknowledging this fund and appreciating this collaboration.

References

Al-Jabri, K., Al-Saidy, A., Taha, R., Al-Kemyani, A.: Effect of using wastewater on the properties of high strength concrete. Procedia Eng. 370376 (2011). doi:10.1016/j.proeng.2011.07.046

ASTM Standard C1602: Standard Specification for Mixing Water Used in the Production of Hydraulic Cement Concrete. ASTM International, West Conshohocken, PA (2012)

Chatveera, B., Lertwattanaruk, P., Makul, N.: Effect of sludge water from ready-mixed concrete plant on properties and durability of concrete. Cem. Concr. Compos. **28**(2), 441–450 (2006)

Chatveera, B., Lertwattanaruk, P.: Use of ready-mixed concrete plant sludge water in concrete containing an additive or admixture. J. Environ. Manag. **90**, 1901–1908 (2009)

FAO, Food and Agriculture Organization of the United Nations: Water Withdrawal and Pressure on Water Resources (2014). http://www.fao.org/nr/water/aquastat/main/index.stm

Ismail, Z., Al-Hashmi, E.: Assessing the recycling potential of industrial wastewater to replace fresh water in concrete mixes: application of polyvinyl acetate resin wastewater. J. Clean. Prod. 197–203 (2010)

Low, G., Ng, K., Ng, W., Tam, C.T., Heng R.B.W.: Use of recycled cement based slurry water for making concrete. Inst. Eng. Malays. **68**(4), 47–55 (2007). http://www.myiem.org/

Sandrolini, F., Franzoni, E.: Waste wash water recycling in ready-mixed concrete plants. Cem. Concr. Res. **31**, 485–489 (2000)

Tsimas, S., Zervaki, M.: Reuse of waste water from ready-mixed concrete plants. Manag. Environ. Quality: Int. J. **22**(1), 7–17 (2011). doi:10.1108/14777831111098444

USEPA Guidelines for Water Reuse. http://www.epa.gov/ (2012)

Hydraulic Conductivity of Layered Compacted Granular Materials Used as Pavement Foundation

Namir K.S. Al-Saoudi[1] and Khawla H.H. Shubber[2(✉)]

[1] Building and Construction Engineering Department, University of Technology, Baghdad, Iraq
namirks@yahoo.com
[2] Engineering College, University of Kufa, Najaf, Iraq
Khawla.shubber@uokufa.edu.iq

Abstract. Base and subbase layers form the structural platform of flexible and rigid pavements. Both layers are commonly made of selected granular material and properly compacted to provide an adequate bearing ability. Improper drainage of these two layers creates various structural defects that ultimately caused failure. Thus, it is of prime importance to evaluate the hydraulic conductivity of these two layers to avoid any structural damage of the pavements.

The current paper is an attempt to evaluate the hydraulic conductivity of several gradation of granular materials in sequences layers usually selected and commonly used as base and subbase course material in pavement construction. The selected materials confirm with SORB specifications (State Commission of Roads and Bridges) issued in Iraq in 2003 for granular base and subbase layers. The tests were performed under constant head condition within the range of validity of Darcy's Law and laminar flow conditions in double layers sequence. Regression analysis of the test results revealed that the hydraulic conductivity is a function of the in place void ratio, mean particle diameter and degree of packing of the granular material. These parameters are presented in the form of equations and can provide approximately quick estimate of hydraulic conductivity of each granular layer.

1 Introduction

Flexible pavement foundation commonly forming from granular material, known as base and subbase courses. In almost cases, these foundation materials should provide adequate drainage to any excess moisture in addition to their function in increasing the structural and load carrying capacity of the subgrade. Highways engineers agree that control of moister and prevention of volume change of subgrade or pumping of fine material in pavement structure is one of the primary and essential tasks to avoid any premature roadway failures. Therefore, the designers started using open-graded, highly permeable layers to overcome the problem of drainage adequacy (NCHRP 1998). So the material specifications should be checked to assure that hydraulic conductivity, strength, load-distribution, and construction stability requirements are met.

L. Mohammad (ed.), *Advancement in the Design and Performance of Sustainable Asphalt Pavements*, Sustainable Civil Infrastructures, DOI 10.1007/978-3-319-61908-8_4

Flow of water through granular material take saturated flow patterns where all voids are filled with water, while in case of presence of water and air in the voids, the patterns will be unsaturated flow. Furthermore, the patterns of flow depend on the source of water and its duration. For practical conditions it is always assumed that completely fully saturated state is not justified and it is not unusual to expect from (1–12)% of air remaining in voids (Reymond and Benno 1975). However, the analysis of water flow through granular materials is easier in terms at fully saturated state.

The aim of the present work is to evaluate and predict the coefficient of hydraulic conductivity of different sequences layers and gradation of granular materials used as base and subbase layers in pavement construction by constant-head hydraulic conductivity tests. The evaluation is based on vertical flow within double layers system, under controlled hydraulic gradient within the range of validity of Darcy's law for laminar flow condition. Hydraulic conductivity tests were carried out in order to demonstrate the effect of stratification, layer thickness, and void ratios of each individual granular materials layer on the overall coefficient of hydraulic conductivity.

2 Materials, Physical Properties, and Methods

In Iraq, the material used in the base and subbase courses must conform to the general specification for road and bridges issued by the Organization of Roads and Bridges, Ministry of Housing and Construction (SORB). Originally (SORB) issued in 2003 and its amendments, these specifications are based on the BS (British Standard), AASHTO (American Association of State Highway and Transportation Officials), or ASTM (American Society for Testing and Materials Standards) specifications, providing a wide choice for selecting materials and equipment. According to SORB there are various types of materials that can be used as base and subbase courses. Such as sand-gravel mixtures, crushed-limestone, crushed-gravel, vibratory-compacted macadam, and stabilized base course with cement, lime, and bitumen, the selection of material types depending on the nature of the site, characteristics of underlying layer-or subgrade materials, available material- equipment, and type of constructional pavement.

The materials used in this Study are sand-gravel mixtures, locally available and currently used in Iraq as pavement construction materials. The mixtures were brought from Al-niba'i quarry near Baghdad capital of Iraq. Table 1 shows the SORB gradation specifications for sand-gravel base and subbase course materials, as specified by ASTM (C-136-84a). Three types of sand-gravel mixtures are selected as materials of base course and other two types are selected as materials of subbase course. The gradation of granular materials used in this research as base and subbase courses compacted layers illustrated in Table 2. In addition to performing grain size distribution to ensure the stratification with SORB specifications, it is essential to specify some physical properties of material necessary to compute the degree of packing of particles within determination effect of granular materials gradation on coefficient of hydraulic conductivity. The degree of packing can range from loose state corresponding to maximum void ratio (minimum dry density or γ_{dry} min.) to very dense state corresponding to minimum void ratio (maximum dry density or γ_{dry} max.). These physical properties are bulk, apparent specific gravity of materials which performed according to the methods described by B.S. (1377) and ASTM (C-126-88) respectively, max. and min. density and corresponding min. and max. void ratio. The results of these tests representing the general physical properties are shown in Table 3.

Table 1. (SORB) specification of base and subbase course material gradation.

Sieve size (mm)	Percent passing by weight for base course materials			Sieve size (mm)	Percent passing by weight for subbase course materials	
	Type A	Type B	Type C		Type A	Type B
50	100			75	100	
37.5	90–100	100		50	95–100	100
25.0	77–95	87–100	100	25	55–90	75–95
19.0	68–90	80–95	92–100	9.5	30–65	40–75
12.5	55–83	70–90	82–95	4.75	25–55	30–60
9.5	47–75	65–85	75–92	2.36	15–42	20–45
4.75	33–65	50–75	60–82	0.30	7–18	14–28
2.0	20–50	33–65	42–70	0.075	2–8	5–15
0.425	10–30	17–40	20–45			
0.180	5–22	10–25	10–28			
0.075	3–10	3–10	3–10			

Table 2. Gradation of material used in study as base and subbase course

Percent passing by weight for base course materials				Percent passing by weight for subbase course materials		
Sieve size (mm)	Type A1 (1)	Type B1 (2)	Type C (3)	Sieve size (mm)	Type A (4)	Type B (5)
50	100			75	100	
37.5	97	100		50	97	100
25	93	90	100	25	85	85
19	83	85	95	9.5	59	46
12.5	79	78	85	4.75	45	35
9.5	70	66	80	2.36	34	28
4.75	59	50	70	0.3	11	14
2.0	48	40	52	0.075	6	5
0.425	20	20	29			
0.18	6	10	22			
0.075	2	4	3			

Table 3. Physical properties of material used as base and subbase course

Test of specific property	Base course materials			Subbase course-materials	
	Type A1 (1)	Type B1 (2)	Type C (3)	Type A (4)	Type B (5)
Bulk specific gravity 23/23 °C	2.649	2.658	2.659	2.663	2.649
Bulk specific gravity (saturated surface dry)	2.671	2.685	2.690	2.681	2.661
App. specific gravity	2.712	2.725	2.731	2.721	2.705
Max. density (gm/cm^3)	2.221	2.239	2.192	2.113	2.130
Min. density(gm/cm^3)	1.529	1.639	1.531	1.570	1.547
Max. void ratio	0.728	0.618	0.731	0.693	0.708
Min. void ratio	0.190	0.185	0.211	0.258	0.241
d_{mean} (Eq. 1)	7.9	8.4	4.9	12.6	12.7

The most reliable equation of mean partial diameter (d) of non-uniform granular materials (such as that used in construction of base and subbase course in pavement structure), is that given as Eq. (1) below which used in calculation of "d" for each gradation (Lindly and Elsayed 1998).

$$d = \frac{1}{100}\sum p_i \cdot D_i \tag{1}$$

where: "P_i" is the percentage of aggregate held between two adjacent sieves; "D_i" is the mean of opening size of two adjacent sieves.

3 Hydraulic Conductivity Principles and Set Up

Ordinarily known that, Darcy's law gives an accurate representation of flow within a porous media for small velocities at laminar flow. Based on constant head principles, Darcy's law donated discharge of fluid through porous media as a function of total head lost per unit length of macroscopic flow path, called hydraulic gradient, "i" across sectional area normal to direction of flow; and coefficient of hydraulic conductivity "K" (Leonards 1962; Bear 1972). While ASTM, 1969, 1986 take coefficient of dynamic viscosity to eliminate the effect of discharge changes resulting from changes in fluid viscosity comply with temperatures changes. In order to ensure the validity of Darcy's low, and to prevention the turbulent flow, the ASTM D2434, and AASHTO T215 specifications suggest employing hydraulic gradients of (0.2–0.3) for materials of loose compactness rating, and of (0.3–0.5) for dense compactness rating, the lower values apply to courser soil. Transition between laminar and turbulent flow is well defined by the dimensionless Reynolds number (Re) (Cedergren 1977). Finally symbol "c" is a parameter depends on the degree of packing (the degree of compaction in highway base and subbase layer) and on the shape of the particles in the porous media

(Roger et al. 1969). Figure 1 below shows testing set up reliant in this work, which is comply with Darcy's law for laminar flow of porous media and agree ASTM D2434 and AASHTO T215 requirements of hydraulic conductivity of the granular materials determination.

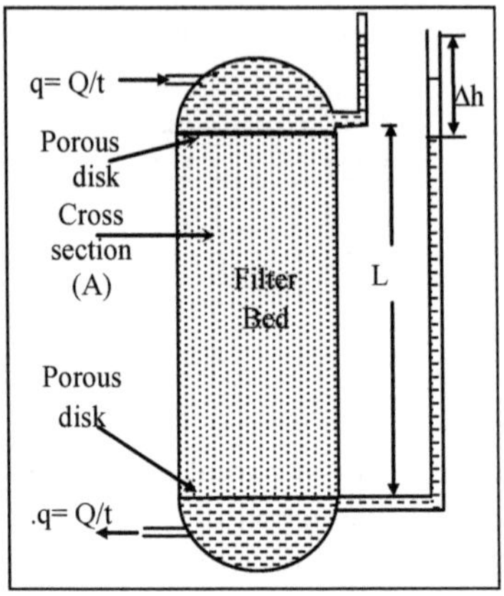

Fig. 1. Test set up

4 Sample Preparation and Testing Procedure

In order to investigate effect of gradation and sequences of compacted layers, samples of different thicknesses ratio (T1/T2) were selected of (0.5, 0.66, 1.0, 1.5, 2.0, 2.5) shown in Table 4. Specific hydraulic gradient selected to be within the validity of Darcy's law for laminar flow (i = 0.2, 0.25, and 0.3) for each different percentage of voids ratio (e_1/e_2) (0.6, 0.75, 1.0, 1.25, and 1.5). Two subbase course materials types A and B and three base course materials types A1, B1, and C compacted provided (T1/T2) ratio listed in Table 5, average mean particle diameter also shown. After the completion of the compaction, water starts to fill gradually and slowly the whole set up by 200 mm height.

Table 4. Selected thickness for layered soil

Thickness of top layer, T1 (cm)	Thickness of bottom layer, T2 (cm)	Ratio of top to bottom (T1/T2)
10	20	0.5
30	20	0.66
20	20	1.0
30	20	1.5
20	10	2.0
25	10	2.5

Table 5. Average particle mean diameter for selected thickness of layered soil

T1 (mm)	T2 (mm)	T1/T2	Layers	D_{ave} (mm)	Layers	D_{ave} (mm)
100	200	0.5	B1/A	11.20	A1/B	11.1
200	300	0.66	B1/A	10.92	A1/B	10.87
200	200	1	B1/A	10.50	A1/B	10.30
300	200	1.5	B1/A	10.08	A1/B	9.82
200	100	2	B1/A	9.80	A1/B	9.50
250	100	2.5	B1/A	9.60	A1/B	9.27
100	200	0.5	C/A	10.03	C/B	10.01
200	300	0.66	C/A	9.52	C/B	9.60
200	200	1	C/A	8.75	C/B	8.80
300	200	1.5	C/A	7.98	C/B	8.02
200	100	2	C/A	7.47	C/B	7.50
250	100	2.5	C/A	7.10	C/B	7.128

To ensure saturation is accrued, the water was left for 24 h prior the beginning of the test. Water from the cylinder outlet was collected at specified interval of 20 s after verifying validity of laminar flow.

5 Results of Experimental Works

The following discussion is based on coefficient of hydraulic conductivity and void ratios obtained for each individual layer of results obtained from previous research on same materials and hydraulic gradients (Al-saoudi and Shubber 2008, 2012). Figure 2 (1–12) show relationship presented in terms of e_1/e_2 ranging from 0.6 to 1.5 versus equivalent coefficient of hydraulic conductivity Ke for different thickness ratios (T1/T2) ranging from 0.5 to 2.5. Charts from No. 1 to 3 in Fig. 2 illustrates the equivalent coefficient of hydraulic conductivity for stratified layer when type A underlies type B1 for hydraulic gradient i = 0.2, 0.25, and 0.3 respectively. When e_1/e_2 ratio approaches 0.75, the equivalent coefficient of hydraulic conductivity controlled by the upper layer (type B1) and the thickness ratio has marginal effect. When e_1/e_2 increases the equivalent coefficient of hydraulic conductivity starts to depend on the coefficient of hydraulic conductivity of the lower layer (type A of low coefficient of hydraulic conductivity (Al-saoudi and Shubber 2008, 2012)). As e_1/e_2 is increase beyond 0.75 the effect of T1/T2 is increased, and the equivalent coefficient of hydraulic conductivity decrease. Otherwise, when type A is underlying type C as shown in charts from 4 to 6 from Fig. 2, for i = 0.2, 0.25, and 0.3 respectively, it can be noted that within the range of e_1/e_2 between 0.6 and 0.75 the equivalent coefficient of hydraulic conductivity increases with decreasing T1/T2 ratio, and the equivalent coefficient of hydraulic conductivity increase with increasing e_1/e_2. Also in that range the equivalent coefficient of hydraulic conductivity (Ke) starts to increase with increasing T1/T2 ratio.

The charts from No. 7 to 9 in Fig. 2 also, demonstrates the coefficient of hydraulic conductivity for case of type B underlying type A1 for i = 0.2, 0.25, and 0.3

respectively. Charts illustrates that when e_1/e_2 is small the effect of the thickness ratios T1/T2 is substantial, and as e_1/e_2 increases gradually, the coefficient of hydraulic conductivity exhibited on increasing trend accompanied by a decrease in the influence of thickness ratio. As the void ratio e_1/e_2 approaches 1.25, the equivalent coefficient of

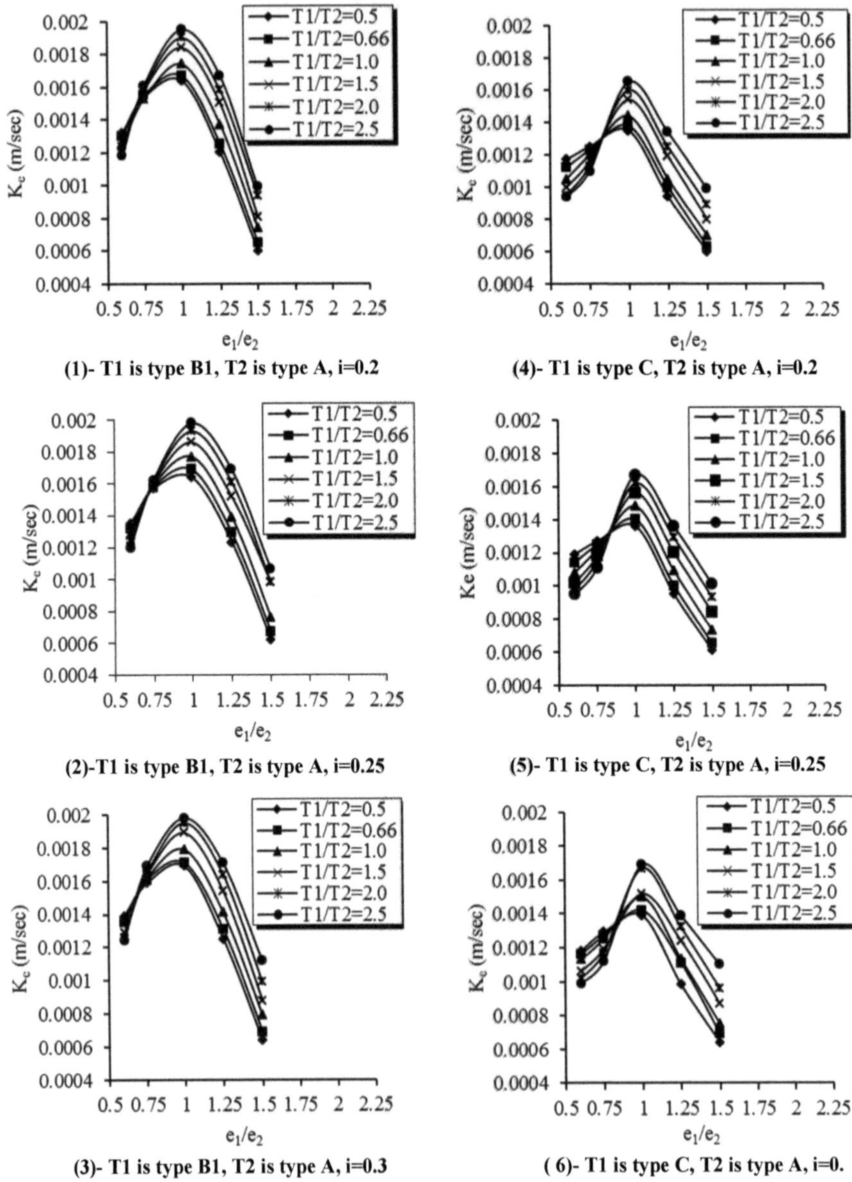

(1)- T1 is type B1, T2 is type A, i=0.2

(4)- T1 is type C, T2 is type A, i=0.2

(2)-T1 is type B1, T2 is type A, i=0.25

(5)- T1 is type C, T2 is type A, i=0.25

(3)- T1 is type B1, T2 is type A, i=0.3

(6)- T1 is type C, T2 is type A, i=0.

Fig. 2. Equivalent hydraulic conductivity for layered granular materials used as base and subbase courses

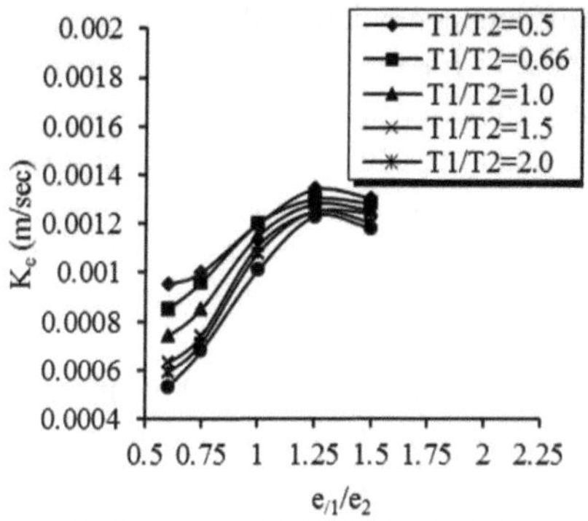

(7)- T1 is type A1, T2 is type B, i=0.2

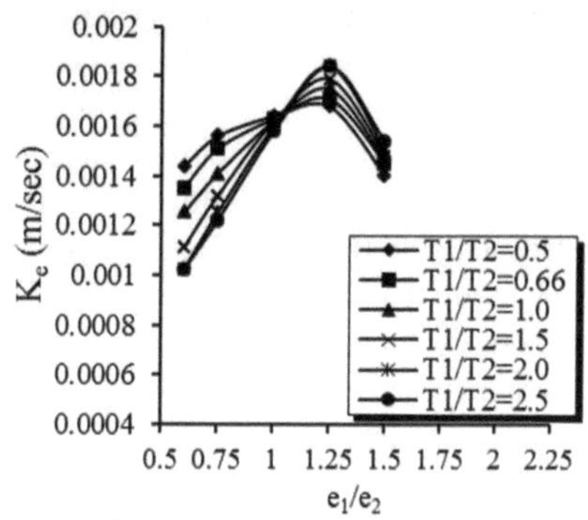

(10)- T1 is type C, T2 is type B, i=0.2

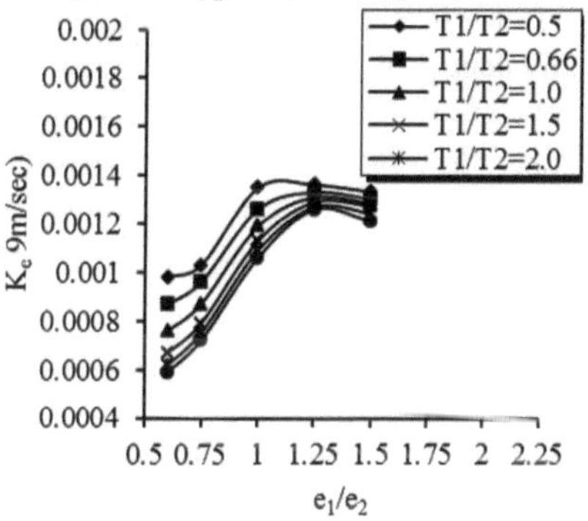

(8)- T1 is type A1, T2 is type B, i=0.25

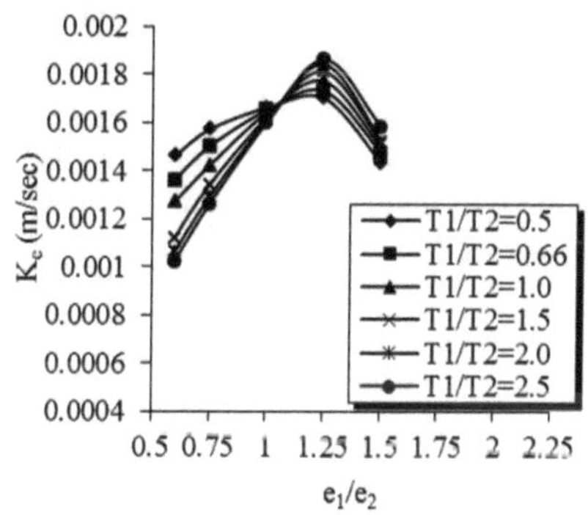

(11)- T1 is type C, T2 is type B, i=0.25

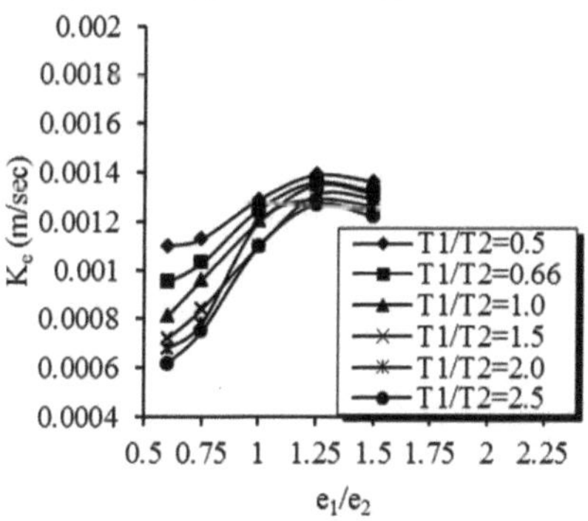

(9)- T1 is type A1, T2 is type B, i=0.3

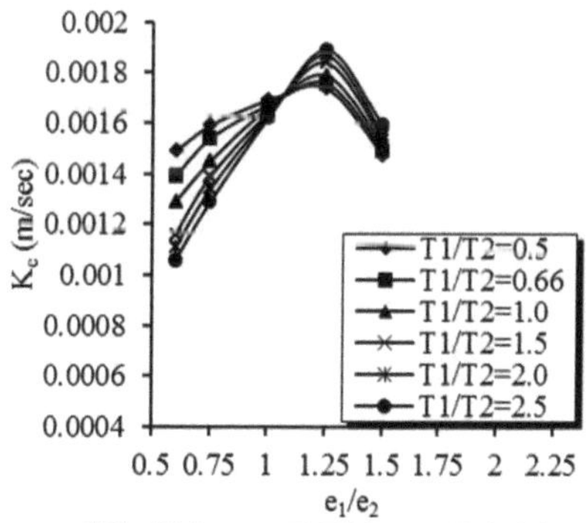

(12)- T1 is type C, T2 is type B, i=0.3

Fig. 2. (continued)

hydraulic conductivity reaches a plateau and any farther increase in e_1/e_2 ratios did not exhibit any further increase in the equivalent coefficient of hydraulic conductivity. These observations were noted for all three hydraulic gradients. The coefficient of hydraulic conductivity for type A1 is less than the values of type B for corresponding void ratios (Al-saoudi and Shubber 2008, 2012).

Charts No. 10 to 12 in Fig. 2 show the equivalent coefficient hydraulic conductivity for stratified soil when type B is underlying type C for i = 0.2, 0.25, and 0.3 respectively, The influence of void ratios take the same trend until $e_1/e_2 \leq 1$. However when $e_1/e_2 > 1$, the behavior of curves appear to have an opposite trend and the equivalent coefficient of hydraulic conductivity starts to decrease with the increasing of e_1/e_2. This can be related to the coefficient of hydraulic conductivity for type C as shown in charts No. 4 to 6, that have low values and narrow ranges between it is maximum and minimum coefficient of hydraulic conductivity.

A specific case of T1/T2 equal to 1 is considered in Fig. 3 with different granular materials stratification for i = 0.2, 0.25, and 0.3. Figure 3 illustrates that maximum equivalent coefficient of hydraulic conductivity is achieved at $e_1/e_2 = 1$ for the cases of B1/A and C/A. when the stratification is change the maximum equivalent coefficient of hydraulic conductivity is achieved at $e_1/e_2 = 1.25$ for C/B and A1/B respectively. This observation is valid for i = 0.2, 0.25, 0.3. Equivalent peak values of equivalent coefficient of hydraulic conductivity are observed at $e_1/e_2 = 1$ and 1.25 for B1/A and C/B stratification respectively. Figure 3 also indicates equivalent coefficient of hydraulic conductivity increases gradually with the increasing e_1/e_2 ratio up to peak values then drop down.

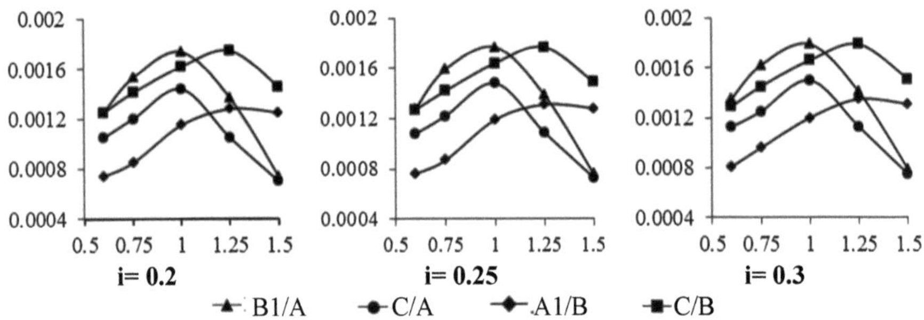

Fig. 3. Influence of different types of granular materials on hydraulic conductivity for thickness ratio (T1/T2) = 1

Linear regression analysis was perform for each stratification system taking equivalent coefficient of hydraulic conductivity (Ke) as dependent variable and the void ratio, thickness ratio, and hydraulic gradient as independent variables. Depending on the data results for the stratification of the higher d_1/d_2 ratio (A_1/B and B_2/A_1) that approach 0.60 for void ratio equal or less than 1 gave the following equation with $R^2 = 0.9275$ for 108 data results:

$$Ke = -0.906 + 0.072i - 0.005(T1/T2) + 0.115(e_1/e_2) + 1.44(d1/d2) \quad (2)$$

Correlation between the actual results and that depended on the Eq. 2 are shown in Fig. 4. Due to the good R^2 that approaching from 1 value, the different between calculated and measured values of coefficient of hydraulic conductivity (Ke) are in

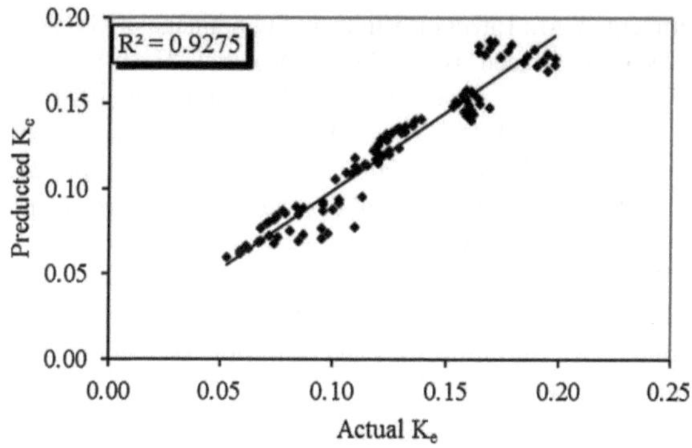

Fig. 4. Actual versus predicted Ke for d1/d2 about 0.6 (Types A1/B and B2/A) depending on Eq. 2.

minimal values. So, in case of same gradation it can be predict the coefficient of hydraulic conductivity (Ke) for any degree of compaction or void ratio, thickness ratio, and hydraulic gradient instead of testing them in laboratory.

6 Conclusions

1. The equivalent coefficient of hydraulic conductivity for the double layer samples depends on the thickness ratio and coefficient of hydraulic conductivity of the individual layers.
2. The equation relating Ke for this case as a function of e_1/e_2, d_1/d_2 and T1/T2 demonstrates that the effect of hydraulic gradient is marginal.
3. In case of type A was underlies type B1 or C, also the hydraulic gradient demonstrates a marginal effect.
4. In general when the void ($e_1/e_2 \geq 1$), Ke starts to decrease gradually.
5. It is difficult to combine the effect of different parameters involved with Ke through the regression analysis. In case of limiting the effect of the mean particle diameter ratio (d_1/d_2) about 0.6 the coefficient of correlation R^2 is raised to 0.927.

References

Bear, J.: Dynamic of Fluid in Porous Media. Elsevier Scientific Publication Company, New York (Chap. 5) (1972)

Cedergren, H.R.: Seepage, Drainage, and Flow Nets, pp. 371, 148–170, 98–100. Wiley, New York (Chaps. 4 & 6) (1977)

Leonards, G.A.: Foundation Engineering. McGraw-Hill Book Company Inc, New York (1962)

Lindly, K., Elsayed, A.: Open-graded highway bases make permeameter set up important. J. Transp. Eng. **124**(2), 144–148 (1998)

Moynahan, J.T., Sternberg, Y.M.: Effect on Highway Subdrainage of Gradation and Direction of Flow Within a Densely Graded Base Course Material "Transp. Res. Bor. 497" "Soil Properties", pp. 50–59. National Research Conical, Washington (1974)

Al-Saoudi, N.K.S., Shubber, K.H.H.: Permeability of some base course materials. In: Proceedings of the First Conference for Pure & Applied Sciences, pp. 223–233 (2008)

Al-Saoudi, N.K.S., Shubber, K.H.H.: Hydraulic conductivity of granular materials. In: 3rd International Conference on New Developments in Soil Mechanics and Geotechnical Engineering, pp. 597–602. Near East University, Nicosia (2012)

NCHRP Synthesis of Highway Practice 96: Pavement subsurface drainage. Road Manag. Eng. J. (1998)

Reymond, N.Y., Benno, P.W.: Soil Properties and Behavior, 1st edn, p. 141. Elsevier Scientific Publishing Company, Amsterdam (1975)

Roger, J.M.D.W.: Flow Through Porous Media, p. 55. Academic Press, New York and London (Chap. 2) (1969)

State Commission of Roads and Bridges (SCRB/R9): General Specification for Roads and Bridges. Republic of Iraq, Ministry of Housing and Construction, Department of Planning and Studies, Baghdad, Revised Edition, Addendum No. 3 (2003)

Extending the Service Life of Bridges Through Proper Compaction of Asphalt Decks

Amir Abd El Halim[1], Ahmed El-Desouky[2], and Abd El Halim[3(✉)]

[1] Stantec, Waterloo, ON, Canada
[2] Department of Civil Engineering, Military Technical College, Kobry El-Kobba, Cairo, Egypt
[3] Program of Infrastructure Protection and International Security, Department of Civil and Environmental Engineering, Carleton University, Ottawa, ON, Canada
AbdelHalim@cunet.carleton.ca

Abstract. Bridges are major components of the highway infrastructure and play important role in the development of societies and their economic growth. The safety and performance of bridges do not depend on design and construction of its structural members alone but also on the degree of protection provided by the asphalt layer added to provide a comfortable ride as well as significant protection against penetration of water and other harming materials to the main structure of the bridge. Compaction of the asphalt deck of the bridge is done only by static rollers while vibratory rollers are not allowed on the deck. However, steel rollers, static or vibratory ones, induce hairline cracks during compaction which allows water, salt and other harmful materials to penetrate through the asphalt deck and reaching the main slab of the bridge which may be made of reinforced concrete or steel members. The presence of water and salt can speed the process of corrosion leading to earlier than expected deterioration of the bridge and in some cases failure of the entire structure. The Ministry of transportation of Ontario (MTO) has used a new asphalt compactor termed Asphalt Multi-Integrated Roller (AMIR) which provided a much tighter asphalt mat with significantly improved permeability performance, higher densities and abilities to compact thicker layers with higher efficiency and less number of passes when compared to current steel rollers. The results led the MTO to include permeability measurements into its Q/A and Q/C requirements for bridge asphalt decks. This paper presents the outline of utilizing AMIR compaction technology on three MTO bridges, the results of field permeability measurements of asphalt sections compacted side by side using AMIR and steel rollers, and laboratory test results. The results showed that the AMIR technology will provide better protection to bridge deck and its steel structure and reinforcement; leading to longer service life with less frequent maintenance. Finally, the paper includes preliminary economic analysis showing the gains and benefits resulting from adopting the new technology to the main parties involved with highway assets.

Keywords: AMIR II compactor · Bridge decks · Permeability · Pavement quality · Economic benefits

L. Mohammad (ed.), *Advancement in the Design and Performance of Sustainable Asphalt Pavements*, Sustainable Civil Infrastructures, DOI 10.1007/978-3-319-61908-8_5

1 Introduction

Since the construction of the early asphalt paved roads, the asphalt industry has become a multi-billion-dollar business and has achieved significant technological advances. However, it has also been experiencing several problems that remain unsolved in spite of decades of intensive research and many attempts to solve them. Early surface cracking of asphalt pavements has been known for decades with the problem of reflection cracking is considered to be one of the most complex and challenging problems in the field (Williams et al. 2015). It is defined as the propagation of existing cracks in the old pavement through the newly constructed asphalt overlay. It is widely assumed that these cracks will propagate from the bottom up to the surface of the new overlay. This problem has been known for several decades and many attempts have been made to solve it. However, despite all these efforts, no reliable solution has been found.

In order to develop a reliable and effective solution to the problem of reflection cracking, one has to deal first with another type of cracking that is initiated at the time of construction (i.e., much earlier than the development of the reflection cracks). This involved a study of the conditions that govern the interface between the new hot asphalt layer being placed and the conventional steel drum roller. The study showed that existing compactors induce hairline surface cracks into the new layer as seen in Fig. 1 and as a result alter the structural behaviour of that layer.

Fig. 1. Construction induced cracks due to steel roller's compaction

The construction-induced cracks on the surface of a newly compacted asphalt layer have been extensively illustrated in Abd El Halim (1985, 1986), Abd El Halim et al. (1987, 1988), and Svec and Abd El Halim (1991). These cracks have also been demonstrated to alter the structural response of the asphalt layer to the applied stresses and to severely reduce the mechanical properties of the new layer. Furthermore, the presence of these built-in cracks at the surface of the asphalt overlay makes any solution to the problem of reflection cracking either ineffective or extremely expensive. Furthermore, the use of vibrating rollers to compact thin lifts of asphalt mats can seriously damage the new layer during compaction. These problems have been well documented by Abd El Halim (1986) and Abd El Halim et al. (1987, 1988). They have also led to the development of a new compaction technology, which overcomes the engineering limitations associated with present rollers, as subsequently described.

1.1 Development of Asphalt Multi Integrated Roller (AMIR)

The concept of relative rigidity was modified and used to explain how construction cracks occur (Abd El Halim 1985). The mathematical formulation and theoretical studies to analyze and explain that phenomenon are described in details in several technical reports summarized in Abd El Halim (1985, 1986) and Abd El Halim et al. (1987). The application of the relative rigidity approach resulted in the design and fabrication of a new compactor, AMIR. It should be mentioned that this compacter was designed, built, and tested with the technical assistance of the Egyptian Ministry of Defence in 1986. It is important to identify the compaction capabilities and limitations of the AMIR compactor. Until recently, the AMIR machine was built as a prototype to prove the main principles of the prevention of construction induced cracks. This AMIR prototype has been upgraded several times since its design and construction in the early 1990's. The AMIR II can provide better degree of manoeuvrability when compared with its earlier version, it has a thicker rubber built, and can complete the required compaction job without sticking to any of the currently employed asphalt mixtures. However, the current compactor has two limitations; the first is due to the width and height of its main frame which sometimes affects its abilities to compact edges of bridge decks which are close to its wall. Figure 2 shows the first self-propelled AMIR I prototype and the current developed AMIR II.

Fig. 2. The Asphalt Multi Integrated Roller (AMIR)

1.2 Review of AMIR Field Trial Results

The AMIR (and subsequently an Australian commercial version of AMIR known as HIPAC) has been used in several field trials and projects in Canada, during 1989–1993 and then in 2003 to 2008, and in Australia during 1997-present. All AMIR compacted sections could be completed using only one AMIR unit, compared with the three rollers currently used by conventional method, and achieves the specifications in less number of passes as reported by several research reports (Vivar and Haddock 2006; Fisher et al. 2012; Abd El Halim et al. 2015). Detailed compaction and performance results observed during the early use of AMIR in Canada compared to conventional compaction equipment are found in numerous technical papers, including Abd El Halim et al. (1993, 1994), Rickards et al. (1999), Abd El Halim and Mostafa (2006) and Abd El Halim et al (2009). However, it is useful to present a brief review of the major conclusions of these field trials.

In general, it was found that the AMIR compacted sections were crack-free with tighter surface texture. Permeability tests completed in Australia indicated that the sections compacted using conventional equipment were between two and four times more permeable than the AMIR compacted sections. Furthermore, AMIR achieved higher specific gravities and better uniformity of the density distribution than those achieved by the static steel vibratory and pneumatic rollers combination. In general, AMIR compactor achieved the desired density after six passes while the combination of vibratory, static steel and pneumatic rollers required 16 to 24 passes. In the field tests carried out in Australia (Rickards et al. 1999), the HIPAC achieved a high compaction only after the first pass, and the required compaction was achieved at a much lower number of passes.

In 2012, the improved AMIR II was used in compaction of two test sections in Ontario, Canada; one section was on the Provincial Highway 28 and the other section was in the contractor firm Tomilson Yard, Ottawa, Ontario. In addition, several tests were performed on MTO paving projects during the summer and fall of 2012 and summer of 2013. Permeability, compaction percentage, surface smoothness and deflection tests were carried out on site while the density tests were performed at Carleton University laboratory on sample cores obtained from the paved sections. The results of the field and laboratory tests showed that the AMIR II compactor was able to achieve a better compacted HMA with less permeability and higher density. Also the deflection measurements of the HMA compacted by the AMIR II displayed an overall average of 10% to 17% lower deflection compared to that of the conventional compacted sections which suggests a longer service life at no extra cost or any changes in the mix design. These better qualities were achieved on the same hot asphalt mixture and with less number of passes. Additionally, AMIR II roller is able to provide surface texture that was tight and crack free leading to significant reduction of the permeability of the asphalt layer. Less permeability will lead to less water penetrating the asphalt layer which will result in less potholing and protection against stripping. The field trials are being monitored on seasonal basis with additional field and laboratory measurements (Abd El Halim et al. 2013). These results encouraged the MTO engineers to try AMIR compactor on a number of bridge projects on major highways in Ontario which is the subject of this paper.

2 The Selected Bridges

To achieve the objectives of this paper, three bridges were included in the testing program. Table 1 shows the details of the selected bridges; Highway 34 Bridge, Holt Road Bridge and Distress River Bridge. The table presents the location of each bridge, number of lanes, date of compaction, and climate conditions. The thickness of asphalt deck of 80 mm was laid and compacted in one lift using the AMIR compactor while the steel roller compacted two-40 mm asphalt lifts on all its assigned lanes. For each bridge, one direction was compacted using the steel roller and the other direction was compacted using the AMIR compactor.

Table 1. Details of the selected bridges

Bridge	Location	No. of lanes	Date of compaction	Climate
Hwy 34	East of Ottawa	2 lanes	Nov. 27, 2014	−5 to −7 °C
Holt Road	Holt Road, Hampton	4 lanes	Sep. 17, 2015	26 °C
Distress River	Magnetawan, North of Ontario	2 lanes	Dec. 7, 2015	0 to −5 °C

3 Bridges Construction and Compaction Process

3.1 MTO Specifications

In Ontario, most asphalt pavements are designed to have an expected service life in the range of 15 to 20 years (Ontario's Transportation Technology Transfer Digest 2010). Ontario Ministry of Transportation specifications recommend the use of static steel rollers only to compact asphalt mixtures on bridge decks. The construction of asphalt pavement on bridge decks is carried out by placing the hot mix asphalt over a water proofed bridge deck by tacking the protection board just before paving. The first operation in the compaction procedure uses a tandem (Class S) roller with a minimum mass of 9 tonnes compacted without vibration immediately after spreading a 40 mm lift of asphalt mix with a temperature not less than 115 °C. The static steel roller passes are increased to induce greater compactive effort to obtain the desired density. Number of passes is often increased because Superpave design method results in mixes which required higher number of passes to meet the required density. The mixes are designed at lower asphalt content with a higher aggregate percentage and aggregate contact compared to traditional mixes. Another reason for the increased number of passes and greater compactive effort is to achieve higher density soon after the initial lay down because asphalt binders 64-34D are much stiffer at a higher temperature. Smoothing out the surface is accomplished with a multi wheeled rubber (class R or V) roller with minimum mass of 18 tonnes followed by a steel roller. A total of 80 mm asphalt thickness is achieved by adding another 40 mm layer after compacting the first lift. The finished product is assumed to be structurally sound and free of defects and surface cracking.

3.2 Highway 34 Bridge Condition

The above procedure was used in the northbound bridge deck of Highway 34 which was paved during the warm temperature of September 2014 as shown in Fig. 3. On November 27th 2014, the Hwy 34 bridge deck paving was carried out using AMIR roller when air temperature at the site was −5 °C to −7 °C. According to Ontario Provincial Standards, during the placing of *Superpave 12.5FC2 mix*, binder paving will not be permitted if the air temperature at the site is less than 2 °C. Surface course ambient air temperature shall be at least 7 °C except for SMA and Superpave *12.5FC2* the ambient air temperature shall be at least 12 °C. Usually bridge deck compaction begins, when the paver spread asphalt mat surface at a temperature of not less than 115 °C. Due to construction staging and racking of asphalt in the joint made AMIR roller to compact South Bound (SB) approach at west ramp end of the asphalt mat surface at 75 °C. AMIR's rolling started from the higher edge of the course and continued towards the lower end as shown in Fig. 4. It was important to realize AMIR ability to compact the mix at that low temperature and achieved a density of 90% in 4 passes only when tested on site with nuclear gauge. Due to asphalt crew laying delays and safety reasons AMIR compacted over all bridge deck and SB approach at east ramp end at temperature of approximately 75 °C instead of compacting at temperature higher than 100 °C throughout the paving project. In spite of lower temperature and cold weather AMIR achieved a density of 90%, tighter asphalt mat surface and well compacted longitudinal joint with less number of passes than observed with the use of the static steel and pneumatic rollers combination.

Fig. 3. The Highway 34 Bridge

Due to the limitations of the current AMIR-II it should be noted that AMIR roller was unable to compact a width of 250 mm due to the projection of barrier wall form work. For this reason a smaller tandem steel roller was used on the edge of the AMIR portion of the bridge deck, Fig. 5. Tandem roller left a trail of broken/ crushed

Fig. 4. AMIR-II compacts Highway 34 Bridge deck

Fig. 5. Steel static roller compacts the edge of AMIR compacted lane, Highway- 34 Bridge

aggregates into AMIR compacted area while compacting lower end, entering and exiting AMIR compacted mat, thus showing the difference of the quality of both compaction techniques side by side on the same asphalt mat. Figure 3 illustrates the lane of which AMIR compactor was used while the lane on the left side of the picture was compacted earlier by the steel roller as mentioned before. The AMIR compacted a thick lift of 92 mm and finished after only 8 passes.

3.3 Holt Road Bridge Condition

Figure 6 shows the Holt Road Bridge on Highway 401 where the asphalt mix "Superpave 12.5 FC2" Traffic Category 2 was used. When AMIR started to compact the asphalt mat good surface texture and no problems were noted as can be noted in Fig. 7. The surface was tight and texture was the familiar AMIR's. The results of AMIR II usage on the Holt Road Bridge will be presented and discussed in more details later.

Fig. 6. The Holt Road Bridge

3.4 Distress River Bridge Condition

The last bridge that AMIR-II was used to compact half of its deck was the Distress River Bridge on Highway 520 near the town of Magnetawan, District of Parry Sound. This bridge is located in the northern parts of Ontario in contrast to the Holt Road Bridge which is considered in southern Ontario. The AMIR compactor finished its share of the project very efficiently and without any problems at all, Fig. 8.

4 Evaluation of the Compaction Quality

On the three bridges, field permeability measurements, in terms of coefficient of permeability (k), were taken on AMIR and Steel compacted lanes, respectively, Fig. 9. To evaluate the permeability of each compaction technology, the National Centre for Asphalt Technology (NCAT) field Permeameter test was used. The NCAT field Permeameter is widely accepted by several researches because of its ease of use, practicality, short time of the test, and non-destructive nature (Tarefder and Ahmad 2015;

Fig. 7. AMIR compacts the first lane, Holt Road Bridge

Fig. 8. AMIR compacts its lane of the Distress River Bridge

Chen et al. 2013). The test is performed by recording the drop in water level in the standpipe over a given time interval. For each of the test sites, five to six measurements were taken and the mean of the measurements are used in the analysis. Also several cores were extracted, Fig. 10, from the approaches to the bridge deck of all compacted bridges for determination of the bulk relative density and relative compaction in the laboratory according to MTO LS-262 (1999) and MTO LS-287 (1996).

Fig. 9. Field measurement of permeability

Fig. 10. Extraction of cores

4.1 Results of Field Permeability Measurements

Table 2 summarizes the results of the field measurements of permeability of AMIR and Steel sections. For Highway 34 Bridge, the overall average field permeability of the AMIR compacted lane was 5.64E−4. The field permeability of the steel finished lane was measured after more than 6 months since construction of the lane. The average permeability on the steel lane was 2.08E−2, which is about 37 times the measured permeability on the AMIR lane. It should be noted that permeability of asphalt pavement is decreased with the pass of traffic as a result of additional compaction due to traffic.

Table 2. Summary of average field permeability measurements (k in cm/sec)

Bridge	Steel	AMIR	Ratio* (%)
Hwy 34	2.08E−02	5.64E−04	3687.9
Holt Road	8.43E−03	4.74E−03	177.9
Distress River	5.60E−03	2.10E−03	266.7

* Ratio of measured permeability of steel and AMIR compacted lanes.

For Holt Road Bridge, the overall average field permeability values were 4.74E−3 and 8.43E−3 for AMIR and steel roller compacted lanes, respectively. It was noted that the ratio of steel and AMIR permeability measurements was 177.9%. For Distress River Bridge, the permeability measurement results showed that the ratio between the permeability of the steel compacted section and AMIR compacted one was 266.7%. The average value of the permeability for the AMIR compacted lane were 2.10E−3. While the corresponding value was 5.60E−3 for the steel compacted lanes. This observation may explain the early deterioration of asphalt roads constructed in regions subject to high precipitation which can also cause early corrosion of steel members of the bridge when exposed to the salt and water penetrating the asphalt layer.

4.2 Results of Densities and Relative Compaction

Table 3 provides the calculated Bulk Relative Density (BRD), and the ratio of the compacted density and the maximum density values for the asphalt (relative compaction) of the asphalt cores extracted from the approaches of the AMIR lane. In

Table 3. Summary of average measured bulk relative density and relative compaction

Bridge	BRD (gm/cm^3)			Relative compaction	
	Steel	AMIR	Ratio* (%)	Steel	AMIR
Hwy 34	N/A	2.293	N/A	N/A	91–93.6 (cores)
Holt Road	2.324	2.377	102.3	92–95 (field)	93–98 (field)
Distress River	2.352	2.368	100.7	N/A	N/A

* Ratio of average densities of AMIR to steel compacted lanes.

addition, during compaction the relative compaction is measured using nuclear gauge as shown in Fig. 11. As compaction completed inspection of the surface texture of the finished sections was completed including taking pictures and videos.

Fig. 11. Measuring of relative compaction behind AMIR

For Highway 34 Bridge, the BRD of AMIR compacted lane was 2.293 gm/cm^3 and the average relative compaction of 92.6% which exceeds the MTO minimum specs. The table shows that AMIR compactor resulted in higher BRD compared to the steel roller. For Holt Road Bridge, the ratio of average BRD of AMIR and steel compacted lanes is 102. 3% while it was 100.7% for Distress River Bridge.

For Holt Road Bridge, the field measurements of the density, conducted using the nuclear gauge, showed that the field relative compaction of the AMIR compacted lanes ranged between 93% to 98% while this range decreased to 92% to 95% for the steel compacted lanes. These readings suggested that while both compaction methods met the minimum MTO requirements, the AMIR compactor provided better compaction with less number of passes and superior surface texture.

4.3 Inspection of Finishing of Asphalt Surface

For Highway 34 Bridge, Fig. 12 shows that the texture of AMIR compacted section was certainly much tighter than steel section with no construction cracks and only few marks left by the edge of the thick rubber belt. Measurements of permeability on top of these marks suggested that the marks were only of a cosmetic nature without any structural effect on the properties of the lane. Figures 13 and 14 show similar findings for Holt Road Bridge and Distress River Bridge, respectively.

It is important to note that the AMIR compactor succeeded not only to meet the minimum requirements as provided by MTO but clearly exceeded them when it was

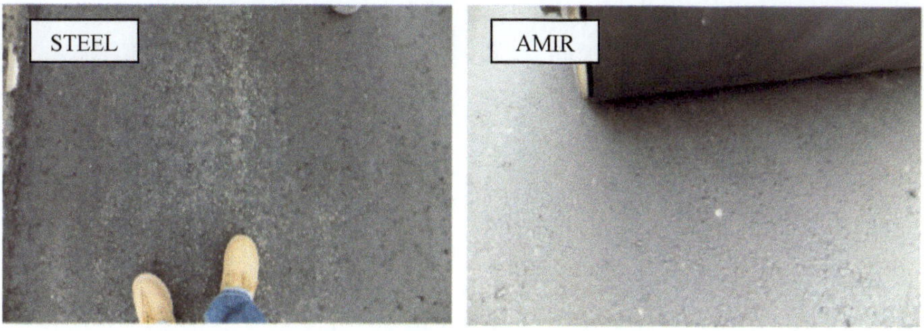

Fig. 12. Finished asphalt deck surface, Highway 34 Bridge

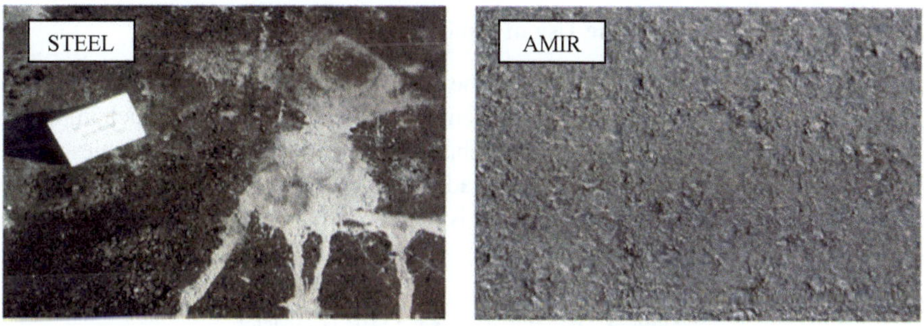

Fig. 13. Finished asphalt deck surface, Holt Road Bridge

Fig. 14. Finished asphalt deck surface, Distress River Bridge

used during the summer months or under cold conditions nearing freezing temperatures during the months of November and December. Several field visits were performed to the three bridges and a number of videos and pictures were taken. The visits and, the videos and pictures suggested that AMIR compacted surfaces maintained superior texture when compared to those compacted with the steel rollers and in some cases the bridge lanes compacted using the steel rollers showed signs of deterioration as the case on Highway 34 Bridge (Fig. 15, taken March 2015).

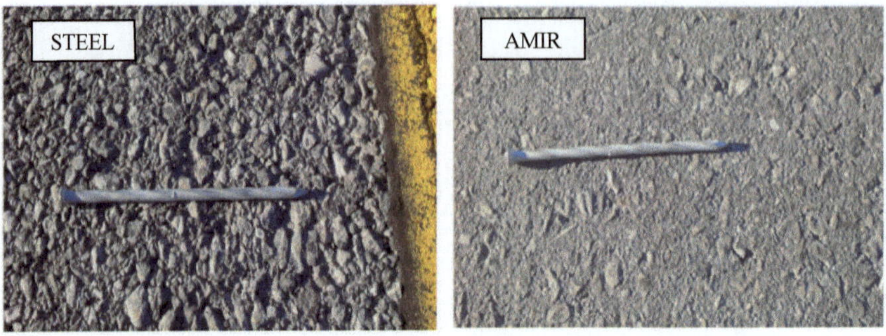

Fig. 15. Close up of surface of Highway 34 Bridge, March 2015

4.4 Economic Benefit of the Use of Amir Technology

The research carried out to date utilizing the AMIR compactor instead of the currently used combination of vibratory steel and pneumatic rollers has results in much improved technical properties of the finished pavements. Clearly, these improved qualities can be of significant economic benefits as presented in Table 4. Firstly, the cost of equipment (asset or rent), operation and maintenance is significantly reduced at least two to three

Table 4. Benefits of using AMIR compactor

AMIR technology	Traditional compaction
Single roller performed all the compaction needs (one operator)	Combination of 3 roller used for compaction (three operators)
Needs one parking spot	Needs three parking spots
Initial purchase cost of one equipment (not more than 40% of the purchase cost of the three compactors)	Initial purchase cost of three equipment
Economical (saves manpower, spare parts, maintenance cost, fuel and time compare to current HMA compaction practice)	Uneconomical compare to AMIR
Less number of passes required to achieve the required field density	Required 5–8 more passes than AMIR to achieve the required field density
Environmental friendly (less carbon emission)	2 times more carbon emission compared to AMIR
Compaction was achieved without using vibration (no vibrating unit)	Vibratory roller was used to achieve required compaction (vibrating unit of vibratory roller is changed every two years)
Better densities and permeability	Equal or lowed densities and much higher permeability
Provided up to 37 times less permeable layer than current HMA compaction practice	Provided up to 37 times more permeable layer than AMIR
Achieved tighter and smoother finished surface	Less tighter and smoother finished surface compared to AMIR

times where one unit of AMIR roller can complete the compaction job instead of the three rollers traditional compaction ones. In addition, AMIR requires one operator instead of three highly trained operators who are required to operate the vibratory or oscillating rollers. Also the cost of spare parts is reduced at least three times when use AMIR compactor. Costs of fuel, oil, and lubricants are reduced five to eight times given that AMIR compactor can achieve the required density in less number of passes, usually six passed compared to 26 passes by the three rollers. These benefits may increase by 100% as AMIR compactor can compact up to 100 mm in one lift while other rollers can compact only up to 60 mm asphalt layer in one lift. Given the number of passes required to finish the asphalt surfaces and the fact that AMIR compactor results in less carbon emissions, AMIR compactor is assumed to be environmental friendly. AMIR compactor achieves tighter (lower permeability coefficient) and smoother surfaces with equal or higher densities than the current compaction rollers. This leads to extended pavement life and less pavement maintenance costs. Tighter pavements finished by AMIR have a great benefit to extend the structural life of steel/concrete bridge elements. Finally AMIR compactor leads to produce asphalt pavements free of surface cracks which will certainly extend the service life of the built pavements and thus preserving the capital investment in road construction.

At the present time the newest development on the AMIR technology has reached a stage by which current steel rollers can be retrofitted with a "kit" that replaces the steel drums of the current roller. The estimated costs of the proposed kit are expected to be a percentage of the total price of a new vibratory roller. Thus, replacing the existing roller with the proposed kit would be more attractive to pavement contractor in terms of initial investment and performance. The preliminary economic analysis showed that benefits will be shared by all stakeholders of the highway system including the Ministry of Transport, the pavement contractors, the road users and of course the tax payers. The benefits clearly will be a function of the quality of the paved road, its added service life, less maintenance and rehabilitation costs and public satisfaction by the end product.

5 Conclusions

The results of the use of the AMIR compactor to compact asphalt decks on three different MTO bridges confirmed the premise of the developed compaction technology and provided a number of advantages which are expected to result in significant improvements of Ontario asphalt pavements. The results presented in this paper support the following conclusions:

1. In all field trials only one unit of AMIR compactor was used to perform the required tasks that at the present time require three different conventional rollers; vibratory, pneumatic, and static steel rollers on highway pavements and two when bridge decks to be compacted.
2. The AMIR provided much tighter surface texture and demonstrated better overall qualities in terms of measured field permeability, density, and crack free surface. And it is important to realize that no vibrations or oscillations are used.

3. AMIR method has achieved the relative compaction with less number of passes and at the same time was able to compact thicker lifts under warmer temperatures.
4. The AMIR compactor provided much better finished longitudinal joints.
5. AMIR compactor has great economic benefits to all highway parties and authorities.

References

Abd El Halim, A.O., Abd El Halim, A.A., Awadalla, M., Hassanin, M.A.: Development of the Asphalt multi-integrated roller field and experimental studies. J. Constr. Eng. Article ID 752674 (2015). doi:10.1155/2015/752674

Abd El Halim, A.O.: Influence of relative rigidity on the problem of reflection cracking. Transp. Res. Rec. **1007**, 53–58 (1985)

Abd El Halim, A.O.: Experimental and field investigation of the influence of relative rigidity on the problem of reflection cracking. Transp. Res. Rec. **1060**, 88–98 (1986)

Abd El Halim, A.O., Mostafa, A.: AMIR and steel drum compactors: evaluating the effect of their compaction on the permeability of asphalt pavements. Transp. Res. Rec. J. Transp. Res. Board. (1967), 173–180 (2006). doi:10.3141/1967-17

Abd El Halim, A.O., Haas, R., Svec, O.J.: Improved asphalt pavement performance through a new method of compaction. In: Proceedings of 17th ARRB Conference, Part. 3, pp. 175–191 (1994)

Abd El Halim, A.O., Phang, W., Haas R.C.: Realizing structural design objectives through minimizing of construction induced cracking. In: Proceedings of Sixth International Conference on Structural Design of Asphalt Pavements, Ann Arbor, USA, July 13–16, vol. I, pp. 965–970 (1987)

Abd El Halim, A.O., Phang, W., Haas, R.C.: Unwanted legacy of asphalt pavement compaction. J. Transp. Eng. **119**(6), 914–932 (1993). doi:10.1061/(ASCE)0733947X(1993)119:6(914)

Abd El Halim, A.O., Phang, W., El Gindy, M.: Extending the service life of asphalt pavements through the prevention of construction cracks. Transp. Res. Rec. **1178**, 1–8 (1988)

Abd El Halim, A.O., Said, D., Mostafa, A.: A protection of the environment through the prevention of surface cracking. Open Civ. Eng. J. **3**, 7–15 (2009). doi:10.2174/1874149500903010007

Abd El Halim, O., Pinder, F., Chelliah, A.R., Abdelalim, O.: Reducing maintenance and rehabilitation costs through the use of AMIR compaction. Civ. Eng. Archit. **1**(3), 51–60 (2013). doi:10.13189/cea.2013.010301

Chen, C., Williams, R.C., Ei, T.A., Lee, H., Schram, S.: Quality control/quality assurance testing for longitudinal joint density and segregation of asphalt mixtures. Constr. Build. Mater. **47**, 80–85 (2013). doi:10.1016/j.conbuildmat.2013.05.007

Fisher, J., Graves, C., Blankenship, P., Hakimzadeh-Khoe, S., Anderson, M.: Factors affecting asphalt pavement density and the effect on long term pavement performance. Kentucky Transportation Center, College of Engineering, University of Kentucky, May 2010 (2012)

Ministry of Transportation Ontario: Laboratory Testing Manual, Test Method LS-287, Rev. No. 16. Method of test for the determination of percent compaction of compacted bituminous paving mixture (MRD Method) (1996)

Ministry of Transportation Ontario: Laboratory Testing Manual, Test Method LS-262, Rev. No. 18. Method of test for bulk relative density of compacted bituminous mixtures (1999)

Ontario's Transportation Technology Transfer Digest. Spring 2010, vol. 16, issue 2 (2010)

Rickards, I., Goodman, S., Pagai, J., Abd El Halim, A.O., Haas, R.: Practical realization of a new concept for asphalt compaction. Transp. Res. Rec. **1654**, 27–35 (1999). doi:10.3141/1654-03

Svec, O.J., Abd El Halim, A.O.: Field verification of a new asphalt compactor, AMIR. Can. J. Civ. Eng. **18**(3), 465–471 (1991). doi:10.1139/l91-057

Tarefder, R., Ahmad, M.: Evaluating the relationship between permeability and moisture damage of asphalt concrete pavements. J. Mater. Civ. Eng. **27**(5), 1–10 (2015). doi:10.1061/(ASCE)MT.1943.5533.0001129

Vivar, E., Haddock, J.E.: HMA pavement performance and durability. Publication FHWA/IN/JTRP-2005/14. Joint Transportation Research Program, Indiana Department of Transportation and Purdue University, West Lafayette, Indiana, pp. 193 (2006)

Williams, R.C., Chen, C., Buss, A. (2015) Reflective crack mitigation guide for flexible pavements. Final report Sponsored by the Iowa Highway Research Board and the Iowa Department of Transportation (IHRB Project TR-641), Institute for Transportation Iowa State University (2015)

Characteristics of Jointed Rigid Airfield Pavement Using Different Material Parameters and Modeling Techniques

Ahmed E. Abu El-Maaty(✉), Ghada M. Hekal,
Eman M. Salah El-Din, and Saad El-Hamrawy

Civil Engineering Department, Menoufia University, Al Minufya, Egypt

Abstract. Rigid pavements have been used broadly in airfield constructions. As the pavement design is expected to deliver acceptable performance along its service time under wide-ranging circumstances. The concept of the load transfer is crucial in pavement design procedures. Many researchers investigated rigid pavement in airfields based on the finite element method. Despite the notable enhancement, significant concerns were overlooked. These simplifications to the developed models may affect the results of the developed models and make them unrealistic.

Sensitivity studies were carried out to investigate the effect of the material parameters of the pavement layers and explore the effect of modeling techniques on the load transfer indicators. These parameters include the influence of the dynamic damping, the joined influence of aggregate interlock and dowel bars at the joint, separation between concrete and base, the bondage of the interface between the dowels and the surrounding pavement and simulation of the gap between the adjacent slabs at the joint.

The development of the three-dimensional model was guided by a set of technical requirements, all of which were met in the final model by using the finite element code ABAQUS (6.13). The verification process presented. Therefore, increases the confidence in its results. Understanding the responses of rigid airfield pavement under such conditions are essential in developing new pavement design procedure and more advanced remedial measure for the present pavements system.

Keywords: ABAQUS · Airfield · Jointed concrete slabs · Load transfer efficiency · Dynamic loading

1 Introduction

Rigid pavements are complex structures that comprised of several concrete slabs where longitudinal and transverse joints are provided between them, which may or may not contain dowel bars. Dowel bars join the slabs to transfer the applied loading across the joint primarily by shear force. The concept of load transfer can be explained by the following: when the load is applied to a concrete slab, stresses and deflections are decreased if a portion of this load is transferred to an adjacent slab. Load transfer is crucial to the Federal Aviation Administration (FAA) pavement design procedure.

L. Mohammad (ed.), *Advancement in the Design and Performance of Sustainable Asphalt Pavements*, Sustainable Civil Infrastructures, DOI 10.1007/978-3-319-61908-8_6

Load transfer is a complex phenomenon that diverges with concrete material, age, environmental conditions, as thermal gradient, shrinkage and moisture content, quality of construction, magnitude and configuration of the wheel load (Ioannides 1997; Jiwon and Keith 2003).

Westergaard established a response model for concrete pavement of a slab-on-grade simulating the paving layer as a thin, infinite or semi-infinite plate founded on a base of springs (Westergaard 1926). It was recommended that a 25% of the load that transfers to the adjacent slab was an appropriate design value (Federal Aviation Administration 2009). The federal aviation administration (FAA) established a new design procedure program named (FAARFIELD). Nevertheless, it resumes considering the 25% of the load transfers to the neighboring slab through the joint as a design constant factor. The measured stress-based load transfer efficiency (LTE) at the National Airport Pavement Test Facility (NAPTF) for the CC2 (construction cycle 2) test items was found significantly higher than 0.25 (Wadkar 2009). Furthermore, the design procedure is based on static analysis assuming that the speed of the wheel is zero. However, load transfer has its effect mainly under moving vehicles. The effect of the dynamic loading on pavement component and the damping effect on the pavement responses are not accounted for in these models. The stress-based load transfer efficiency may not remain constant at 0.25 as it can be influenced by these parameters (Wadkar 2009).

The finite element method was considered an effective method, which is used for rigid pavement analysis since the early 1970s. In spite of the outstanding enhancement, some serious sides of the pavement modeling have been ignored. Previously developed models either have ignored modeling the dowel bar or modeled their action by a beam or spring elements and, consequently, the interface between the dowel and the surrounding concrete is not modeled. Most of the earlier studies used low-grade meshing element, using poor grade element without an appropriate type and integration affect the precision of the obtained outputs and the time which the model takes until it converges. (Abu El-Maaty et al. 2016). Such conditions may disrupt the results obtained at the joint to be unrealistic. The foundation modeling usually as Winkler foundation, the effect interaction with the slab or base was not accounted in the simulation. Also, simulation of the lift-off of the pavements, especially when curling or warping due to thermal gradients was not accounted for. Load transferred through the joints by aggregate interlock effect used to be modeled by shear spring elements; this simplification does not simulate the realistic performance of aggregate interlock action (Joshi 2012; Han Jin Oh 2014; Mackiewicz 2015).

2 Objectives

Sensitivity studies were carried out to investigate the effect of the material parameters of the pavement layers and explore the effect of modeling techniques on the load transfer indicators on pavement responses. These parameters include the influence of the dynamic damping, the joined influence of aggregate interlock and dowel bars at the joint, separation between concrete and base, bondage at the interface between the dowels and the surrounding pavement and simulation of the gap between the adjacent. A 3D, dynamic, finite element analysis was performed using ABAQUS (6.13). Calibration of model parameters is essential for procurement an accurate dynamic behavior of rigid pavements. For developing such efficient model, parameters such as element type, mesh size, interactions between foundation layers, boundary conditions, joint stiffness value and damping parameters, were accomplished in a sequence of steps.

3 Model Description

The pavement system was selected based on a typical rigid pavement constructed at the airfields in Egypt. The developed model comprises two dowels-jointed concrete slabs based on layers of base, subbase and subgrade layers as shown in Fig. 1. To avoid problems related to boundary conditions, the pavement slabs were represented by their full widths of 5.0 m with full lengths of 5.0 m. The base, subbase, and subgrade are formed marginally wider than the slab to permit a better distribution of the stresses and broadened by 0.5 m for each edge of the slab. The two adjacent slabs are linked with 14 dowel bars placed at 350 mm spacing center to center, at mid-height of the slab. The dowel bars are 32 mm in diameter and 500 mm in length. The slab thickness is 340 mm. The slabs lie on top of a 150 mm-thick of the base layer. The subbase thickness is 250 mm, and the subgrade is 2.5 m to guarantee better simulation of subgrade behavior as an approximation of the infinite foundation (Abu El-Maaty et al. 2016).

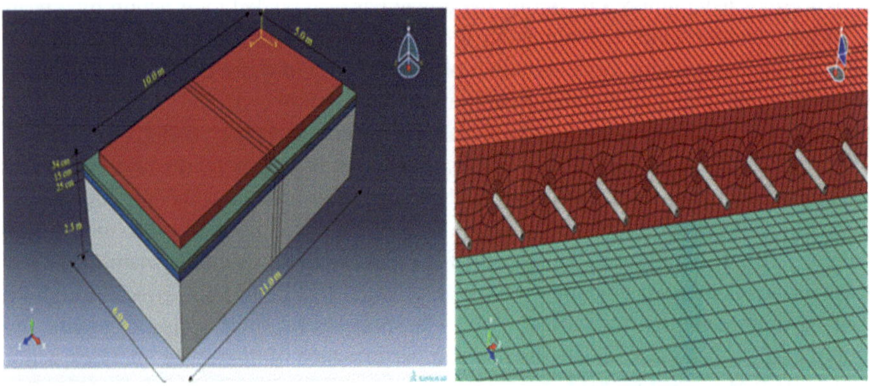

Fig. 1. Three-dimensional model assembly and the detailed section at the joint

3.1 Pavement Material Models

The nonlinear behavior of concrete pavement can be modeled in ABAQUS using many constitutive models available Such as the smeared crack, the brittle cracking, and the concrete damaged plasticity model. For the purpose of this study, the concrete damaged plasticity model is used as it set for applications of dynamic loading and can be in combination with material damping (ABAQUS User's Guide 2013). In ABAQUS, the concrete damaged plasticity model is defined by the following parameters; Dilation angle, Eccentricity, the ratio of initial equibiaxial compressive yield stress to initial uniaxial compressive yield stress (F_b/f_c), the ratio of the second stress invariant on the tensile meridian (K), Viscosity parameter. The values of these parameters were 38°, 0.1, 1.16, 2/3 and zero respectively (ABAQUS User's Guide 2013). Furthermore, both concrete compression hardening and concrete compression damage were used in defining the model in addition to the concrete tension damage and concrete tension stiffening (Jankowiak 2005).

Federal Aviation Administration functions a full-scale pavement test facility concerned with airfield pavement only for research at National Airport Pavement Test Facility (NAPTF). A construction cycle 6 (CC6) includes test pavement and instrumentation layout and materials testing data. Items used in the test are set by the 3-letter sign of MRS (medium-strength subgrade, rigid pavement, stabilized base), then a number and a letter. The numbers (1, 2, and 3) are linked to the objective strength of the concrete pavement (500, 750 and 1000 psi respectively). The density is used to apply the self-weight loading on the concrete (NAPTF-Databases). FAARFIELD program is developed by FAA to help in designing rigid pavement thickness. The program comprises three base courses types of Item P-306 of Econocrete Subbase, P-304 of Cement-Treated Base and P-301 of Soil-Cement Base (FAA 2010). The material of the base layer was represented as elastic isotropic models. Subbase and subgrade layers were simulated as solid brick elements. The input data required to define Subbase and subgrade layers are obtained from FAA report on developing FEDFAA program (Edward and Izydor 2007). Dowel bars were represented as elastic isotropic material models. The properties constants used are listed in Table 1.

Table 1. Material properties of the developed finite element model.

Cases	Parameter	Value
PCC slab (MRS-1)	Modulus of elasticity	3,800,000 psi
	Poisson's ratio	0.15
PCC slab (MRS-2)	Modulus of elasticity	5,700,000 psi
	Poisson's ratio	0.15
PCC slab (MRS-3)	Modulus of elasticity	7,600,000 psi
	Poisson's ratio	0.15
Item P-306	Modulus of elasticity	700,000 psi
	Poisson's ratio	0.2
Item P-304	Modulus of elasticity	500,000 psi
	Poisson's ratio	0.2
Item P-301	Modulus of elasticity	250,000 psi
	Poisson's ratio	0.2
Steel bar	Modulus of elasticity	30,458,000 psi
	Poisson's ratio	0.3
Foundation		
Item p-209	Modulus of elasticity	14,474 psi
	Poisson's ratio	0.35
Subgrade	Modulus of elasticity	4,500 psi
	Poisson's ratio	0.4

3.1.1 The Development of the Finite Element Model

The interface between half of the dowel bar in a slab and concrete has been demonstrated as a perfect bond and the other half in the neighboring slab can move along the dowel bar's axial direction. The tangential behavior of the dowel is demonstrated using Coulomb frictional contact between the surfaces. The friction coefficients were taken as 0.3 for the perfectly bonded side and 0.05 for the coated side of the dowel. Separation is permitted between the surfaces (Samir et al. 2003). The normal behavior of the load transfer device is displayed by using hard contact pressure definition between the two surfaces. The tangential behavior of surface between the slab and base was demonstrated as Isotropic Coulomb friction. The friction coefficient between the slab and base interface was assumed 1.5 (Huang 2003). Loss of contact between slab and base is modeled using normal hard contact that allows the surfaces to separate after coming in contact. No separation is permitted between foundation layers. The interaction between the upper of the subbase and the bottom of the base, the lower surface of the subbase and the upper surface of subgrade were simulated by the use of hard contact interface elements and isotropic Coulomb friction model, the coefficient of friction was assumed 1.5. The sides between the two neighboring pavement slabs at the joint are considered having a zero gap with a Coulomb friction of 0.01 imitating a restricted aggregate interlocking.

Meshing irregularities in the developed model in a non-uniform manner can generate stresses does not exist in actual assemblies. Wedge elements with very fine meshing were designated for the dowel bars to guarantee the consistency in mesh distribution around dowel bar; the fine mesh allows accurate calculation of the interaction stress evolution around the dowels. First order 6-node linear triangular prism elements are used for the dowel bars meshing. Eight-node linear continuum three-dimensional brick element (C3D8R), reduced integration, and hourglass control available in ABAQUS (6.13) are used for meshing the concrete slabs. Realizing that the joints, the area around the dowels and loading path are critical stress regions that can initiate pavement failure, a refined mesh was developed in these areas, to capture the flow of stresses precisely around the dowel bars. All layers of the pavement (base, subbase, and subgrade) are imitated with the identical element type to conserve the continuity of nodes between consecutive layers. Figure 2 shows the cross section at the pavement joint and its meshing details (ABAQUS User's Guide 2013).

The bedrock was assumed deep enough to mimic semi-infinite boundaries. All degrees of freedom were constrained at the bottom of the subgrade layer. The edges of subgrade were restricted in their Y-direction and also at the edges of the base as well as side surfaces of the subbase. As the pavement may lose the contact with the base, no external constraints are used at the concrete pavement whose stability is conserved by its interaction with the base and initiation of the slab's own-weight. The dowel bars were linked to the slab by the interaction properties and stabilized by their weight; no further boundary conditions were applied.

3.2 Aircraft Loading

The moving tire is to be demonstrated as tire imprint area to characterize a smooth pavement surface. Traditionally, the contact stresses at the wheel–pavement interface is assumed to be equivalent to the tire pressure and is distributed on a regular basis in a

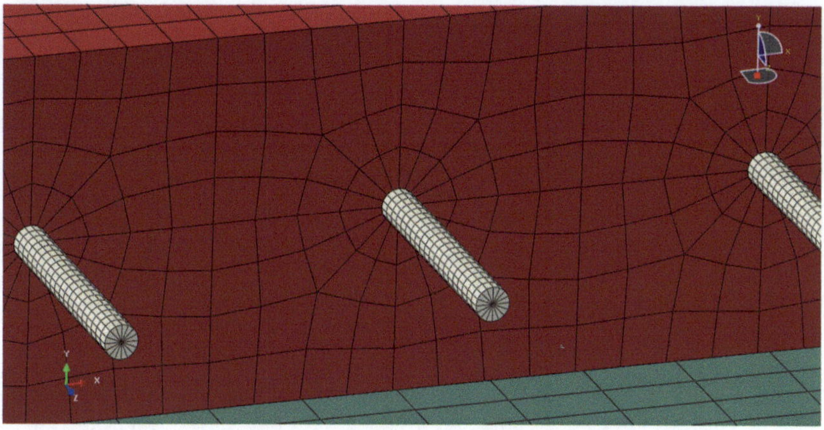

Fig. 2. Meshing details at the joint.

rectangular contact zone. The contact area used in this study was considered using the Portland Cement Association technique (Huang 2003). This technique assumes that the wheel-pavement contact area "Ac," that considered by dividing the load on each tire by the tire pressure, to be a rectangular figure with 0.8712 L length and 0.6 L width. The next equation calculates the length of the contact area (L):

$$L = \sqrt{\frac{A_c}{0.5227}} \quad (1)$$

The F-15 is used for the parametric cases in this study. The main characteristics of the aircraft are shown in Table 2 and Fig. 3 (FAA 2010). In this study, transient fidelity approach, which founded on the notion of moving the tire print patch at successive locations along the pavement for each step time, are used. Damping effect is considered in both analysis types. ABAQUS offers "Rayleigh" model to simulate the damping effect. The pavement damping is mainly stiffness proportional and hence the first term is neglected. Here, the Rayleigh coefficient β is considered 0.2. Previous studies show that dynamic stress-based LTE is not sensitive foundation damping and hence not used in this model (Xinhua and Xiaochun 2010).

Table 2. Main features of the aircraft.

Aircraft	Tire pressure (MPa)	Footprint area (mm^2)	Tire contact length (mm)	Tire contact width (mm)	Dual spacing (mm)	Tandem spacing (mm)
F-15	2.344	61290	353.4	220.8	0.0	0.0

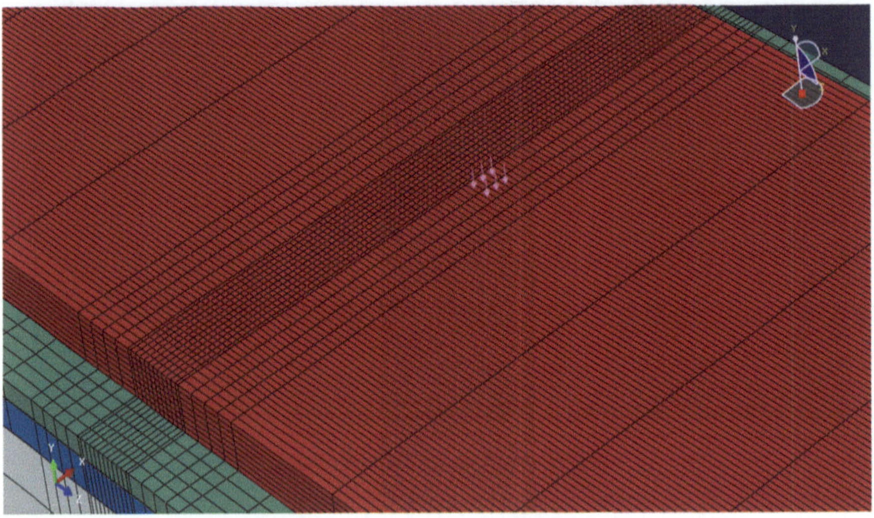

Fig. 3. Single loading of F-15 aircraft.

3.3 Verification of the Model

With the advances in programs' user interface, most finite element software can be used as a "black box" by the operators without wide knowledge of FEM. Therefore, the FEM packages could be misused resulting in what is termed "garbage in, garbage out" simulations (Liu 2014). The most reliable method of investigating the accuracy of the theoretical model is to compare its results with field measurements for the same structure under identical loading conditions. The developed model verification process was achieved using NAPTF field data. A comparison was made between the effects of load transfer at the joint in a full-slab finite element model and the full-scale tests data available from testing of the CC2 test strip at FAA's NAPTF facility. (NAPTF-Databases). The dual tandem wheel configuration of the National Airport Pavement Test vehicle (NAPTV) is simulated in ABAQUS by applying the load to a set of elements covering the loaded footprint area in two-step movement across the joint as shown in Fig. 4. The maximum principal stress responses in wheel position 1 and position 2 were considered in this analysis and shown in Fig. 5. The average stress based LTE of transverse joints from test data was found 0.47 for MRS (Wadkar 2009). The stress-based LTE is calculated when the load is located at the edge of each slab:

$$\begin{aligned} \text{Stress-based LTE position 1} &= \frac{\text{stress at loaded slab}}{(\text{stress at unloaded} + \text{stress at loaded})} \\ &= \frac{2.89655}{(2.89655 + 3.08708)} = 0.4841 \end{aligned} \tag{2}$$

$$\text{Stress-based LTE position 2} = \frac{2.4541}{(2.4541 + 2.729)} = 0.4734 \tag{3}$$

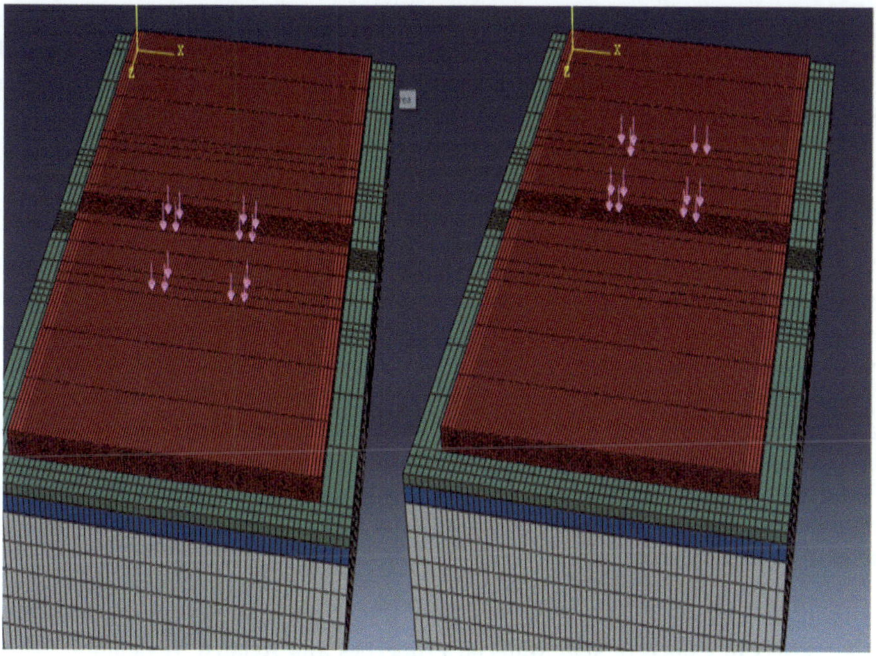

Fig. 4. The two-step movement of the load across the joint.

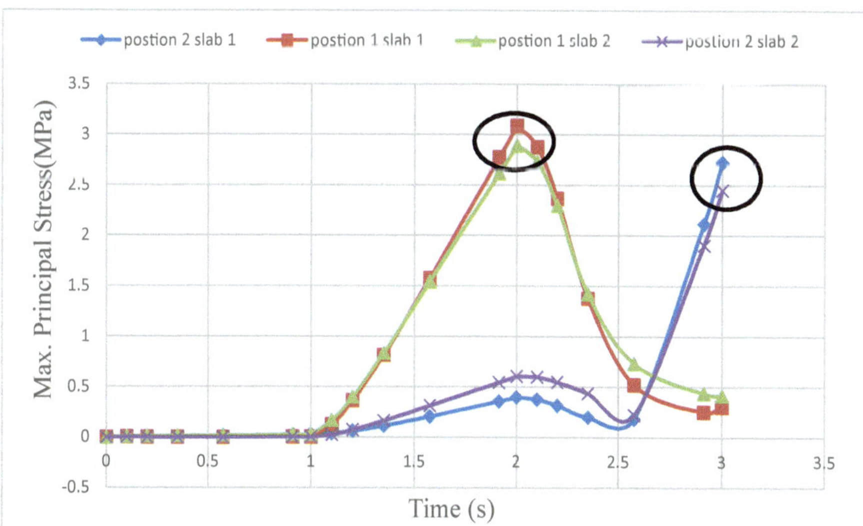

Fig. 5. The maximum principal stresses across the joint during load moving from position 1 to position 2.

The computed stress-based LTEs were found to be similar to the average stress-based LTE from the test data, which was 0.47 for MRS section under 4-wheel dynamic gear configuration and the difference between the model results and field data is in an acceptable range.

4 Analysis of Results

The results of this study primarily concentrate on viewing stresses around deformed dowel hole at the location of maximum stresses at the end and beginning of load application for each step of the entire analysis history. They also concentrate on viewing stresses and deformation at the critical edge of the loaded and unloaded slab. The history of a specific variable was plotted against the time of the moving loads from a specific point to the joint edge. Sensitivity studies were made to evaluate the effect of different modeling and material parameters on the response of rigid airfield pavement. These effects were mainly assessed using load transfer indicators. This section describes the Investigation of pavement response under the impact of some modeling related parameters, those are:

4.1 Effect of Concrete Slab Material Properties and Related Modeling Parameters

In this section, the pavement response is examined for concrete moduli of elasticity values of MR1, MRS2 MRS3 and plastic case that were described previously. The base used for this case was Item P-306, and the foundation setting was the "very low" case, which allows the responses of pavement to appear more visible and easier to observe and analysis. The slabs are loaded using a single wheel aircraft at the joint and the load transfer efficiencies are calculated using the stresses and strains at the loaded and unloaded slabs. A single wheel aircraft, F-15, with an edge loading case is used for the analysis.

The analysis of the results shows that the increasing in the pavement modulus of elasticity increases the stresses but significantly reduces the strains, which follows Hooke's law:

$$E = \sigma/\varepsilon \tag{4}$$

where E is the modulus of elasticity, σ is the stress, and "ε" is the strain. In the plastic case, the deflection is increased and its load transfer efficiency is decreased significantly. The deflection based load transfer is decreased with the increasing the slab modulus of elasticity and the stress based load transfer is increased with the increasing the slab modulus. However, the plastic case, which has the lowest modulus of elasticity, was found to be the highest value in stress LTE. This result may be attributed to the plastic behavior itself. Cracks may increase the interlock effect, which increases the stress-based LTE. The results show that the stress based load transfer efficiency is not sensitive to surface layer modulus of elasticity. On the other hand, the impact of the modulus of elasticity of the PCC layer on deflection based load transfer efficiency is quite notable. Figures 6 and 7 summarize the behavior indicators observed in these models.

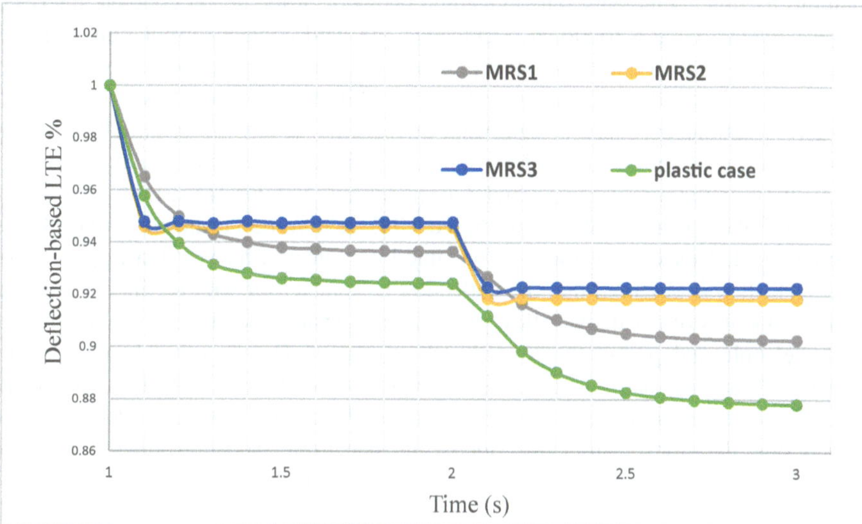

Fig. 6. Deflection-based LTE across the joint with concrete material change

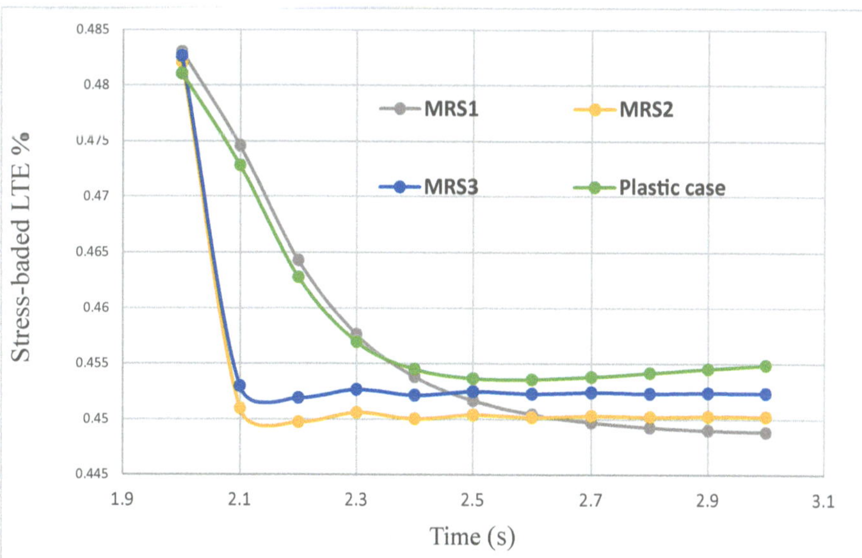

Fig. 7. Stress-based LTE across the joint with concrete material change

4.1.1 Bondage Condition at the Interface Between the Dowels and the Surrounding Concrete

In this case, the focus is on the transverse interaction between the dowel and surrounding concrete, which subjected to extensive contact or bearing stresses and additional friction bond in the longitudinal direction due to the normal contacts and Coulomb friction coefficients. The tangential behavior of the dowel-concrete interface can be modeled using a frictional coefficient. The hard contact formulation is a rule used in ABAQUS to define the contact pressure. Separation is not allowed between the surfaces on the tied side. The change in the bond of dowel bars model using "not allow separation" option, produces a significant stress drop across the joint. This finding establishes the efficiency of dowel bars in holding joint cracking, which may develop to joint faulting. In the plastic case, which has the lowest modulus of elasticity was found to be the highest value of the stress based LTE. It may be attributed to the plastic behavior itself. Cracks may increase the interlock effect, which increases the stress based LTE. It was observed that the deflection LTE increases with the increase of the concrete modulus of elasticity. Allowing separation enable the dowel to slide through the dowel hole and permit the loaded slab to deflect. On the other hand, for "no separation" case, the slab is slightly restricted form deflection. The restrictions of the dowel bar increase the dowel slip in the unbounded side and restrict the dowel slip in the tied side, therefore, restrict this side of the slab from deflection. Figures 8 and 9 summarize the behavior indicator observed in these models.

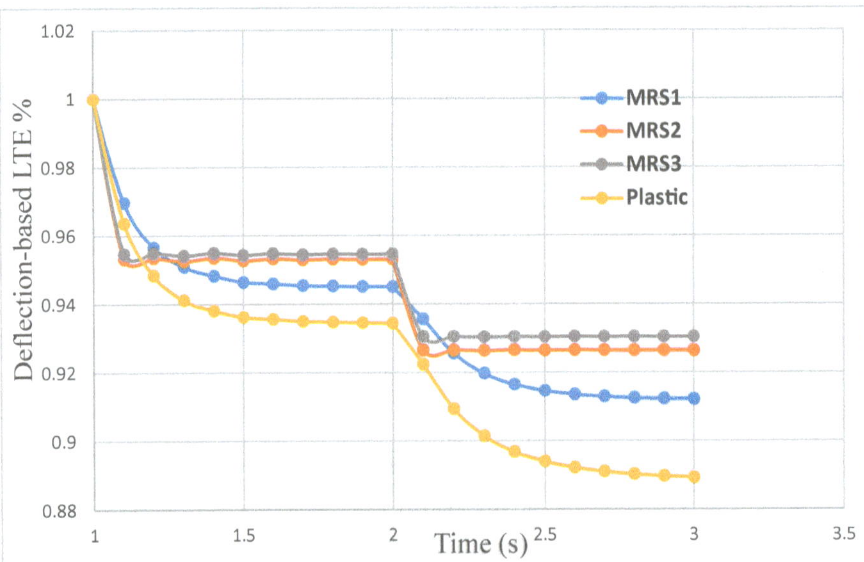

Fig. 8. Deflection based LTE according to dowel bondage change

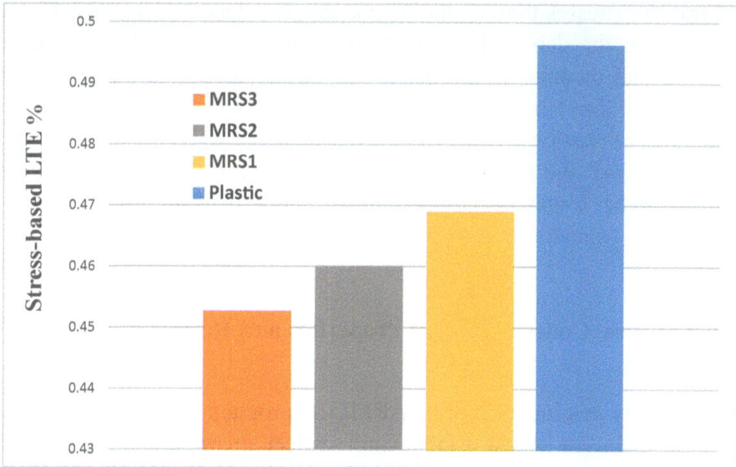

Fig. 9. Stress based LTE according to dowel bondage change

4.1.2 Combined Effect of Aggregate Interlock and Dowel Bars

A modified model of the concrete pavement has been developed to investigate the impact on rigid pavement load transfer indicators due to the combined effect of aggregate interlock and the dowel-concrete interaction as load transfer mechanisms. The interlocking mechanism has been modeled by a zero-gap opening and a Columb friction of 1.5 to simulate a restricted aggregate interlocking.

Results show a significant decreasing in the stresses and strains across the joint. Zero gap effect with the interlocking bond makes the adjacent slab response slightly as

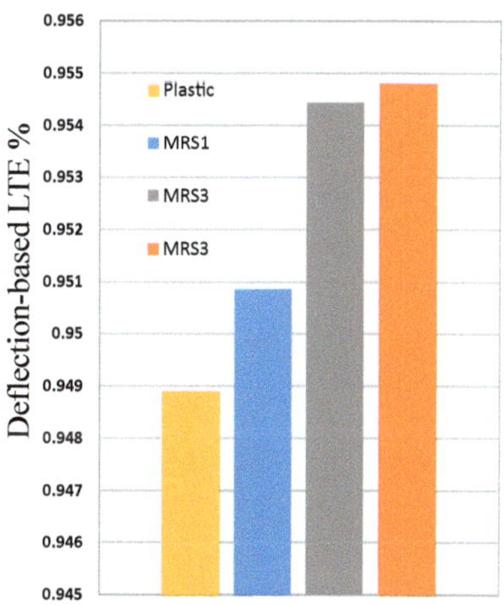

Fig. 10. Deflection based LTE across the joint due to joint stiffness change with pavement material

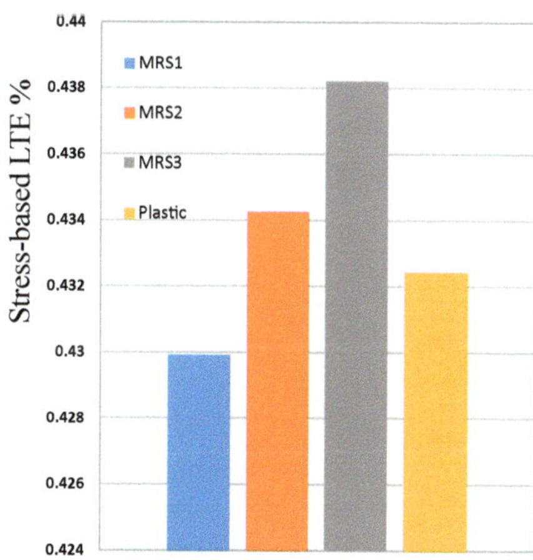

Fig. 11. Stress based LTE across the joint due to joint stiffness change with pavement material

a one and transfer stresses through the base layer. The deflection-based LTE increased with the increasing the slab modulus of elasticity although the deflection reduced with the rise of the slab modulus of elasticity. The stress-based load transfer is increased with the increasing the slab modulus. Also, the stresses values increase with the rise of the slab modulus of elasticity. However, the plastic case, which has the lowest modulus of elasticity However, the stress based LTE, was found to be in a higher value from what supposed to be. It may be attributed to the plastic behavior itself. Figures 10 and 11 summarize the behavior indicator observed in these models.

4.2 Effect of Base Course Material Properties and Related Modeling Parameters

Three items usually that included in FAARFIELD program are used as base courses in this study. The slabs are loaded using a single wheel aircraft at the joint, and the load transfer efficiencies are calculated using the stresses and strains at the loaded and unloaded slabs. A single wheel aircraft, F-15, with an edge loading case is used for the analysis. Figure 12 the behavior indicator observed in these models. The changing in the base course modulus of elasticity has an insignificant effect on either stresses or strains magnitudes or distributions. Therefore, the deflection based load transfer efficiency is not sensitive to base layer modulus of elasticity. It is observed that stress-based LTE is not sensitive to the modulus of elasticity of the base layer. As seen from the figure, the LTE indicators almost remain the same with the change in elasticity modulus of the base layer.

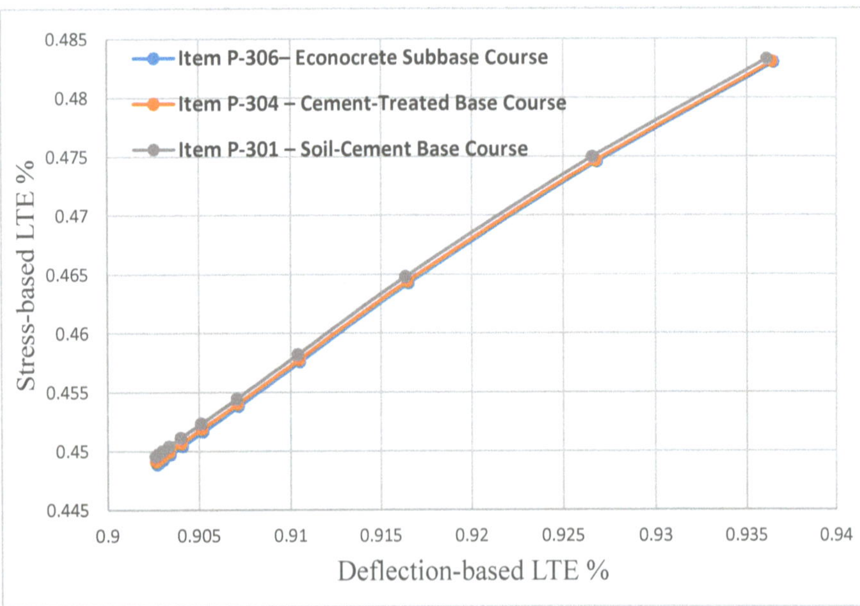

Fig. 12. Stress and deflection-based LTE across joint with base material change.

4.2.1 Separation Condition Applied Between Concrete and Base (Allowed, Not Allowed)

The effect of interaction with the slab or base is important to understand the rigid pavement responses; especially taking into account the impact of the lift-off of the pavements due to the thermal gradient. The allowance of the surfaces to separate after coming in contact, for this sensitivity study, is restricted. Three different bases stated used earlier are the concern of this case. The Columb friction coefficient between concrete and base was fixed. The observations showed that both of the stresses and of strains across and at the top of the joint is reduced. Also, the stresses on the lower side of the slab and upper side of the base layer are reduced. The base layer plays in this case as a device of load transfer, which increases the bond between adjacent slabs. Therefore, there was a decrease in the stress based LT across the joint. It was observed that the stress based LTE is significantly reduced compared to changing the base material only case and decreased with the increase of base layer stiffness. With the rise of base layer's modulus of elasticity, the deflection based LTE increases, which is attributed to the increase in layer resistance to slab deflection because of the bondage between them. Figure 13 summarizes the behavior indicator observed in these models.

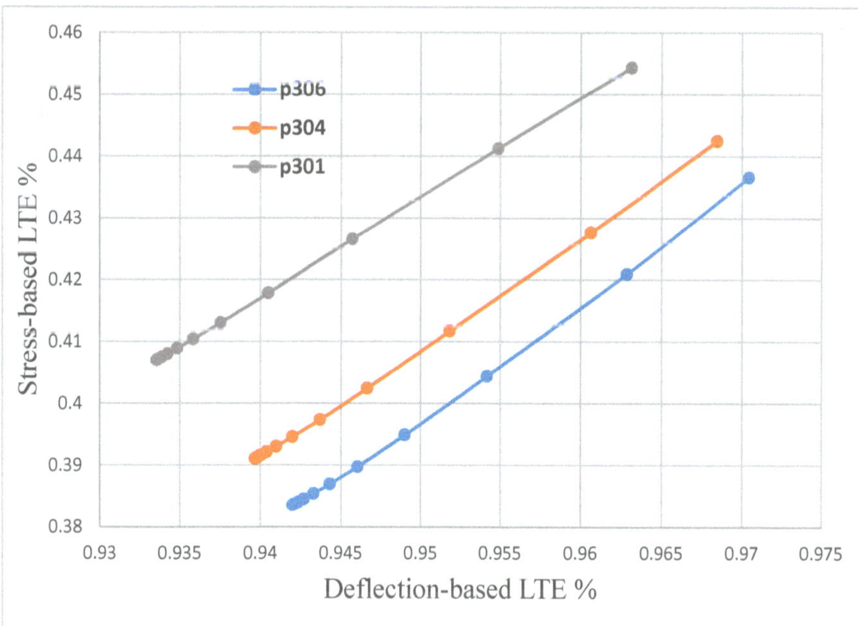

Fig. 13. Relation between LT indicators with base "separation" condition change.

4.3 Modeling the Spacing Between the Adjacent Slabs at the Joint With/Out a Gap

Many factors affect the alignment of dowels assemblies; one of them is the location and width of saw cut in contraction joints about dowel location. The wider joint gap permitted within pavements that containing dowel bars would consequence in dowel misalignment. It is necessary to determine the effect of varying joint opening spaces and to predict the joint opening behavior. A refined model of dowelled joints with different openings is set for this sensitivity study. Four model cases in are considered for calculating LTEs efficiencies. Those openings are zero, four, eight, 12 mm, which are used in this case.

The change in the joint gap has a minimal effect on the magnitudes or the distributions of the stresses and strains values induced in the slab. Deflection-based LTE is decreased with the increase of the joint opening. It can be attributed to the little allowance that opening permits the slab to deflect due to the reduction of the joint stiffness. This decrease is credited to the loss of the aggregate interlock effect across the joint. The zero-gap case was found the highest in both stress and deflection based LTE. However, the change between zero gap case and other cases can be attributed to the change of modeling techniques from non-gap interaction to the interaction with a gap. It is observed on the other indicator, stress-based load transfer, which significantly decreased from the zero-gap case compared to with-gap cases. However, the stress based load transfer slightly increases with the rise of the increasing gap. In a previous study (Wadkar 2009) suggest that the dynamic effect and the lag of load transferred through the joint due to the gap may be behind the increase of the LTE. Figures 14 and 15 summarize the behavior indicator observed in these models.

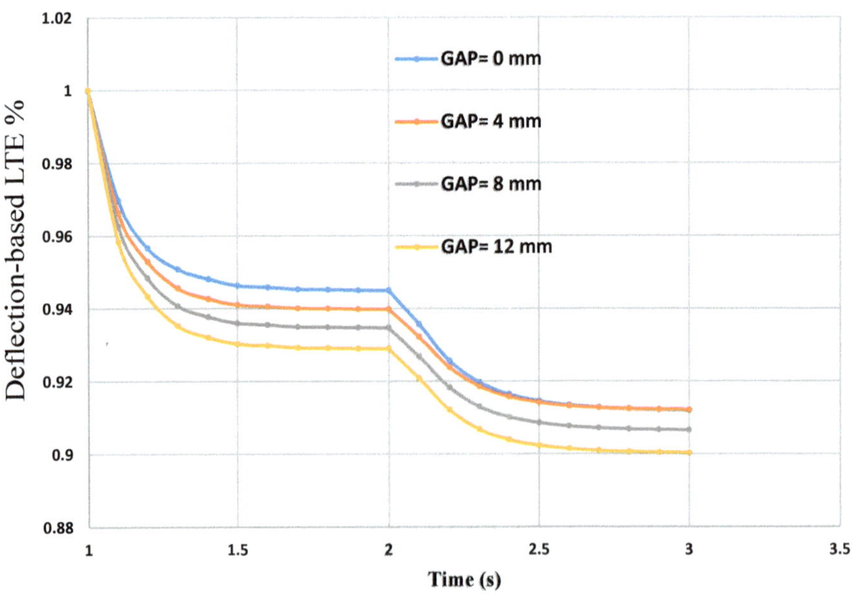

Fig. 14. Deflection based LTE according to gap opening change

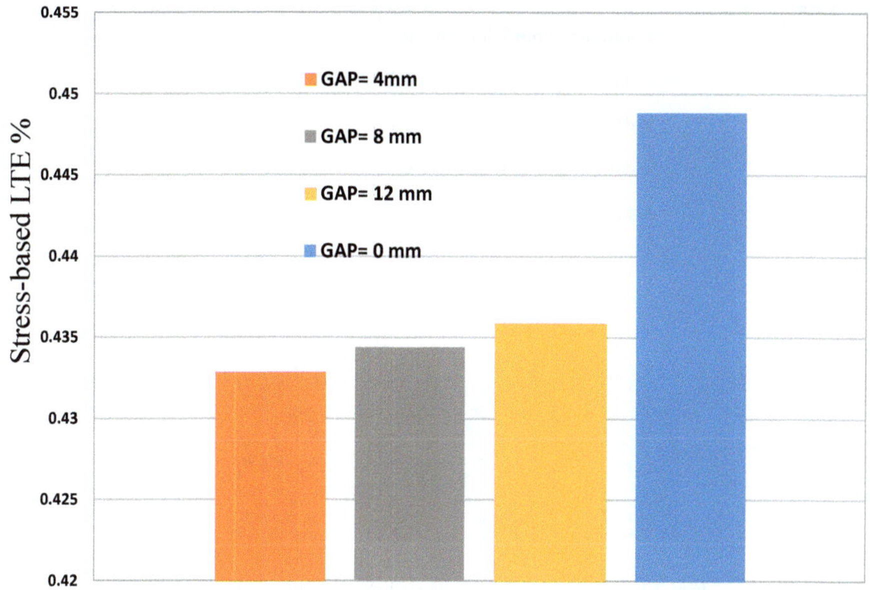

Fig. 15. Stress based LTE according to gap opening change

4.4 Effect of the Dynamic Damping Factor of the Pavement

A sensitivity study is performed to understand the impact of dynamic damping on load transfer indicators. Ranges of stiffness proportional damping coefficient ‘β’ values from

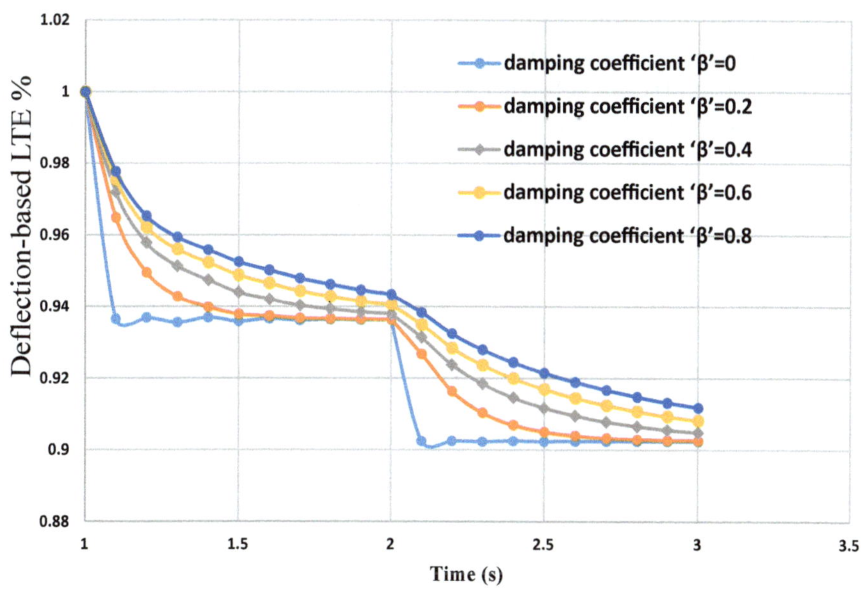

Fig. 16. Deflection based LTE according to dynamic damping

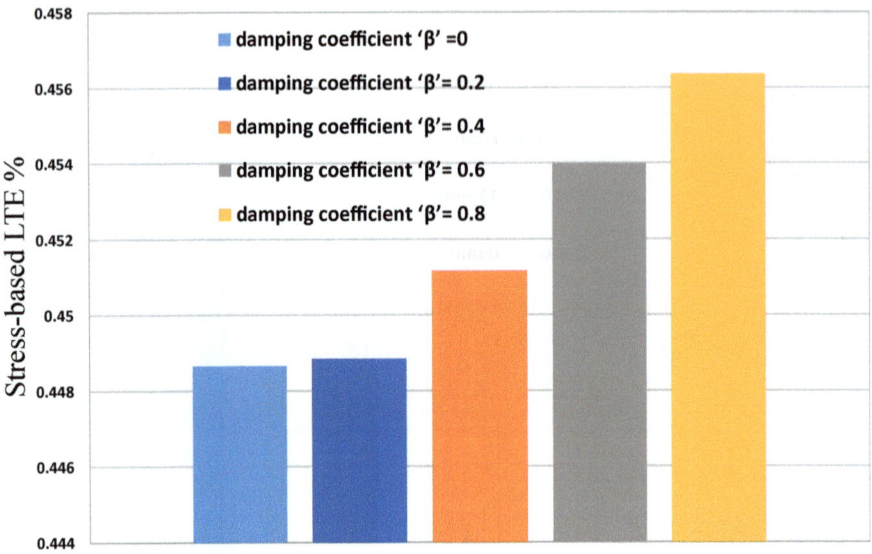

Fig. 17. Stress based LTE according to dynamic damping

0.0 to 0.8 s are selected. Results show a significant decreasing in stresses with the increase of the dynamic damping factor at the same aircraft speed and configuration. In addition, a decreasing in pavement deflection occurs with the rise of the dynamic damping factor. However, the decreasing in pavement deflection was less significant compared to the effect on the stresses and even can be neglected. The dynamic the stress based LTE at the joint increases with pavement damping. Deflection-based LTE is decreasing with the increase of the dynamic damping factor at the same aircraft speed and configuration. The stress-based LTE values were found to be closer to 0.5 for higher pavement damping values. Figures 16 and 17 summarize the behavior indicator observed in these models.

5 Conclusion

This research focused on the influence of material parameters and modeling characteristics on the load transfer efficiency indicators of rigid jointed pavement in the airfield. The following deductions were drawn based on the analysis and observations of the developed 3D finite element model which present valuable insights on the behavior of the slab-dowel system when subjected to loading:

1. At the same aircraft speed and configuration, a significant decrease in pavement stresses and deflections were observed with the increase of the dynamic damping factor. The dynamic stress based LTE at the joint increased with pavement damping. The dynamic stress based LTE values were closer to 0.5 for higher pavement damping values.

2. The change of base layer material has an insignificant effect on LTE indicators. However, Separation condition applied between the pavement and the base layer has a significant impact on LTE indicators and makes changing base's material matter.
3. Changing joint-gap size had a negligible impact on the magnitudes or the distributions of the stresses and deflections values on the pavement. The deflection based load transfer efficiency was decreased with the increase of the joint opening. The stress-based load transfer was significantly decreased from the zero-gap case compared to with-gap cases.
4. Using an unbounded interface, where separation condition allowed, using surface-to-surface combined with Columb friction between the pavement and base permitted a better imitation of the pavement system, especially the gap formulation under the effect of the thermal gradient.
5. The change in the bond of dowel bars model using "not allow separation" option, produce a significant stress drop across the joint. This finding establishes the efficiency of dowel bars in holding joint cracking, which may develop into joint faulting. Allowing separation enable the dowel to slide through the dowel hole and permit the loaded slab to deflect.
6. Using wedge elements for modeling ensured uniform meshing around dowel bars to allow the model to capture the dowel responses accurately. Meshing the pavement parts using solid brick elements had numerous benefits especially in tracing the zones of high stresses and allow the simulation of the interfaces between the model parts.
7. Using surface-to-surface contact to model aggregate interlocking alongside the joint, permits simulation of a realistic load transfer devices' behavior as relative motion, gap origination. Zero gap effect with the high interlocking bond makes the adjacent slabs response slightly as a one and transfer stresses through the base layer.

References

Abaqus Documentation User's Guide: ABAQUS User's Guide: Dassault Systèmes. Simulia Corp, Providence (2013)

Abu El-Maaty, A.E., Ghada, M.H., Eman, M.S.: Modeling of dowel jointed rigid airfield pavement under thermal gradients and dynamic loads. Civ. Eng. J. **2**(2), 38–51 (2016)

Advisory Circular AC Report No: 150/5320-6E: Federal Aviation Administration U.S. Department of Transportation, Airport Pavement Design and Evaluation (2009)

Joshi, A.: Influence of moving load, structure, temperature gradient, and wheel configuration on load transfer efficiency. M.S., Civil and Environmental Engineering, Henry Rowan College (2012)

Edward, G., Izydor, K.: FAA finite element design procedure for rigid pavements. Report for U.S. Department of Transportation Federal Aviation Administration (2007)

Federal Aviation Administration: FAARFIELD 1.305 program database (2010). http://www.faa.gov/airports/engineering/design_software/

Huang, Y.: Pavement Analysis and Design, 2nd edn. Prentice Hall, Upper Saddle River (2003)

Ioannides, M.: Advanced pavement design: finite element modeling for rigid pavement joints. Report I, U.S. Department of Transportation Federal Aviation Administration (1997)

Jiwon, K., Keith, H.: Three-dimensional finite element analysis of doweled joints for airport pavements. Transp. Res. Rec. J. Transp. Res. Board **1853**, 100–109 (2003)

Mackiewicz, P.: Analysis of stresses in concrete pavement under a dowel according to its diameter and load transfer efficiency. Can. J. Civ. Eng. **42**(11), 845–853 (2015)

NAPTF-Databases: National airport pavement test facility. U.S. Department of Transportation, Federal Aviation Administration (2016)

Samir, S., William, G., Mourad, Y., Motamarri, S.: Effect of bonding force on stresses in concrete slabs. West Virginia University, Department of Civil and Environmental Engineering, Morgantown, TRB (2003)

Wadkar, A.: Study of load transfer efficiency of airfield rigid pavement joints based on stresses and deflections. M.S., Civil and Environmental Engineering, Henry Rowan College (2009)

Westergaard, H.M.: Analysis of stresses in concrete pavements due to variations of temperature. Highw. Res. Board **6**, 201–215 (1926)

Xinhua, Y.U., Xiaochun,W.: Considering joint load transfer efficiency of rigid pavement dynamic effects under a single moving load. In: 2nd International Conference of Information Engineering and Computer Science (ICIECS), pp. 1–4. IEEE Publisher, Wuhan (2010)

Effect of Construction Induced Cracks on Tensile Strength and Bonding Between Asphalt Concrete Layers of Pavement Under Different Temperatures

Graziela Girardi(✉), Mohammad Ramezani, and A.O. Abd El Halim

Department of Civil and Environmental Engineering,
Carleton University, Ottawa K1S 5B6, Canada
GrazielaGirardi@cmail.carleton.ca

Abstract. Pavement designers consider traffic loads and climate conditions to be the most important variables governing the long term performance of asphalt pavement. However, research conducted at Carleton University found that construction methods can also affect the long term performance by creating surface hairline cracks that allow water to penetrate the asphalt layer, which destroys bonds in the asphalt and between pavement layers. This paper presents the results and major findings of an experimental investigation performed on large scale asphalt slabs extracted from newly laid pavement. For the first time, our research included evaluating the bonds between asphalt layers built in the field. The laboratory study included testing the slabs under direct tensile stress, and subjecting them to temperatures ranging from room temperature to −20 °C. The testing included a total of 100 asphalt samples 300 mm long and 100 mm wide, with a thickness of 50 mm for single layer and 100 mm for double layers.

The results showed that construction induced cracks affect the tensile strength of new asphalt roads, and can weaken the bond between the upper and lower layers when overlays are constructed. These issues impact maintenance and rehabilitation investment, and waste the new overlay.

Keywords: Asphalt concrete · Low temperature · Tensile strength · Bonding · Cracks

1 Introduction

The maintenance and repair of highway network systems are major road agencies' expenditures. More than 14 million tonnes of asphalt is produced annually in Ontario, Canada alone. Though researchers have created Superpave and other mix designs intended to increase the lifespan of pavement, studies show that these efforts improved the product, but not enough to delay its early deterioration. Researchers have found that traffic, moisture, subgrade, construction quality and maintenance all influence the performance of pavement, but the root cause is still unknown. The combined effects of environmental conditions and traffic loading contribute to pavement deterioration, particularly in Canada and the northern United States with their low winter temperatures.

L. Mohammad (ed.), *Advancement in the Design and Performance of Sustainable Asphalt Pavements*, Sustainable Civil Infrastructures, DOI 10.1007/978-3-319-61908-8_7

Research by Tabatabaee et al. (2011) showed low temperature cracking is a major distress in many cold regions, and proposed that excessive brittleness due to increased stiffness and diminished capability to adapt to stress causes a buildup of thermally induced stress and, ultimately, the cracking of mixtures in the pavement. Raab et al. showed that these cracks, which typically appear after severely cold weather events or a number of cold thermal cycles, are transverse to the longitudinal direction of the pavement due to the thermal contractions in roadways. A parameter study by Shen Weixin found that material homogeneity, asphalt-concrete ductility, frictional constraints on the interface and the rate of cooling, significantly influence the thermal cracking of asphalt concrete pavement.

To this point, asphalt compaction is the only aspect of the process that has not been changed or improved; steel rollers have been used since its discovery. Compaction of the asphalt mix has been recognized as an important factor in the degradation and long-term performance of pavement. Problems with compacting asphalt mixes have generally been ascribed to the mix alone. Halim et al. showed that the surface cracks initiated by a steel wheel roller could explain the frequently reported early deterioration of asphalt pavements. They also found that the construction cracks significantly reduce the resistance of pavement to thermal stresses and accelerate deterioration.

Researchers have been studying thermal cracking for years, and have tried a variety of methods to evaluate the low temperature behavior of asphalt mixes. Several have attempted to describe the primary mechanism of various types of thermal cracking (Haas et al. 1987), and to predict the formation of thermal cracks. Empirical models (Fromm and Phang 1972; Chang et al. 1976; Haas et al. 1987) are based on field investigations and laboratory tests, and associate the spacing of the cracks with major environmental factors and material properties. Most analytical models (Carpenter and Lytton 1975; Lytton et al. 1983; Hiltunen and Roque 1994) employ fracture mechanics to predict the propagation of thermal cracks. Harold Timm et al. (2001) investigated several empirical and mechanistic-based models, and found those models do not fully explain the mechanism leading to the regularly spaced cracks observed in the field. Tests to provide data for the models include indirect tension, direct tension, direct tensile creep, flexural bending, thermal stress restrained specimen, and coefficient of thermal expansion and contraction.

Pablo E. Bolzan and Gerald Huber conducted research to determine if running the Direct Tensile Test (DTT) routinely would provide suitable data for Highway Agencies. However, the temperatures selected ranged from 4 °C to 25 °C; thus no testing was done in sub-zero conditions. Halim et al. (1990a, b) performed DTT on asphalt-concrete using rectangular specimens, and found that they prevented stress buildup and eliminated the errors in the results of the original DTT.

The objective of this paper is to study the tensile strength of asphalt pavement in temperatures from +20 °C to −20 °C and assess the condition of the bonding between asphalt layers, using special Direct Tensile Strength Test (DTT) designed for asphalt concrete samples extracted from newly constructed pavement. We also study the impact of microcracks created on asphalt surfaces during the compaction process.

2 Test Objectives

The objective of the selected testing program is to simulate actual field conditions and replicate field loading and boundary conditions as closely as possible. In a controlled laboratory environment, direct tensile testing was performed on asphalt concrete samples in the simulated thermal conditions that lead to cracking. Tests were divided in two phases: Phase A studied the stress in the asphalt specimens, and Phase B focused on the bonding of the samples.

Test samples were saw-cut from field extracted slabs, and loaded in direct tension. Both single and double layer asphalt slabs were tested under different temperatures, and the effects of asphalt tensile strength and bonding in lower temperatures was evaluated. Failure was induced by monotonic loading to simulate extreme low-temperatures, and material properties were examined using tensile strength, which is also affected by temperature, as well as the loading rate.

3 Direct Tension Strength Test (DTT)

The Direct Tension Strength Test (DTT) for asphalt is known for its ability to determine asphalt bituminous, and not asphalt concrete, failure stress and strain at low temperatures. The original direct tension test is performed by applying an axial tensile force to a round specimen, and measuring the stress strain characteristics of the material. Holding the specimen is one of the major problems, as the test is executed under the assumption that only pure tension is applied to the specimen. Bending stresses are produced due to misalignments, and this is the source of errors in test results. However, the use of rectangular specimens prevented the buildup of stresses in specimens, and eliminated the errors in the results of the original DTT. The tensile force and corresponding deformation are recorded, then used to calculate the tensile strength and the maximum elongation of the specimen. Another assumption in this test is that the stress is uniformly distributed across the central section. This method of DTT is used to determine the tensile strength, stiffness and the limiting strain of a given mix. A uniform displacement rate is applied to the base of a moving plate at one end of the sample, and the other end is restrained. The displacement is controlled with a Linear Variable Displacement Transducer (LVDT), and a load cell is used to measure the tensile load on the sample. The test is performed in the temperature controlled chamber, and can be applied to field extracted slabs.

4 Test Facility

To perform the Direct Tensile Strength Test (DTT) on rectangular asphalt concrete samples, a test facility (Figs. 1 and 2) was designed by Halim et al., and assembled in the Civil and Environmental Engineering Laboratory of Carleton University inside an environmental chamber.

Fig. 1. Specially designed testing table, for DTT on asphalt concrete

Fig. 2. Specially designed testing table, for DTT on asphalt concrete

The facility consisted of (Raab and Parti 2009):

- an environmental chamber with a temperature range of +60 ° to −40 °C;
- a hydraulic actuator with 100 kN capacity and 250 mm stroke;
- a direct tensile strength testing table for asphalt concrete samples, equipped with fixed and moving mounting plates;
- a moving plate connected to a load cell and a positioning system consisting of a linear variable displacement transducer (LVDT) that feeds back to a data acquisition computer;
- a moving plate on a Teflon sheet to reduce friction forces, with the load cell adjusted accordingly to eliminate the loads required to move the plate and resist friction; and,
- a computer, data acquisition system and programs for monitoring and control of the test equipment.

The specially designed testing table consists of two sections: one is fixed, and the other can move horizontally by sliding over Teflon guides. An asphalt concrete slab is fixed to the surface of the table, and the movable section is then pulled at a constant rate of displacement, with a data acquisition system monitoring the applied load. The corresponding deformation, (i.e. the horizontal displacement of the movable table section) is monitored with an LVDT (Linear Variable Differential Transformer). Figure 3 shows the horizontal outline of the testing table.

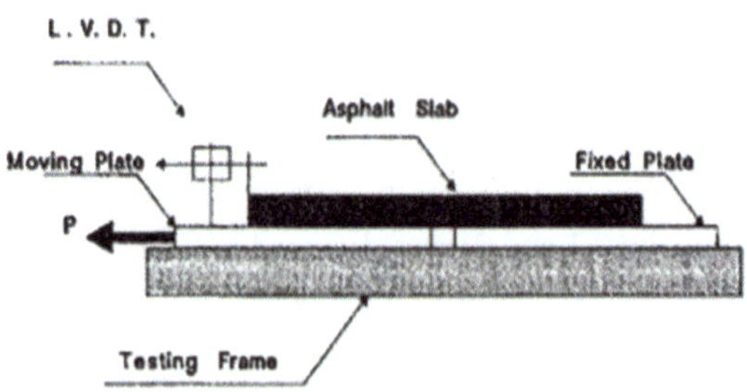

Fig. 3. Outline of the DTT testing table (Halim et al.)

5 Material

The asphalt mixes used in this study were Superpave 12.5 mm FC2 PGAC 70-34 and Superpave 19 PGAC 64-34. Both mixes are designed to meet the aggregate gradations specified by the Ontario Ministry of Transportation. The mixes' main differences are aggregate source, density and gradations, as shown in Table 1.

Table 1. Design material properties of Mix A and Mix B

Property	Mix A - top course	Mix B - base course
Asphalt cement content	SP 12.5 FC2 70-34	SP 19 64-34
Binder content (mass%)	4.8	4.6
Voids in mineral aggregates (VMA %)	15.1	14.36
Air voids	4.0	4.0
Bulk relative density (BRD)	2.378	2.447
Air temperature at construction site	+20 °C	+20 °C
Date of construction	September, 2015	Sep, 2015

6 Experimental Procedure (Testing Program)

6.1 Mix Preparation and Compaction - In the Field

The sampling lot was constructed on access road near Highway 417, on Lees avenue, East of Ottawa, Canada (Fig. 4). Samples were taken from this access road while asphalt was asphalt material was from the actual highway paving project under construction at the same day and compaction was done by the same crew and machinery. No interference was allowed in the contractors work during the construction of the major highway project. Ontario Provincial Standard Specifications (OPSS) for the placement and compaction of hot mix asphalt designed using the Superpave method were followed as the original road project was for the Ministry of Transportation of Ontario (MTO).

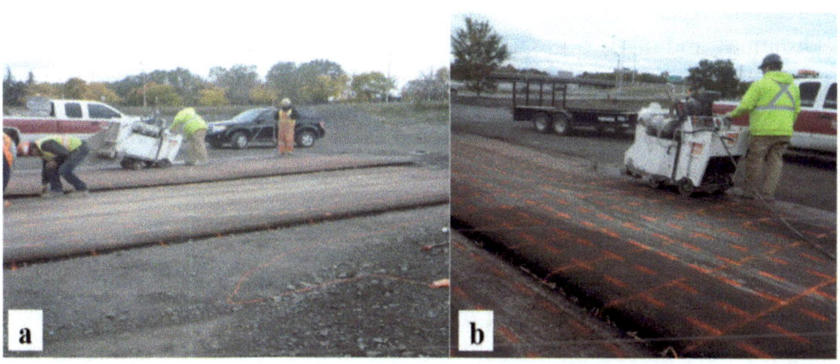

Fig. 4. Sampling lot (Ottawa, Canada, 2015), marking and saw-cutting the slabs

The asphalt and placement of the asphalt Mix design for the top course was 12.5FC-12 mmSP mix with A/C type (70-34D), and mix design for the base course was 19mmSP mix with A/C type (64-34D). Tack Coat was applied at the rate of 0.35 kg/m^2 to all base course pavement surfaces. As required by OPSS 310, the tack coat consisted of SS-1 emulsified asphalt diluted with an equal volume of water (the undiluted material was according to OPSS 1103.) The lot for the sampling was compacted as in similar main projects, with three rollers: a breakdown roller, a compaction roller and a finish roller as below:

- Class S, Self-propelled steel-drum, tandem, or three-wheel rollers
- Class R, Self-propelled pneumatic-tired rollers
- Class V, Self-propelled vibratory rollers

The test samples were obtained from asphalt concrete mixed and compacted in the field. The mixes were laid out in 10 m × 30 m slabs at the paving site, and compacted with eight passes of half-ton steel, pneumatic and vibratory rollers, to achieve a pavement thickness of 50 mm (for single layer samples) and 100 mm (for double layer samples). After 30 days the slabs were saw-cut to 30 × 60 cm, and carefully marked. The specimens were extracted from the pavement by cutting the asphalt concrete through the new overlay and the old pavement, then transported to the laboratory. Great care was taken to lift the samples from the road, to avoid damage deformation or crack propagation.

6.2 Preparation of Test Slabs in the Lab

The 60 cm × 30 cm asphalt slabs were received from the field, then saw-cut to the test sizes of 300 mm × 100 mm in the lab. The marking scheme kept track of the sample locations during compaction, and the rolling direction was marked on each sample. The thickness was 50 mm for single layer samples, and 100 mm for double layer samples. Sample transportation, storage, handling and preparation techniques (e.g. gluing, alignment) are very important and can affect test results.

6.3 Test Procedure

The test itself is relatively simple and quick, whereas sample preparation requires expert technicians and takes quite a long time, as the cut samples must be handled with great care. Following are the steps for preparation and testing of the slabs:

1. The 300 mm × 100 mm samples were cleaned with metal and a plastic brush, and all loose aggregates were removed from the cut surface.
2. Gluing:
   a. Samples were placed on a table with the surface to be glued face up (Fig. 5a).
   b. Epoxy adhesive (Hilti HIT-RE 500 V3) was applied at both ends of the sample, and the middle area at 50 mm did not receive any adhesive (Fig. 5b).
   c. Samples were placed on the centre of metal plates, and the centre of the sample was aligned with the joint between the plates (Fig. 5c).

Fig. 5. Preparation of test slabs in the lab

   d. Samples were put under low pressure by placing weight on top, and prevented from sliding (Fig. 5d).
   e. The adhesive was cured for 24 h (minimum cure time at 72 °F is 6.5 h).
3. Conditioning the samples:
   a. The plates (with samples glued on them) were placed inside an environmental chamber (EC), which was set at the planned test temperature.
   b. The samples remained in the chamber for 24 h, to allow them to be conditioned and ready for the testing.
4. The sample was placed on the pulling table, and bolted to the plates on the table (Fig. 6a, b).
5. The pulling table was activated, and the sample was pulled apart at a constant strain rate of 0.06 mm per second until it failed, at which point the strain at failure and stroke in millimetres and Load (kN) is recorded (Fig. 6c, d). The test was repeated at room temperature, 0 °C, −10 °C and −20 °C, with a minimum five samples for each temperature set. Table 2 summarizes the completed tests at different temperatures for both the single and double-layer asphalt concrete samples.
6. The above test was performed for samples glued to plates on their roller-compacted face, and for samples glued to plates on their rough side, the face that was connected to the lower layer in the field.

Samples are conditioned at the desired test temperature for 24 h prior to testing. The test temperatures are −20 °C, −10 °C, 0 °C and +20 °C. Direct tensile tests are performed at a loading rate 0.6 mm/min. The test continues until visible failure or separation has occurred, or the total displacement is greater than 15 mm.

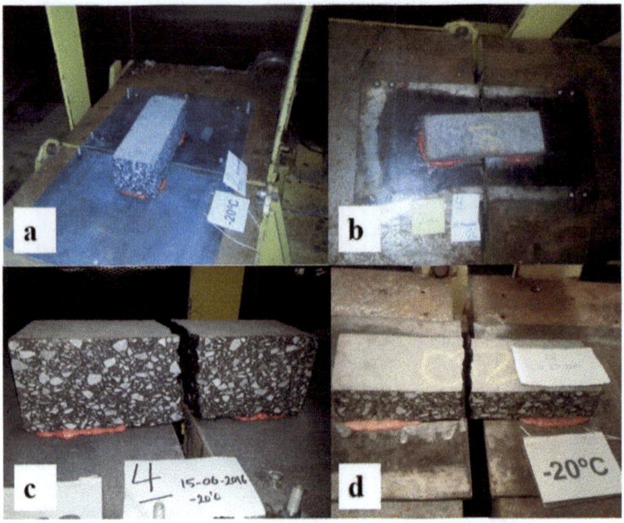

Fig. 6. Testing slabs in the lab

Table 2. Asphalt slabs with cracked bottom layer and disconnected top layer

Date	Temperature (°C)	Compacted face	Max displacement average (mm)	Max load average (kN)
11/12/2015	20	Down	−3.6172	−0.7932
11/05/2016	20	Up	−9.9851	−1.3179
20/05/2016	0	Up	−15.5102	−5.8564
02/06/2016	0	Up	−19.6701	−8.0441
09/02/2016	−10	Up	−13.0803	−8.6139
01/04/2016	−10	Up	−20.1534	−13.3477

7 Summary

Laboratory test methods were developed that simulate the field conditions of pavement subjected to thermal stresses. The tests examined the effects of 'roller checking' on the tensile strength and strain energy of asphalt paving mixtures at different temperatures, and established an accelerated test methodology. Tests were conducted on asphalt slab specimens with dimensions of 100 × 300 × 50 mm (single layer) and 100 × 300 100 mm (double-layer), in temperatures from −20 °C to +20 °C. Each sample was affixed to a set of moving and fixed plates with high strength epoxy glue. A forced monotonic was applied in the horizontal direction only, at a rate of 0.060 mm/sec until sample failure.

8 Results and Analysis

Direct tensile testing was performed on saw-cut slabs, which were conditioned in advance at the test temperature for at least 24 h. Table 2 summarizes the average results of testing single and double slabs. Each group was tested with the compacted face up and the compacted face down, and the asphalt bonding between the layers was also analysed. Figures 7, 8, 9 and 10 are representations of the raw data recorded by the data acquisition program.

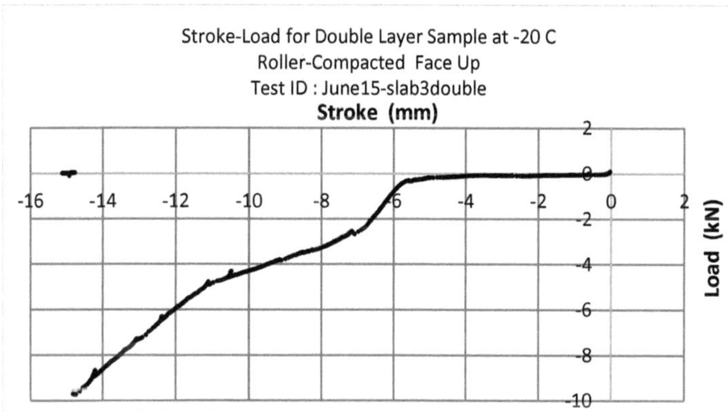

Fig. 7. Test result stroke vs. load at −20 °C

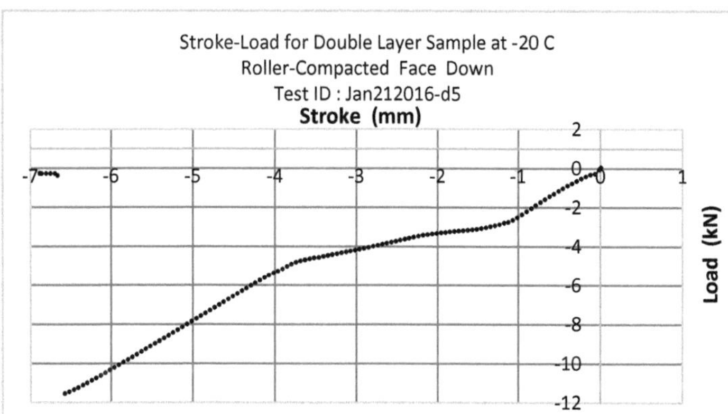

Fig. 8. Test result stroke vs. load at −20 °C

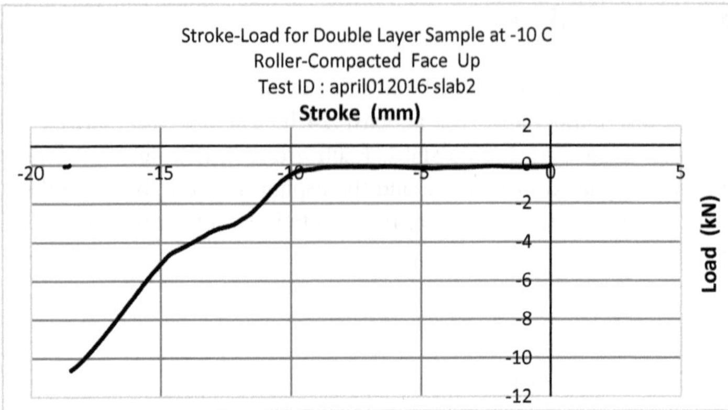

Fig. 9. Test result stroke vs. load at −10 °C

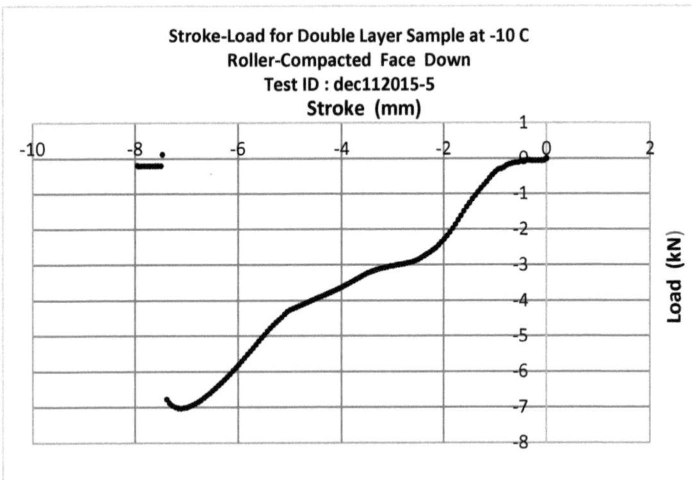

Fig. 10. Test result stroke vs. load at −10 °C

The calculated stress was lower for roller-compacted face down samples for:

- Double-layer samples tested at −20 °C. (by 9%)
- Single-layer samples tested at −20 °C. (by 37%)
- Single-layer samples tested at −10 °C. (by 11%)
- Double-layer samples tested at +20 °C, (by 194%)

The review of the test results indicates that calculated stress was smaller for roller-compacted face down double-layer samples tested at −20 °C by 9%, and 194% for double-layer samples tested at +20 °C. On single-layer samples tested at −20 °C it presented 37%, and on single-layer samples tested at −10 °C the stress was smaller by

11% when compared to roller-compacted face down. However, at −10 °C, double-layer samples presented higher stress for roller-compacted face up, 10%.

Results indicated that roller-compacted face down samples were less strong, which the authors attribute to the formation of microcracks during the compaction process. These construction induced cracks can significantly reduce the tensile strength of the asphalt layers, and accelerate the development of cracks when the sample is subjected to tensile forces. Samples tested with the roller-compacted face up were stronger, and required more energy to initiate and develop cracks.

The results show a significant difference between identical pairs of samples, based on which compaction face is tested; the pavement exhibited higher strength with the compacted face up. The decrease in strength can be explained by the presence of construction induced cracks, which form when the pressure and curvature of the rollers cause a deflection basin and create construction cracks in the transverse direction, thus reducing the load carrying capacity in the longitudinal direction.

Furthermore results show that slabs at −20 °C broke at higher loads than −10 °C. At +20 °C failure occurred gradually and only through the binder due to material softening, while failure occurred abruptly and through the aggregates in samples at −20 °C.

The bonding between the two layers of 50 asphalt slabs was also tested, at temperatures of +20 °C, 0 °C, −10 °C and −20 °C. The results showed that six slabs (12%) of the 50 slabs had bottom layer cracking, and the top layer was disconnected from the bottom layer at the tack coat line. All the other slabs both layers cracked at the same time. Table 2 shows the test results.

As expected, tack coat did not display resistance at room temperature 20 °C. The samples began to slide as soon as the load was applied (with 0.1 kN load).

At the colder temperatures (−10 °C and −20 °C) the tack coat bonding did create resistance. At −10 °C the maximum load observed was 7.7 kN, and at −20 °C it was 22.5 kN. The test results of the laboratory tests on double slabs indicated that the tack coat gluing the two samples to each other performs better in cold temperatures.

It was also observed that, of the 103 double-layer, 600 mm × 300 mm slabs received from the sampling lot, 27 (26%) showed separation of the top layer (top course) from the bottom layer (base course) after approximately five months storage at room temperature (+20 °C). These samples were excluded from the testing program. This finding is important because it shows that, even without any load or exposure to low temperatures, tensile stress asphalt layers can still separate. It could be due to the lack of bonding, mishandling or difference in temperature. Figure 11 shows examples of these slabs.

This paper and related laboratory tests explains how cracks initiated during construction process is reducing tensile strength of the newly paved asphalt layers. The authors' point of view is that current asphalt compaction equipment and processes are the main cause of crack initiations in asphalt layers. These cracks induced during construction will grow rapidly under the effect of the traffic and cold temperatures. These cracks will not be prevented unless construction methods are adjusted and compactors are modified to avoid creating the construction-hairline cracks. This view is in contrast with those who believe that asphalt cracks are initiated due to thermal stresses.

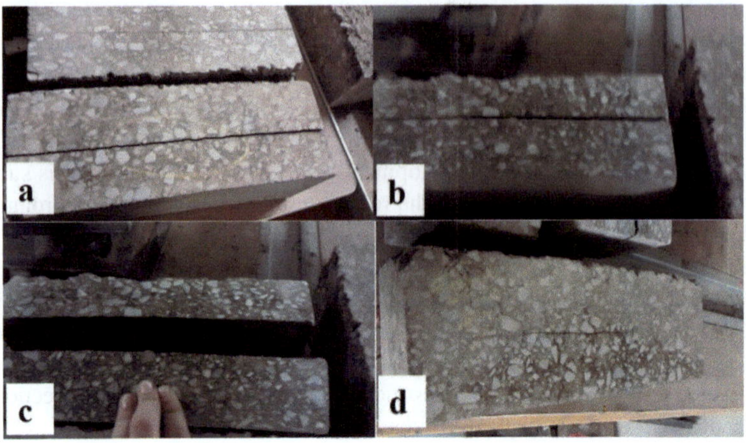

Fig. 11. Asphalt slabs disconnect at the bond line prior to test.

9 Conclusion

Damages by microcracks, originated during the construction, in an asphalt concrete were assessed by using special DTT test in a range of temperatures (−20 °C, −10 °C, 0 °C, +20 °C).

Based upon the work presented herein and summarized above, the following conclusions have been drawn:

- Tack coat is stronger in cold temperatures (below 0 °C) compared to ambient temperature (22 °C). Tack coat provides better bonding in lower temperatures than room temperature (+22 °C).
- The different samples' width, 5 cm to 30 cm, did not affect the results.
- The length of the samples should not be smaller than 30 cm.
- Asphalt samples of the same dimensions and mixture failed at a higher load than 4 kN in cold temperature (−10 °C and −20 °C) compared to ambient temperature (20 °C) that failed around 1.2 kN.
- There is a meaningful and important difference between the samples when tested with the compacted surface down and up. Slabs with compacted surface up demonstrated higher stress values when compared with similar samples tested with the compacted surface down. This observation suggests the construction cracks play a role in the tensile strength of asphalt pavements.

This paper demonstrates the negative impacts of the compactors' induced microcracks on structural behavior of the new asphalt layer. The results show how these cracks created during construction of asphalt can alter the structural response of the asphalt layer to applied stresses and to reduce the mechanical properties of the newly laid asphalt pavement layer.

It is the authors' opinion that reducing the amount of early cracks, especially construction induced cracks which under the effect of cold temperature these cracks will propagate, could results in higher strength of asphalt layers and eventually extending the service life of the asphalt pavement.

Acknowledgments. The authors would like to acknowledge NSERC, Carleton University for their financial support, and from RW Tomlinson Limited for providing the asphalt specimens. The authors would also like to thank Stanley Conley, and Jason Arnott for their assistance in operating the environmental chamber and assistance during the testing.

References

Adlinge, S.S., Gupta, K.A.: Pavement deterioration and its causes. IOSR J. Mech. Civil Eng. (IOSR-JMCE). In: Second International Conference on Emerging Trends in Engineering (SICETE), pp. 9–15. ISSN: 2278-1684

Raab, C., Parti, M.N.: Long-term interlayer shear performance of base and subbase layers in asphalt pavements. In: 6th International Conference on Maintenance and Rehabilitation of Pavements and Technological Control, Mairepav6, Torino, Italy, vol. II, pp. 646–653 (2009)

Hass, R.G.G., Phang, W.A.: Relationships between mix characteristics and low temperature pavement cracking. Paper presented to AAPT (1998)

Shalaby, A., Abd El Halim, A.O., Easa, S.M.: Influence of thermal stresses on construction-induced crack. In: Beuving, E., Franched Rilem, L. (eds.) Third Conference on Reflection Cracking, pp. 30–39. E & FN Spon, Masstricht, The Netherlands (1996)

Tabatabaee, H.A., Velasquez R., Bahia, H.U.: Predicting low temperature physical hardening in asphalt binders. Submitted for publication in the J. Constr. Build. Mater., Elsevier (2011)

Abd El-Halim, A.O., Svec, O.J.: Field and laboratory evaluation of a new compaction technique. In: Proceeding of the RILEM Symposium on Bituminous Mixes, Budapest, Hungary, pp. 37–54, October 1990a

Abd El-Halim, A.O., Svec, O.J.: Influence of compaction techniques on the properties of asphalt pavements. In: Proceedings ot the Canadian Technical Asphalt Association Annual Meeting, Winnipeg, Man, pp. 18–33, November 1990b

Kirkner, D.J., Shen, W.: Numerical Simulation of Thermal Cracking of Asphalt Pavements. Transportation Research Board Preprint, Washington, DC (1999)

Turos, M.I.: Determining the Flexural Strength of Asphalt Mixtures using the Bending Beam Rheometer (2010)

Kandhal, P.S., Dongre, R., Malone, M.S.: Prediction of low-temperature cracking using superpave binder specifications. NCAT Report No. 96-2, February 1996

Easa, S.M., Shalaby, A., Abd El-Halim, A.O.: Reliability based model for predicting pavement thermal cracking. J. Transp. Eng. **122**, 374 (1996)

Effect of Wet-Dry Cycle on Durability of Cement-Stabilized Recycled Pavement Base Aggregates

Saif Bin Salah[1(✉)], Sahadat Hossain[2], Mohammad Faysal[1], Mohammad Sadik Khan[3], and Carla Maria Flores[1]

[1] Department of Civil Engineering, The University of Texas at Arlington, 416 Yates Street, NH 119, Arlington, TX 76019, USA
{saif.salah,md.faysal,carla.flores}@mavs.uta.edu

[2] Department of Civil Engineering, The University of Texas at Arlington, 416 Yates Street, NH 404, Arlington, TX 76019, USA
hossain@uta.edu

[3] Department of Civil and Environmental Engineering, Jackson State University, 1400 J.R. Lynch Street, Jackson, MS 39217-0168, USA
mohammad_sadik.khan@jsums.edu

Abstract. Resilient Modulus (M_R) is one of the most important stiffness parameter to determine the thickness of a pavement layer (AASHTO 2003 pavement design guideline). At present, the design procedure does not consider the effect of deterioration of pavement layers due to seasonal variations. Recent studies conducted by researchers on the variability of M_R with wetting-drying (WD) and freeze-thaw (FT) cycles show that long-term durability is an important criterion to be considered in designing pavement base with recycled materials. However, limited study has been conducted to evaluate the effect of environmental deterioration on the reclaimed asphalt pavement (RAP) and recycled crushed concrete aggregate (RCCA) mixtures stabilized with cement. The objective of this research is to evaluate the durability of 50% RAP-50% RCCA mixtures stabilized with 4% and 6% cement subjected to 0, 4, 8, 16 and 30 wetting-drying cycles. Wetting and drying cycles were applied to the specimens after curing them for 7 days. Resilient modulus tests were then conducted on the specimens at the end of these specified number of cycles. Environmental tests were also conducted to assess the effect of WD cycles on the leachate quality. These tests included total suspended solids (TSS), total dissolved solids (TDS), turbidity, chemical oxygen demand (COD) and pH. Tests results indicated that with the increase in the number of WD cycles, the value of M_R decreased by about 11% of the initial value at the end of 30 cycles. Results obtained from the environmental tests after 30 WD cycles were found to be within the permissible values provided by EPA guidelines.

1 Introduction

Reclaimed asphalt pavement (RAP) and recycled crushed concrete aggregate (RCCA) are waste materials very abundantly produced owing to the demolition and rehabilitation projects carried out all over the United States. According to a report from the

L. Mohammad (ed.), *Advancement in the Design and Performance of Sustainable Asphalt Pavements*, Sustainable Civil Infrastructures, DOI 10.1007/978-3-319-61908-8_8

National Asphalt Pavement Association (Hansen and Copeland 2015), the amount of RAP produced in 2014 was 71.9 million tons. The amount of RCCA generated from the construction and demolition is expected to be 123 million tons per year (USDOT 2004). These non-biodegradable materials when discharged to landfills can pose significant threat to the environment. The Texas Department of Transportation (TxDOT) has already started using RCCA as flex-base material for pavement base layer construction (Faysal et al. 2016b). Recent studies conducted on these materials suggest that they can be reused for flexible base construction of newer pavements and for rehabilitation of older roadways. Most of these studies were conducted to assess the strength and stiffness properties of the recycled materials in terms of compressive strength (Faysal et al. 2016b), resilient modulus response (Faysal et al. 2016b; Puppala et al. 2011), geophysical response (Mahedi 2016), and accelerated load testing (Jones et al. 2014).

However, pavement engineers are nowadays more concerned about the long-term performance of pavement structures that is significantly affected by seasonal variations (Khoury and Zaman 2007). Wetting and drying (WD) actions induced by these seasonal variations result in episodic moisture movement in various pavement layers. Moisture variation in pavements causes pavement distresses like edge cracking, edge drop and longitudinal cracking (Hedayati and Hossain 2015; Hossain et al. 2016). This indicates possible changes in engineering properties that are associated with the pavement materials. Various studies have been conducted in recent times on the effect of wetting-drying (WD) cycles on performance of pavement materials. Miller et al. (2003) and Tao and Zhang (2006) used unconfined compressive strength to assess durability of stabilized aggregates under WD cycles. Khoury and Zaman (2005) evaluated the effect of WD cycles on resilient modulus of virgin aggregates stabilized by lean stabilizers. Regression models have also been developed to predict resilient modulus values from aggregate properties and WD cycles (Maalouf et al. 2012). This research area is however not fully explored and additional studies are still needed (Little et al. 2005).

The objective of this study is to evaluate the durability of RCCA and RAP mix materials in terms of structural competency and environmental soundness when subjected to repeated wetting and drying cycles. A comprehensive experimental program was undertaken to characterize strength and long term durability these material mixes. For this study, a 50–50% mix of RAP and RCCA materials were used. This combination was selected with a view to ensure maximum use of RAP materials. Previous studies reported RAP material to be structurally incompetent when mixed by more than 50% with RCCA and stabilized with up to 6% cement content (Faysal et al. 2016a). The material mix was stabilized with 4% and 6% cement content. Subsequent wetting and drying (WD) cycles were then applied to specimens prepared from these materials. The laboratory procedures mentioned in most of the previous studies followed the ASTM D 559 test method that is developed specifically for soil-cement specimens. Kalankamary and Donald (1963) considered this method to be severe and not a good indicator of actual field conditions. No established laboratory method for conducting WD cycles on stabilized pavement aggregates have been found in previous studies. The method followed by Khoury and Zaman (2002) was considered most acceptable and is adopted in this study. Resilient modulus tests were then conducted to assess the stiffness properties of the RAP and RCCA mixes. Furthermore, environmental soundness was determined by performing pH, total dissolved solids (TDS), total suspended solids (TSS), chemical oxygen demand (COD) and turbidity tests. These test results were compared with Environmental Protection Agency (EPA) guidelines and recent literature.

2 Sources of Materials

In this research, two types of recycled materials were used. One of them is the recycled crushed concrete aggregate (RCCA) which was collected from Big City Crushed Concrete stockpile located in Dallas, Texas. The other is the reclaimed asphalt pavement (RAP) material collected from specified stockpiles of Rockwall County, Texas. Both site locations are TxDOT approved stockpile sites in the Dallas-Fort Worth area.

3 Experimental Program

The experimental program undertaken in this study aimed at evaluating the structural durability and environmental viability of using combinations of the RCCA and RAP materials in flexible pavement base construction. The RCCA and RAP materials used for this study contained particle size ranging from 1 in. (25 mm) to No. 200 (75 μm). For this research, a 50–50% mix of RAP and RCCA materials was utilized. The material mix was then stabilized at cement dosage of 4% (MIX4) and 6% (MIX6). Portland Type I/II type cement was used to treat the base materials. Structural competency of the material was determined using the resilient modulus (M_R) test. Durability tests include wetting-drying (WD) cycles followed by M_R tests on the prepared specimens (Table 1). The environmental tests include total suspended solids (TSS), and total dissolved solids (TDS), chemical oxygen demand (COD), turbidity and pH which were conducted on the leachate samples. A tabular presentation of these environmental tests is given in Table 2. Repeatability of the tests were ensured by performing the same test on three identical specimens for each type of mixes.

Table 1. Experimental program for resilient modulus (M_R) tests

Material	Cement content (%)	Mix ID	Tests	Test samples					Control samples				
				Wetting-drying (WD) cycles					Curing (days)				
				0	4	8	16	30	7	15	25	40	70
50% RCCA + 50% RAP	4	MIX4	M_R	√	√	√	√	√	√	√	√	√	√
	6	MIX6	M_R	√	√	√	√	√	√	√	√	√	√

4 Particles Size Distribution

Varying particle size has a significant effect on the strength characteristics of granular materials (UFGS 2010). Sieve analysis tests were conducted on the collected RCCA and RAP materials following Standard TxDOT Specifications (Tex-110E). Figure 1 shows the particle-size distribution curves for both the materials. It is evident that for both RCCA and RAP, less than 1% of the material passes through the No. 200 sieve. Therefore, as per TxDOT specification (Item 276), hydrometer analysis was not deemed necessary for this study.

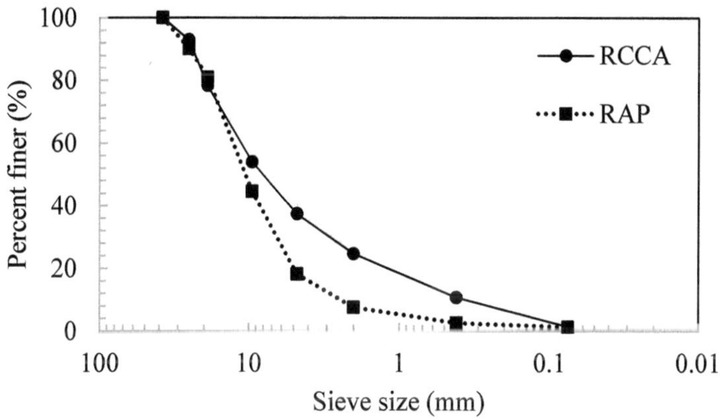

Fig. 1. Particle size distribution for RAP and RCCA material

5 Optimum Moisture Content (OMC) and Maximum Dry Density (MDD)

The moisture content at which compaction of a certain material yields its maximum dry density (MDD) is the optimum moisture content (OMC). In this study, OMC and MDD tests were conducted following the Tex-113 E guidelines on each of the material combinations at different cement contents as shown in Table 1. The value of the optimum moisture content was found to be 8.5% with a maximum dry density value of 1950 kg/m^3 (122 pcf).

6 Test Methods

6.1 Resilient Modulus (M_R)

Resilient modulus tests were conducted on 152 mm (6 in.) × 305 mm (12 in.) cylindrical specimens prepared by compaction at 6 lifts with each layer subjected to 50 blows to achieve the required compaction at OMC level. After curing for 7 days, the control samples and test samples were subjected to curing and WD cycles respectively. At the end of these specified curing and WD cycles, resilient modulus tests were conducted in accordance with the AASHTO T 307-99 test procedure using the resilient modulus testing machine manufactured by Geocomp Inc. The confining and deviator stresses were used as given in the AASHTO guidelines for base materials.

6.2 Wetting-Drying (W-D) Cycle

The effect of successive wetting-drying (W-D) cycles on strength properties of cement stabilized RCCA-RAP mix materials was investigated in this study. Due to unavailability of standard procedures for wetting-drying of stabilized base materials, experimental methods reported by researchers in recent times (Khoury and Zaman 2002; Faysal et al. 2017b) has been adopted in this study. RCCA and RAP materials mixed in 50–50% proportions and stabilized with 4% (MIX4) and 6% (MIX6) cement content have been used for this purpose. For each of these material combinations, three "control samples" and three "test samples" were prepared, cured for 7 days and then tested for resilient modulus as per AASHTO T 307-99 test procedures. The test samples were then subjected to wetting-drying (W-D) cycles. Each WD cycle consisted of drying the sample in the oven (71 °C/160 °F) for 24 h followed by submerging it in potable water for 24 h. For this study, the numbers of WD cycles considered were 0, 4, 8, 16 and 30 cycles. After completing a specified number of cycles, the samples were tests for resilient modulus (M_R). On the other hand, the control samples were cured following conventional process. After curing for up to 15, 25, 40 and 70 days, the samples were tested again for M_R. These curing periods represent the time corresponding to 4, 8, 16 and 30 wetting-drying (WD) cycles. In this study, each sample was subjected to multiple resilient modulus tests after specific times. This approach was considered reasonable since resilient modulus tests involve very low levels of strain (Khoury and Zaman 2007).

6.3 Environmental Tests

Environmental tests were conducted on the test samples prepared from the RCCA-RAP material mix as shown in Table 2. These samples were subjected to 0, 4, 8, 16 and 30 WD cycles. At the end of these specified cycles, the samples were soaked in deionized water for 24 h and leachate samples were collected. Environmental tests such as total suspended solids (TSS), and total dissolved solids (TDS), chemical oxygen demand (COD), turbidity and pH tests were conducted on the collected leachate samples as per the ASTM standard test methods.

Table 2. Experimental program for environmental tests

Material	Cement content (%)	Mix ID	Environmental tests	Test samples Wetting-drying (WD) cycles				
				0	4	8	16	30
50% RCCA + 50% RAP	4	MIX4	COD	√	√	√	√	√
			TDS	√	√	√	√	√
			TSS	√	√	√	√	√
			Turbidity	√	√	√	√	√
			pH	√	√	√	√	√
	6	MIX6	COD	√	√	√	√	√
			TDS	√	√	√	√	√
			TSS	√	√	√	√	√
			Turbidity	√	√	√	√	√
			pH	√	√	√	√	√

7 Results and Discussion

7.1 Resilient Modulus Tests

7.1.1 Data Analysis and Model Fitting

For each of the MIX4 and MIX6 materials two separate sets of Control and Tests samples were prepared. Each set consisted of three identical samples on which resilient modulus tests were conducted. For all the cases, the coefficient of variation (COV) of the M_R values was found to be within 0.15–8.63% which shows good repeatability of the performed tests. According to AASHTO test procedure requirements, the obtained resilient modulus data were used to develop prediction models. One of them is the "k-θ model" proposed by Moosazedh and Witczak (1981): $M_R = k_1\theta^{k2}$, where k_1 and k_2 are model parameters and θ is the bulk stress expressed as a combination of confining (σ_c) and deviator stresses (σ_d) in the form $3\sigma_c + \sigma_d$. Figures 2 and 3 shows the k-θ models developed for Control and Test samples of MIX4 and MIX6. Another model used in this study is the improved three-parameter model (Puppala et al. 1996): $M_R = k_3\sigma_c^{k4}\sigma_d^{k5}$, where k_3, k_4 and k_5 are model parameters. Statistical analysis was conducted to examine the accuracy of these models. The model parameters along with the calculated statistical parameters are presented in Tables 3 and 4.

7.1.2 Effect of Curing

The k-θ models (two-parameter) developed for the control samples of MIX6 is plotted in Fig. 2(a). With the increase in curing time, the resilient modulus (M_R) curves for the samples shifts resulting in higher values of resilient modulus. The M_R values increased by about 30% after 25 days of curing beyond which the curves of 40 and 70 days almost coincide with each other. This indicates that the MIX6 material gains almost all of its stiffness within the first 25 days.

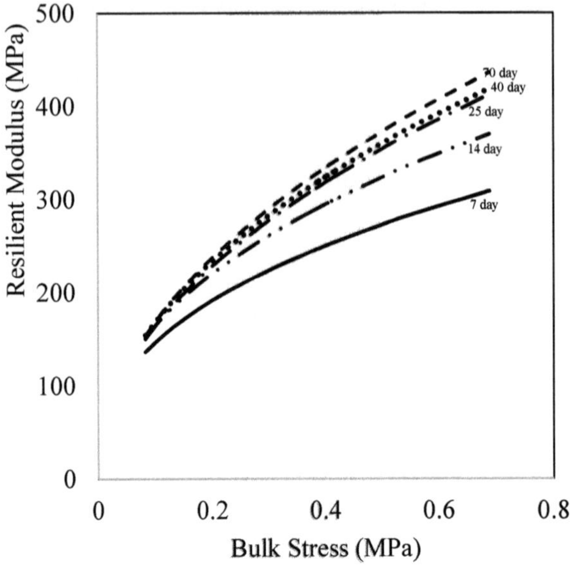

(a)

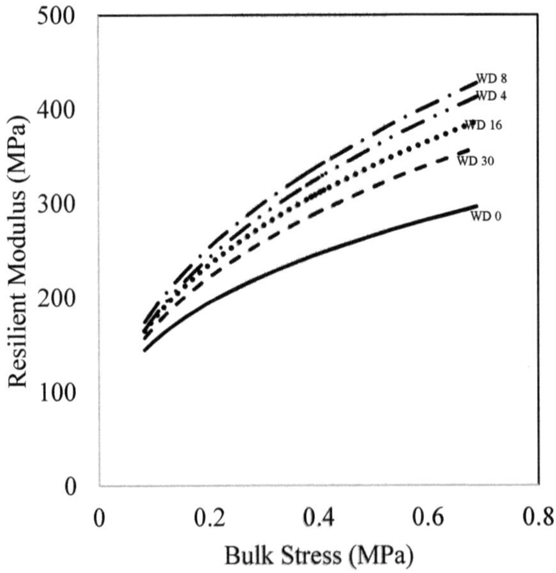

(b)

Fig. 2. Graphical plot of k-θ models developed for (a) Control samples and (b) test samples of MIX6 materials

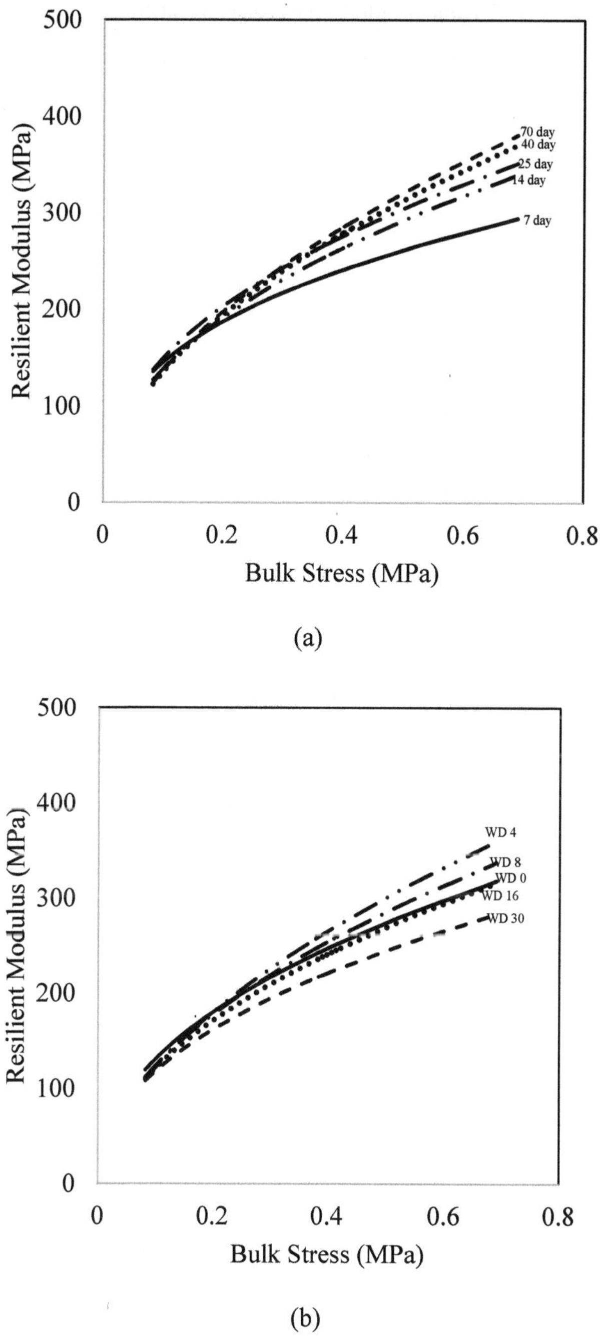

Fig. 3. Graphical plot of k-θ models developed for (a) Control samples and (b) test samples of MIX4 materials

Table 3. Model parameters for MIX4

	Three parameter model			Two parameter model			
	k_1	k_2	R^2	k_3	k_4	k_5	R^2
W-D cycles							
0	377.8	0.46	0.83	625.7	0.04	0.41	0.89
4	443.7	0.56	0.92	949.7	0.27	0.31	0.95
8	409.6	0.52	0.94	836.9	0.28	0.24	0.95
16	380.5	0.49	0.91	762.5	0.39	0.11	0.91
30	336.1	0.45	0.91	635.7	0.40	0.03	0.91
Curing							
7 days	337.5	0.37	0.82	554.7	0.03	0.37	0.97
15 days	403.9	0.46	0.83	759.5	0.08	0.41	0.94
25 days	414.6	0.44	0.97	774.8	0.25	0.20	0.98
40 days	450.3	0.52	0.95	924.9	0.28	0.25	0.96
70 days	464.3	0.53	0.94	985.4	0.28	0.27	0.96

Table 4. Model parameters for MIX6

	Three parameter model			Two parameter model			
	k_1	k_2	R^2	k_3	k_4	k_5	R^2
W-D cycles							
0	335.6	0.34	0.82	531.4	0.04	0.32	0.97
4	484.1	0.43	0.84	859.6	0.09	0.37	0.97
8	500.1	0.42	0.90	771.5	0.11	0.39	0.96
16	448.9	0.40	0.89	785.7	0.16	0.26	0.96
30	414.9	0.39	0.94	679.4	0.16	0.22	0.96
Curing							
7 days	356.2	0.39	0.82	602.5	0.06	0.36	0.95
15 days	432.2	0.42	0.82	783.1	0.10	0.35	0.93
25 days	492.3	0.48	0.87	957.1	0.11	0.40	0.97
40 days	498.9	0.47	0.85	956.7	0.09	0.42	0.96
70 days	523.7	0.49	0.81	1074.2	0.09	0.46	0.96

Plots for MIX4 material as shown in Fig. 3(a) also reflect similar trends. In this case, the M_R values increased to about 20% at the end of 25 days. Curing up to 70 days only increased the stiffness of the material by further 10%. All these percentage variations have been calculated with respect to the initial 7-day stiffness values for both MIX4 and MIX6 materials.

7.1.3 Effect of Wetting-Drying (WD) Cycles

The Test samples of both MIX4 and MIX6 materials were first cured for 7 days and then subjected to wet-dry (WD) cycles. Resilient modulus (M_R) tests were conducted on the samples after 0, 4, 8, 16 and 30 WD cycles. The k-θ models developed for these Test samples of MIX6 is plotted in Fig. 2(b). It is observed that M_R increases by about 42% up to 8 WD cycles. This is because moisture intrusion resulting from the first 8 WD cycles contributes towards cement hydration rather than weakening the materials. This induces higher stiffness properties into the samples which increases resilient modulus. Additional WD cycles caused a reduction in M_R values. A cumulative drop of about 21% in resilient modulus was observed from 8 to 30 WD cycles.

MIX4 materials exhibited a comparatively weaker response to the WD cycles. Values of resilient modulus were found to increase by up to 16% at the end of 4 WD cycles (Fig. 3(b)). A further increase in WD cycles causes a decrease in M_R values.

At the end of 30 WD cycles, the M_R of the Test samples was found to be lower than the 7-day cured value by about 11%. This can be attributed to the fact that repeated wetting-drying processes have adverse effects on binding properties of cement (Khoury and Zaman 2007).

7.2 Pavement Layer Coefficient a_2

Resilient modulus (M_R) is a key parameter to determine the structural layer coefficient 'a_2' that is used for designing pavement base layer thickness. The layer coefficient a_2 was calculated in this study from M_R values using the correlation (AASHTO 2003a, b):

$$a_2 = 0.249 * \log\left[M_R(\text{psi})\right] - 0.977 \tag{1}$$

The resilient modulus (M_R) at a specific bulk stress of 30 psi (0.207 MPa) was used for this purpose. This stress level is considered ideal for pavement base layers in accordance with NCHRP 1-28A guidelines. For flexible pavement base, a minimum value of 0.10 for layer coefficient a_2 is considered acceptable (Faysal et al. 2016b). Figure 4 shows a plot of the M_R and layer coefficient a_2 calculated at a bulk stress of 30 psi (0.207 MPa) for both MIX4 and MIX6. It is evident from Fig. 4(a) that with increase in curing time, the a_2 values of both the materials increases. These values remain above the threshold value of 0.10 even after 70 days of curing. Thus, under ideal conditions both MIX4 and MIX6 can exhibit good long-term performance.

Effect of wetting-drying (WD) cycles on a_2 values have been shown in Fig. 4(b). At the end of 30 WD cycles, the MIX6 material has an a_2 value of about 0.15. However, a_2 values of MIX4 decreased with increasing WD cycles. The value decreased to about 0.115 at the end of 30 WD cycles. This is marginally above the minimum permissible value for flexible pavement base. A further reduction in a_2 values from addition WD cycles might cause the MIX4 material to be structurally incompetent as flexible pavement base.

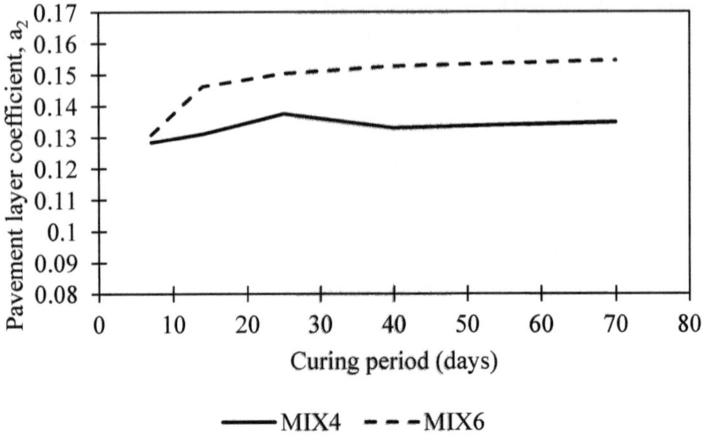

(a)

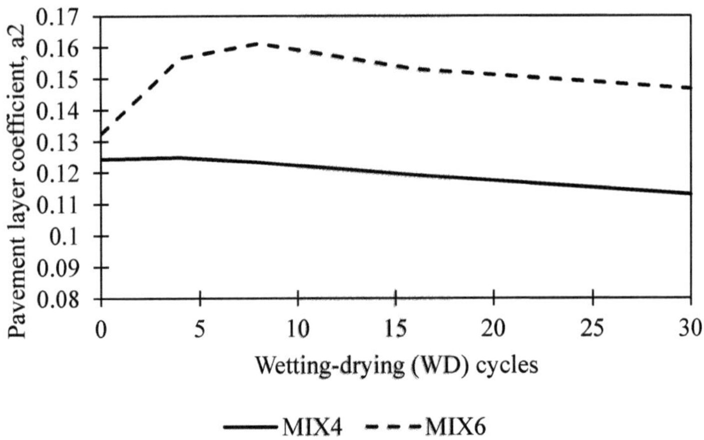

(b)

Fig. 4. Variation layer coefficient 'a_2' values with (a) curing and (b) wetting-drying (WD) cycles for MIX4 and MIX6 at a bulk stress of 30 psi (0.207 MPa)

8 Environmental Tests

8.1 Chemical Oxygen Demand (COD)

The COD concentration in the leachate samples was determined using Spectronic 20D+ model Spectrometer. Changes in COD with different number of wet and dry cycles is shown in Fig. 5. Test results indicate that the value of COD increases until 4 wet-dry cycles and COD decreases from 4 to 8 wet–dry cycles. MIX4 samples release higher

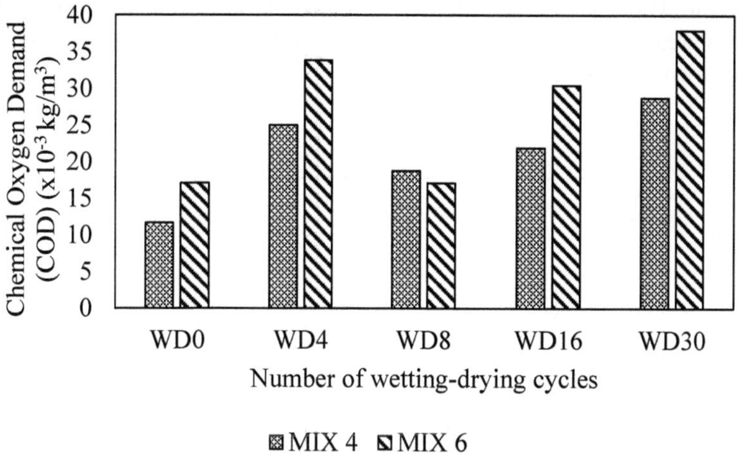

Fig. 5. Change in chemical oxygen demand (COD) with wetting-drying (WD) cycles

concentration of chemical when subjected to 4 wet-dry cycles after that there is a reduction in the released chemical compound. The amount of chemical compound releases into the leachate for MIX6 samples up to 4 wet-dry cycles is higher than the cycles from 4 to 8 wet-dry cycles which might have occurred because of the reduction in the rate of hydration. The value of COD obtained for MIX4 samples is less than the MIX6 samples which indicates that specimens stabilized with higher cement content releases more chemical compound in the leachate.

The value of resilient modulus decreases with increase in the number of WD cycles. Khoury and Zaman (2007) reported that this reduction in strength is due to the adverse effects of repeated wetting-drying processes on binding properties of cement. The value of resilient modulus decreases after the completion of 8 WD cycles for the specimens stabilized using 4% and 6% cement. The decrease in the resilient modulus shows the degradation of the specimens which led to the increased value of COD. From 8 to 30 wet-dry cycles the value of COD increased by 1.5 and 2.2 times for MIX4 and MIX6 samples respectively. The value of COD obtained is less than 120×10^{-3} kg/m^3, which is within the EPA guidelines.

8.2 Total Suspended Solids (TSS)

Total suspended solids (TSS) test was conducted on the leachate samples in accordance with ASTM D5907-13 standard method for non-filterable matter. TSS test results show that the value of TSS in leachate samples decreases by 80% and 83% for MIX4 and MIX6 respectively with the increase in the number of wet and dry cycles from 0 to 30. This might have occurred because of the improved inter particle bond and well developed matrix due to hydration of cementitious materials within the specimens. According to the test results, the value of TSS is higher for the leachate samples obtained for MIX4 than that for MIX6 which might have attributed to the fact that the specimens prepared using 4% cement content is weaker than the prepared specimens at 6% cement content as shown in Fig. 6.

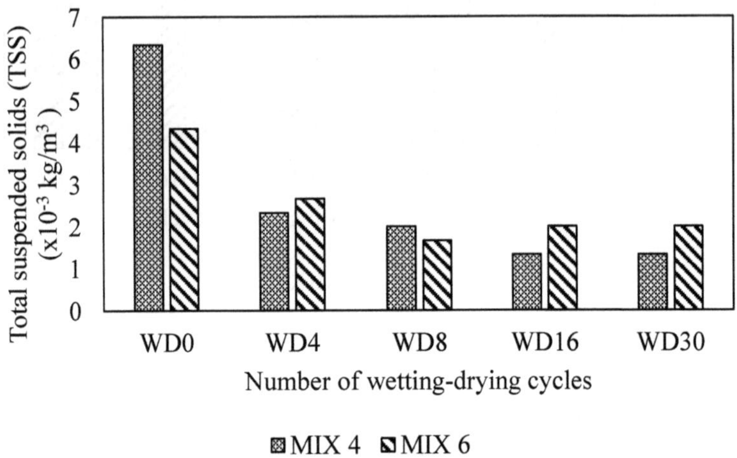

Fig. 6. Change in total suspended solids (TSS) with wetting-drying (WD) cycle

8.3 Total Dissolved Solids (TDS)

Total dissolved solids (TDS) or filterable parameter is one of the important parameters for the treatment of raw water, wastewater and in monitoring of streams. TDS tests were performed in accordance with the ASTM D5907-13 standard method for filterable matter. Filtrate obtained after passing the leachate through the glass fiber filter paper of 1.5 μm nominal pore size. Dissolved solids represent the amount of cementitious materials washed out from the specimen due to its reaction with water (Faysal et al. 2017a). TDS tests were conducted on the collected leachate samples after selected number of wet and dry cycles such as 0, 4, 8, 16, and 30.

The value of TDS changes with the cement content and number of wet and dry cycles (Fig. 7). The lower cement content results in lower value of TDS than the specimens contain higher cement content. TDS value obtained at different cycles follows similar trend as COD. TDS values increase with increase in wet-dry cycles until 4 after that it reduces after completion of 8 wet-dry cycles. During the first 4 wet-dry cycles the value of TDS increase due to rapid hydration process. The value of TDS increases by 75% and 83% for MIX4 and MIX6 samples respectively with the increase in WD cycles from 8 to 30 cycles. The increased value of TDS denotes the degradation of strength of specimens, which complies with trend obtained from resilient modulus test results at different number of wet and dry cycles. TDS test results are well within the limit of 500×10^{-3} kg/m^3 as per EPA guidelines (EPA 2005).

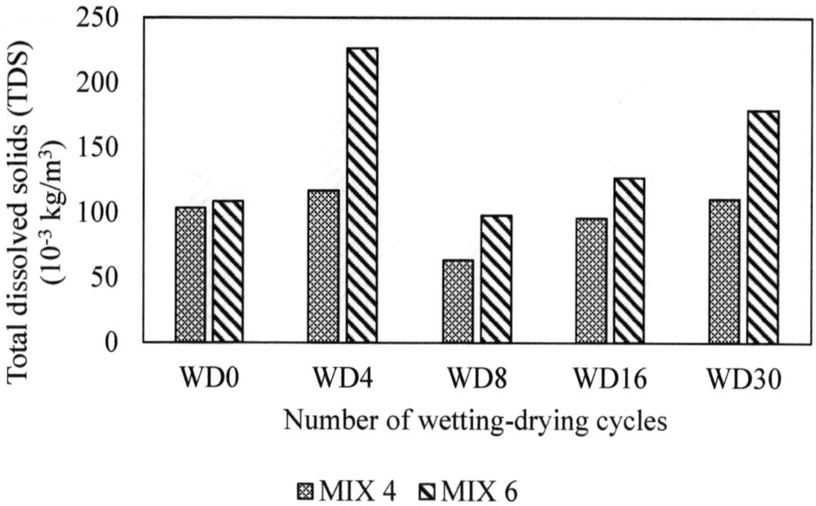

Fig. 7. Change in total dissolved solids with wetting-drying (WD) cycle

8.4 Turbidity

Turbidity is a parameter to determine the amount of suspended matter such as soil particles, different types of organic and inorganic matter and microorganisms present in water. Turbidity was measured using a HACH 2100P Model portable turbidimeter that operates on the nephelometric principle of turbidity measurement in Nephelometric Turbidity Unit (NTU). This equipment measures the optical property of water such as the amount of light scattered and absorbed while passing through the water sample. The variations in turbidity test results due to different combinations of cement-stabilized materials are included in Fig. 8. It can be inferred that the value of turbidity decreases with increase in cement content. The maximum value of turbidity is 1 NTU that is less than 5 NTU. This satisfies the EPA guidelines.

8.5 pH

pH is the measure of the acidity or alkalinity of water or leachate samples. pH test was performed in on the obtained leachate samples in accordance with ASTM D1287. The value of pH increases with increase in cement content in the prepared specimens. This might have occurred because of the hydration reaction taking place between water and calcium carbonate to form soluble calcium hydroxide (Faysal et al. 2017a). Test results are shown in Fig. 9. The value of pH is well within the range of 6–9 as per EPA guidelines for storm water sampling (EPA 2005).

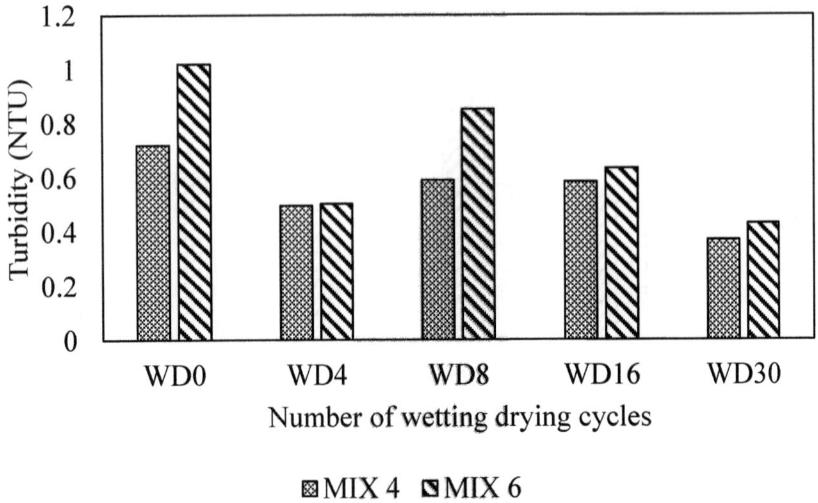

Fig. 8. Change in turbidity with wetting-drying (WD) cycle

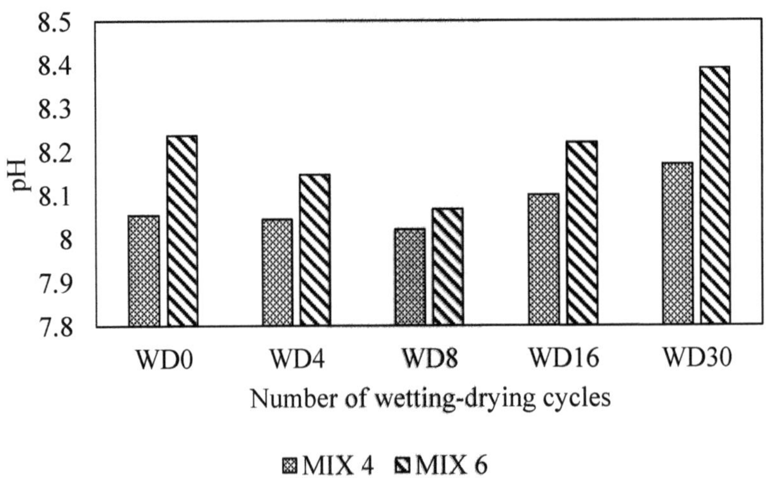

Fig. 9. Change in pH with wetting-drying (WD) cycles for MIX4 and MIX6

9 Conclusion

This study was conducted to evaluate the long-term performance of stabilized pavement base containing high percentage of RAP when subjected to repeated wetting-drying cycles. Certain conclusions drawn from this study can be summarized as follows:

- Conventional curing did not have any detrimental effects on stiffness properties of 50–50% RCCA-RAP materials when stabilized with 4% and 6% cement content. Resilient modulus (M_R) values for both the materials increased until the end of 70 days.
- Resilient modulus values of 50–50% RCCA-RAP materials at 6% cement content after 30 WD cycles was found to be higher by about 10% than the initial value obtained at 0 WD cycles. On the other hand, M_R values of the of 50–50% RCCA-RAP materials at 4% cement content after 30 WD cycles was found to be lower by about 11% than the initial value obtained at 0 WD cycles. This deterioration is an indication of loss in binding properties due to the applied wetting-drying (WD) cycles in the 50–50% RCCA-RAP material stabilized with 4% cement.
- Pavement layer coefficient (a_2) for both material combinations were found to be above the minimum required value of 0.10 for pavement base.
- At the end of 30 WD cycles, values of COD, TDS, TSS, Turbidity and pH for 50–50% RCCA-RAP materials mixed with 4% and 6% cement content were found to be within the permissible values for storm water (EPA 2005).

Therefore, RAP and RCCA materials if mixed at 50–50% and stabilized with 4% or 6% cement content was found to be structurally competent and environmentally sound up to 30 wetting-drying (WD) cycles. However, increased number of WD cycles might have more detrimental effects on the strength of these materials. Thus, further studies on the effect of higher number of wetting-drying cycles on RCCA and RAP materials should be conducted.

References

AASHTO T307-99: Standard Method of Test for Determining the Resilient Modulus of Soils and Aggregate Materials. American Association of State Highway and Transportation Officials, Washington, D.C. (2003a)

AASHTO T307-99: Standard Method of Test for Determining the Resilient Modulus of Soils and Aggregate Materials. AASHTO Part II Test Methods, American Association of State Highway and Transportation Officials, Washington, D.C. (2003b)

ASTM D5907-13: Standard Test Methods for Filterable Matter (Total Dissolved Solids) and Nonfilterable Matter (Total Suspended Solids) in Water. ASTM International, West Conshohocken, PA (2013)

Faysal, M., Hossain, M.S., Salah, S.B., Bhattacharjee, S., Thain, B., Khan, M.S.: Characterization of Geo-Environmental Properties of Untreated or Cement Treated Recycled Base Materials in Pavement Base Layer Applications. Accepted for publication in the Proceedings of Geotechnical Frontiers, Orlando, FL (2017a)

Faysal, M., Hossain, M.S., Salah, S.B., Khan, M.S., Flores, C.M.: Characterization of Strength and Long-Term Durability of Recycled Flex Base Materials. Accepted for presentation in the proceedings of Transportation Research Board, Washington, D.C. (2017b)

Faysal, M., Mahedi, M., Aramoon, A., Thian, B., Hossain, M.S., Khan, M.S.: Strength characterization of untreated and cement-treated recycled flex-base materials. In: Geotechnical and Structural Engineering Congress 2016, pp. 1233–1244 (2016a)

Faysal, M., Mahedi, M., Aramoon, A., Thian, B., Hossain, M.S., Khan, M.A., Khan, M.S.: Determination of the structural coefficient of different combinations of cement-treated/untreated recycled base materials. In: Geotechnical and Structural Engineering Congress 2016, pp. 1198–1208 (2016b)

George, K.P., Davidson, D.T.: Development of a freeze-thaw test for design of soil-cement. Highw. Res. Rec. (36) (1963)

Hansen, K.R., Copeland, A.: Asphalt Pavement Industry Survey on Recycled Materials and Warm-Mix Asphalt Usage: 2014 (No. Information Series 138) (2015)

Hedayati, M., Hossain, S.: Data based model to estimate subgrade moisture variation case study: low volume pavement in North Texas. Transp. Geotech. **3**, 48–57 (2015)

Hossain, S., Ahmed, A., Khan, M.S., Aramoon, A., Thian, B.: Expansive subgrade behavior on a state highway in North Texas. In: Geotechnical and Structural Engineering Congress 2016 (2016)

Jones, D., Wu, R., Louw, S.: Comparison of full-depth reclamation with foamed asphalt and full-depth reclamation with no stabilizer in accelerated loading test. Transp. Res. Rec. J. Transp. Res. Board **2462**, 126–135 (2014)

Khoury, N., Zaman, M.: Effect of wet-dry cycles on resilient modulus of class C coal fly ash-stabilized aggregate base. Transp. Res. Rec. J. Transp. Res. Board **1787**, 13–21 (2002)

Khoury, N., Zaman, M.: Durability effects on flexural behavior of fly ash stabilized limestone aggregate. J. Test. Eval. **34**(3), 1–9 (2005)

Khoury, N., Zaman, M.M.: Durability of stabilized base courses subjected to wet-dry cycles. Int. J. Pavement Eng. **8**(4), 265–276 (2007)

Little, D.N., Males, E.H., Pursinski, J.R., Stewart, B.: Cementitious stabilization. http://www.nationalacademies.org/trb/publications/millennium/00016.pdf. Accessed Jan 2005

Maalouf, M., Khoury, N., Laguros, J.G., Kumin, H.: Support vector regression to predict the performance of stabilized aggregate bases subject to wet-dry cycles. Int. J. Numer. Anal. Methods Geomech. **36**(6), 675–696 (2012)

Mahedi, M.: Potential applicability of stress wave velocity method on pavement base materials as a non-destructive testing technique (2016)

Miller, G.A., Zaman, M.M., Rahman, J., Tan, N.K.: Laboratory and field evaluation of soil stabilization using cement kiln dust (2003)

Puppala, A.J., Hoyos, L.R., Potturi, A.K.: Resilient moduli response of moderately cement-treated reclaimed asphalt pavement aggregates. J. Mater. Civ. Eng. **23**(7), 990–998 (2011)

Puppala, A., Mohammad, L., Allen, A.: Engineering behavior of lime-treated Louisiana subgrade soil. Transp. Res. Rec. J. Transp. Res. Board **1546**, 24–31 (1996)

Unified Facilities Guide Specifications: Concrete Pavement for Airfields and Other Heavy-Duty Pavements (2010)

Tao, M., Zhang, Z.: Durability of cement-stabilized low-plastic soils. In: Transportation Research Board 85th Annual Meeting (No. 06-1255)

Tex-110-E: Particle size analysis of soils. Texas Department of Transportation, Austin, TX (1999)

Tex-113-E: Laboratory compaction characteristics and moisture-density relationship of base materials. Texas Department of Transportation, Austin, TX (1999)

Transportation Applications of Recycled Concrete Aggregate. State of the Practice National Review. U.S. Department of Transportation Federal Highway Administration, Washington, D.C. (2004)

Full Depth Reclamation (FDR) to Repair Road Damaged in the Energy Sector

Dar Hao Chen[1(✉)], Tom Scullion[2], and Kun Li[3]

[1] Texas Department of Transportation, 4203 Bull Creek #39, Austin, TX 78731, USA
darhao2008@gmail.com
[2] Texas Transportation Institute, Texas A&M University, College Station, TX, USA
T-Scullion@tti.tamu.edu
[3] College of Civil Engineering and Architecture, China Three Gorges University, University Avenue 8, Yichang 443002, Hubei Province, People's Republic of China
likun@ctgu.edu.cn

Abstract. Rehabilitating the roadways in the energy sector is one of Texas Department of Transportation's (TxDOT) top priority activities and is also a high profile activity attracting a lot of legislative oversight. Adding to this problem is that many of these roadways are often in remote areas where there is no access to quality paving materials. In addition, there is no detour and the roadway must be opened at end of work day. The FDR offers great potential to stabilize roadways in-place making use of existing materials and determining the optimal stabilizing agent to make these roadways structurally adequate. Although TxDOT has utilized the FDR for several decades, the performance results vary. Some FDR projects last over 10 years without any distress. However, there were several premature failure occurred within one month of trafficking. This paper presented a successful FDR project using foamed asphalt. The main reason of using foamed asphalt is its one-pass operations that can open to traffic immediately after compaction. Lab mix design was conducted to determine the optimum asphalt and cement contents. FWD tests were conducted to evaluate the change of foamed asphalt base. The determined modulus from foamed asphalt base can be used for future design purpose. The backcalculated moduli for foamed asphalt base increased from 493 ksi (2 weeks after construction) to 800 ksi (6 months after construction). District personal is very pleased with the performance of FDR using foamed asphalt.

1 Background

Texas Department of Transportation (TxDOT) now faces major pavement rehabilitation and funding problems with the roadways all over Texas which have been severely damaged by energy development activities. These are often low volume roadways with thin weak pavements with often marginal materials which were never designed to carry these large numbers of very heavy trucks. Rehabilitating the roadways in the energy sector is one of TxDOT's top priority activities and is also a high profile activity

L. Mohammad (ed.), *Advancement in the Design and Performance of Sustainable Asphalt Pavements*, Sustainable Civil Infrastructures, DOI 10.1007/978-3-319-61908-8_9

attracting a lot of legislative oversight. Many districts including Corpus Christi, Laredo, San Antonio, San Angelo, Odessa and several others have hundreds of miles of totally destroyed roadways. Adding to this problem is that many of these roadways are often in remote areas where there is no access to quality paving materials. In addition, there is no detour and the roadway must be opened at end of work day.

Rehabilitation of highway pavements, particularly, low volume roads through full-depth reclamation (FDR) is a cost-effective option that reduces the use of existing base materials and eliminates the effort on disposal of the base materials (Garibay et al. 2008; Franco et al. 2009). The process of FDR usually consists of pulverizing a predetermined amount of the existing unbound granular base and if desired asphalt layer, optionally stabilizing the material with additives, and compacting the new layer to a proper density, as shown in Fig. 1. FDR can be used to treat a wide range of problems, particularly problems related to weak base courses. If designed and constructed properly, FDR is capable of rectifying rutting and fatigue cracking problems, deterioration of pavements due to maintenance patching and deterioration of ride quality caused by depressions and heaving.

Fig. 1. (A) Deteriorated roadway (B) full depth reclamation (FDR) (C) opening the rehabilitated roadway to traffic in the same day

The FDR offers great potential to stabilize roadways in-place making use of existing materials and determining the optimal stabilizing agent to make these roadways structurally adequate. Although TxDOT has utilized the FDR for several decades, the performance results vary (Chen 2009; Franco et al. 2009). Some FDR projects last over 10 years without any distress. For example, One TxDOT District has used emulsion FDR in 2004 to rehab SH87 and the pavement condition is excellent as evident in Fig. 2 with IRI in the range of 80 in/mile (1.26 m/km). The same district also used emulsion FDR in 2004 to rehab the frontage road of IH35 (Chen et al. 2008). The premature failure occurred within one month of trafficking, as shown in Fig. 3. Figure 4 shows another recent premature failure (deep rutting) due to insufficient early strength to handle heavy loads from oil field truck traffic. District indicated that it would cost additional $2 million dollars for traffic control to allow FDR section to cure for 1 week. It means the pavement section would need traffic control personal at the job site for 24/7 during the 3-month construction period. The maximum FWD deflections (normalized to 9000 lb) in the deep rutting area are in the range of 50 mils (1.27 mm).

Base on the empirical experience, it needs the maximum deflection in the range of 20 mils (0.508 mm) to provide sufficient structural capacity to handle such heavy truck traffic load. It means the constructed roadway has deflection 2.5 times higher than the desirable.

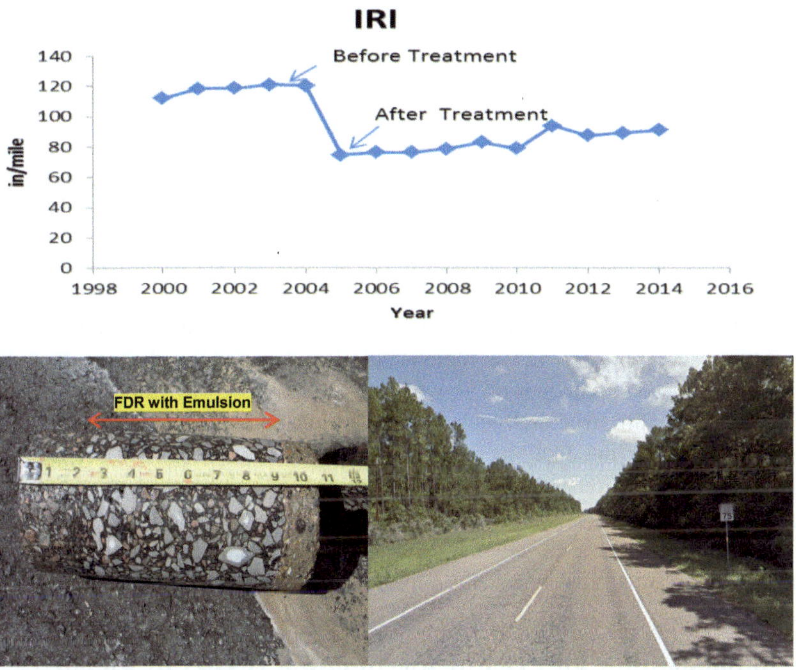

Fig. 2. Successful FDR project that maintain low IRI for the last 10 years

Fig. 3. Premature failure-occurred within one month of trafficking

Fig. 4. A recent premature failure (deep rutting) due to insufficient early strength from FDR layer

Texas is a big state with regions of different climates and local materials. Districts in wet regions expressed concerns for cement only FDR because frequent rains disrupt constructions before FDR being compacted to shape. Emulsion FDR has been successful for several projects in wet regions but premature failures do occur from time to time. Districts need guidelines on how to take existing roadways thru the FDR process to make long lasting structurally adequate roadways able to carry overloaded vehicles. Correctly applying the FDR techniques to these roadways can save TxDOT many millions of dollars. Recently, over $500 million dollars were appropriated for conducting these repairs and more money is urgently needed. It is estimated that applying these FDR techniques will generate a 20% cost saving over traditional methods in terms of reduced haul distances and improved pavement life.

2 Implementing Full Depth Reclamation (FDR) to Repair Road Damaged in the Energy Sector

To accelerate the rehabilitation of energy-impacted pavements, TxDOT is evaluating different FDR options which can be constructed in one pass of the stabilizer. This evaluation of one-pass operations led to the evaluation of foamed asphalt.

FM 99 in the Corpus Christi District is a 5 mile section which was designed to use the cement treated subbase approach described above. To evaluate an alternative design the District decided to construct an experimental foamed asphalt section. This paper provides the design, construction and initial field monitoring of a one mile section placed on FM 99. Construction was completed in June of 2014.

3 FM 99

FM 99 is an extremely heavily trafficked energy development roadway. The typical section in the design plan consists of 1 in. of asphaltic materials as a surface layer and 6 in. of flexible base material. However, with years of maintenance activities, the asphalt thickness has been found to be in the range of 3 to 5 in. The existing condition is shown below in Fig. 5 with deep rutting and severe alligator cracking in many places. Table 1 presents the results from the laboratory mixture design tests. Based on the results in Table 1, it was decided to treat the materials on FM 99 with 1.5% cement and 2.4% foamed asphalt. The current design easily passes both the wet and dry strength requirement and the Unconfined Compressive Strength (UCS) after soaking shows that this is a strong water proof base material. The materials fails the retained strength requirement, it is recommended that this 70% retained strength be waived for this project. This criteria was recommended by Wirtgen (Franco et al. 2009) for areas with substantial rainfall. A subsequent revision was proposed to permit the retained strength to vary from 50% to 70% based on rainfall amounts based on location specific annual rainfall amounts.

Fig. 5. Existing pavement conditions on US99 before FDR

Table 1. Lab results for foamed asphalt design (FM 99)

	TxDOT test method	1.5% cement and 2.4% foam
Dry IDT	Tex 226-F	75 psi
Wet IDT	Tex 226-F	41 psi
Min UCS	Tex 117-E	171 psi*
Dry density		121.5 lb/ft^3
Opt. % Mois		7.3%

After completion of each section, the contractor turned traffic onto the section no more than 1 h after completing compaction, as show in Fig. 6. At the end of the day the section was primed with a light application of MC 30 and grade 5 rock. The foam asphalt section was subject to normal traffic until placement of a surface treatment. The application of the MC 30 prime and the final surface seal was completed one week after construction.

The FWD has been widely used to measure the structural strength and uniformity of the pavement structure (Chen and Scullion 2007). FWD was used in this study to measure the foamed asphalt treated base. FWD data was collected 2 weeks after the foamed asphalt section was completed, as shown in Fig. 6. The average maximum

Fig. 6. Traffic condition during FDR construction and FWD testing on top of foamed asphalt treated base

deflection with close to 10,000 lb. loading is 10 mils. The backcalculated modulus of the 11 in. foamed asphalt treated base by using TxDOT's MODULUS 6 software was found to be 493 ksi. This value is excellent and reflects highly effective stabilization. FWD tests were conducted in December 2014 which was approximately 6 months after the construction. The average maximum deflection with close to 10,000 lb. loading is 6 mils and the backcalulated foamed asphalt base modulus increased to 800 ksi. Condition survey was conducted in August 2015 that was one year after the intense trafficking. No visible distress can be found and district personal is very pleased with the performance.

4 Conclusions

The data generated from the sections on FM 99 support the following conclusions:

- Foamed asphalt treatment appears a viable option for rehabilitating the deteriorated pavements. The main advantage of foamed asphalt treatment is the ability to allow traffic on the section almost immediately after compaction.
- After 6 months of heavy truck trafficking, field FWD data from FM 99 indicated the modulus actually increasing from 493 ksi to 800 ksi.

References

Chen, D.-H., Scullion, T.: Using nondestructive testing technologies to assist in selecting the optimal pavement rehabilitation strategy. J. Test. Eval. **35**(2), 211–219 (2007)

Chen, D.-H., Scullion, T., Lee, T.-C., Bilyeu, J.: Results from a forensic investigation of a failed cement treated base. J. Perform. Constr. Facil. **22**(3), 143–153 (2008)

Chen, D.-H.: Investigation of a pavement premature failure on a weak and moisture susceptible base. J. Perform. Constr. Facil. **23**(5), 309–313 (2009)

Franco, S., Moss, S.P., Yuan, D., Nazarian, S.: Design, Constructability Review and Performance of Dual Base Stabilizer Applications. Report FHWA/TX-09/0-5797-1, Texas Department of Transportation, 172 p. (2009)

Garibay, J.L., Yuan, D., Nazarian, S., Abdallah, I.: Guidelines for Pulverization of Stabilized Bases. Report FHWA/TX-08 0-5223-2, Texas Department of Transportation (2008)

Prediction of Strength and Stiffness Properties of Recycled Pavement Base Materials Using Non-destructive Impact Echo Test

Masrur Mahedi[1(✉)], Sahadat Hossain[2], Mohammad Faysal[2], Mohammad Sadik Khan[3], and Asif Ahmed[2]

[1] Department of Civil, Construction and Environmental Engineering, Iowa State University, Ames, IA 50011, USA
mmahedi@iastate.edu

[2] Department of Civil Engineering, The University of Texas at Arlington, Arlington, TX 76019, USA
hossain@uta.edu, {md.faysal,asif.ahmed0}@mavs.uta.edu

[3] Department of Civil and Environmental Engineering, Jackson State University, Jackson, MS 39217, USA
mohammad_sadik.khan@jsums.edu

Abstract. Reclaimed Asphalt Pavement (RAP) and Recycled Crushed Concrete Aggregates (RCCA) are being used increasingly as an alternative to the conventional base materials in pavement construction. However, product variability and lack of strength-stiffness characteristics are limiting the use of recycled materials in pavement application. A non-destructive evaluation technique could provide a better quality control tool for the highway officials during the construction. Therefore, the objective of the current study is to develop a correlation between the non-destructive impact echo and Unconfined Compressive Strength (UCS) test for RAP and RCCA materials. Both impact echo and UCS tests were performed on specimens prepared with 100% RAP and RCCA. The RAP and RCCA were separately treated using four (0, 2, 4 and 6%) different dosages of Portland cement. Based on the experimental results, the range of P-wave velocity was found between 175 m/s to 475 m/s, the compressive strength varied between 67 kPa to 2860 kPa and tangent modulus range was 19 MPa to 280 MPa. Dynamic modulus of elasticity was also calculated from the P-wave velocity, density, and Poisson's ratio. At 4% and 6% cement contents, dynamic modulus of elasticity was within 10% of the tangent modulus found from UCS test. At 0% and 2% cement content, higher variation was observed. Inadequate fines to fill the voids might be the reason of lower P-wave velocity at 0% and 2% cement contents. This could eventually have predicted lower values of dynamic modulus.

Keywords: Recycled pavement materials · Non-destructive testing · Strength · Resilient modulus

L. Mohammad (ed.), *Advancement in the Design and Performance of Sustainable Asphalt Pavements*, Sustainable Civil Infrastructures, DOI 10.1007/978-3-319-61908-8_10

1 Introduction

Pavement is a layered system which typically consists of a surface layer, base course, (optionally) subbase course, and the subgrade layers. Among all the layers, base layer plays the most important role in transferring the induced stress to the underlying layers. A properly designed base layer provides drainage to water entering the pavement system, provides the insulation to frost susceptible subgrade, prevents the intrusion of fine grained particles into the surface layer, and provides a working platform for the construction operation. The rate of load distribution is also significantly affected by the quality of the base course materials (Potturi 2006). Therefore, the base layer must have sufficient strength to meet the design specifications without any trace of failure.

Natural aggregates have been traditionally used for pavement base construction. In recent times, depletion of natural resources, collection costs, and environmental concerns have constrained their use (Hoyos et al. 2011). According to the report of USDOT 2004, the estimated use of aggregates is expected to be more than 2.5 billion tons per year by the year 2020. This is due to the rapid increase of building and pavement construction (Gonzalez and Moo-Young 2004). Moreover, waste generation from the pavement rehabilitation and structural demolition projects are declining landfill spacing. These causes have raised the importance for the pavement industry to find an alternative way of reusing these materials (Ordonez 2006). Considering these facts, Reclaimed Asphalt Pavement (RAP) and Recycled Crushed Concrete Aggregate (RCCA) have become a potential alternative to natural limestone aggregates. Maintenance of flexible pavements and rehabilitation projects generate huge amounts of RAP as by product. According to the National Asphalt Pavement Association (Hansen and Copeland 2013), approximately 350.7 million tons of plant mix asphalt were produced in the United States of America. The total reported RAP generation was around 76.1 million tons. Similarly, demolition of concrete pavements, bridges, curbs, and buildings are the main sources for of RCCA. Research on compressive strength (Faysal et al. 2016a), resilient modulus response (Faysal et al. 2016b), compression wave velocity (Mahedi et al. 2017), flexural strength, unrestrained shrinkage (Euch Khay et al. 2014), accelerated loading test (Jones et al. 2014), and physical–mechanical properties (Medina et al. 2014) are the recent works on RAP and RCCA materials.

Properties of recycled materials are source dependent and highly variable, which requires extensive evaluation before using in construction (Wilburn and Goonan 1998). For large scale investigation, traditionally practiced methods might be unreasonable in terms of time, cost, reliability, and applicability leading to the necessity of further research on other potential methods (Graveen 2001). It would be beneficial to establish an economic, reliable, and less time-consuming method to determine the mechanical properties for adequate QA/QC during the pavement construction for recycled aggregate materials. Based on these requirements, the objective of this study is to establish a non-destructive test method to utilize in determining the strength and stiffness properties of recycled pavement base materials. Focusing this goal, Reclaimed Asphalt

Pavement (RAP) and two types (Grade-1 and Grade-2) of Recycled Crushed Concrete Aggregates (RCCA) treated with four different dosage levels (0, 2, 4 and 6%) of Portland cement were assessed based on UCS and impact echo test. Compressive strength found from UCS test and P-wave velocity from impact echo test were utilized to evaluate the materials. Dynamic modulus of elasticity was determined using the P-wave velocity, density, and Poisson's ratio. The stress vs. strain curves obtained from the UCS test were also used to determine the tangent modulus of elasticity for the samples. Finally, P-wave velocity was applied to characterize the strength and stiffness properties of cement treated recycled base materials.

According to the test results, inclusion of cement has a significant effect on P-wave velocity, strength, and stiffness properties. P-wave velocity, strength, and modulus of elasticity increase with an increase in cement content. P-wave velocity ranged between 175 m/s to 475 m/s, compressive strength varied between 67 kPa to 2860 kPa, and tangent modulus range was 19 MPa to 280 MPa. Elastic modulus obtained from UCS test results were within the 10% range of the dynamic modulus obtained from impact echo test results.

2 Material Collection

Two different types of recycled materials (RAP and RCCA), which are commonly used in Texas as the pavement base materials, were selected for the test program of this study. RAP materials were collected from the TxDOT stockpile situated in Dallas County, Texas. Grade-1 and Grade-2 type RCCA were collected from a commercial concrete crushing facility located in Dallas. This facility is one of the TxDOT approved companies supplying recycled flex base materials in Dallas–Fort Worth (DFW) area. Grade-2 type RCCA was derived from two different sources (Source-1 and Source-2). Portland Type II cement was used as the binder in this study, which has a 28-day compressive strength greater than 50 MPa.

3 Basic Engineering Tests

Standard test method specified in TxDOT guidelines (Tex-110-E) were followed for sieve analysis and the gradation curves, both shown in Fig. 1. Sieve analysis shows that about 99% of the materials are retained on No. 200 sieve. According to TxDOT specification Item 276, no hydrometer analysis was performed. Atterberg limits were also not determined because of the same reason. Coefficient of curvature and uniformity were calculated for both materials, reported in Table 1. All the materials were classified as well graded gravel according to the Unified Soil Classification System (USCS). Initial moisture content and bulk specific gravity were also determined and are reported in Table 1.

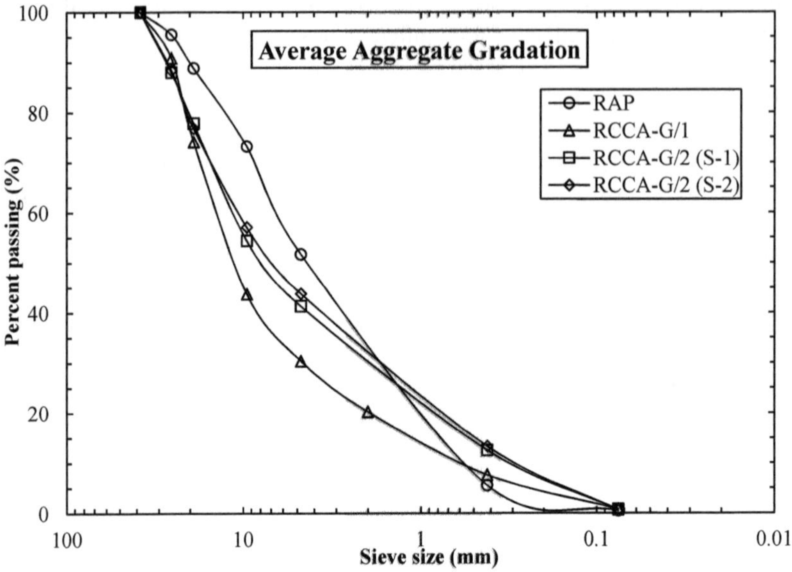

Fig. 1. Sieve analysis result of RAP and RCA

Table 1. Basic materials properties used for this study

Parameters	RAP	RCCA-G/1	RCCA-G/2 (S-1)	RCCA-G/2 (S-2)
Coefficient of curvature (C_C)	1.33	2.51	2.28	2.33
Coefficient of uniformity (C_U)	7.84	23.21	34.09	24.85
Moisture content (%)	0.23	1.12	0.93	0.87
Dry bulk specific gravity	1.90	1.88	2.02	1.93

Optimum Moisture Content (OMC) and Maximum Dry Density (MDD) were determined in accordance with Tex-113-E: Laboratory Compaction Characteristics and Moisture-Density Relationship of Base Materials. The summary of OMC and MDD have been shown in Fig. 2, indicating that optimum moisture content varies within the range of 6.5–11%. The maximum dry density ranges between 18–21 kN/m^3 (Table 2).

4 Experimental Setup

RAP and RCCA from Source-1 were considered for UCS and impact echo test. Typically cement treated base layer consists of 3–10% cement of the total weight of the mixture. Test samples for each combination were treated by 4 different cement contents: 0%, 2%, 4%, and 6%. Three identical samples were prepared for each combination to check the repeatability of the test results. Samples were prepared at optimum

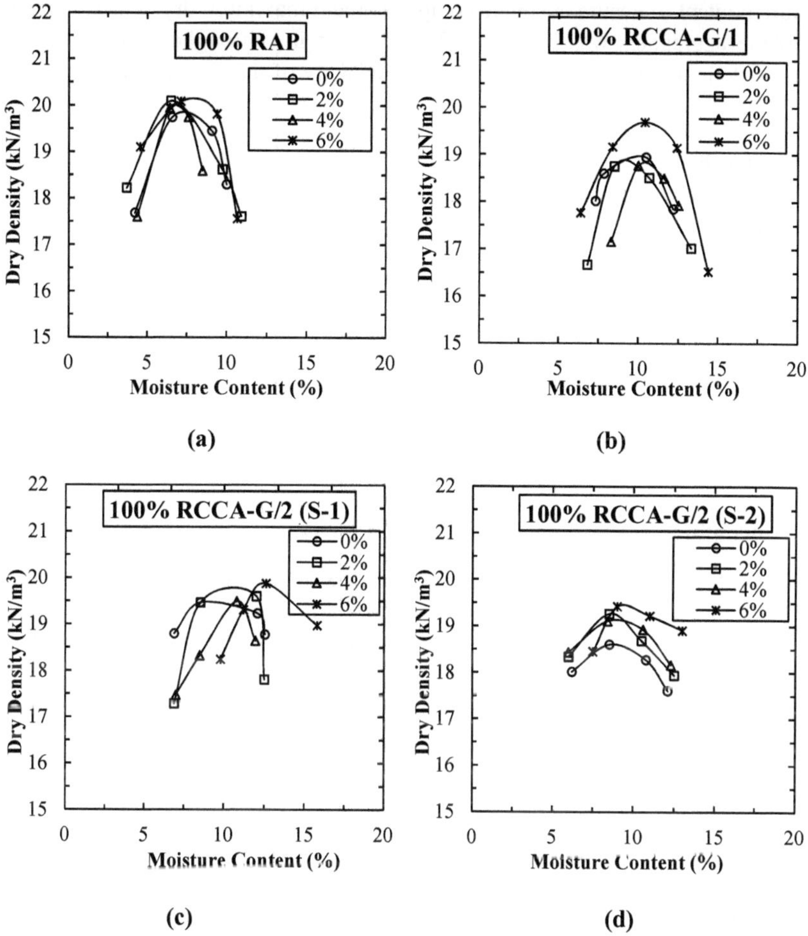

Fig. 2. Moisture-density relationship of cement treated (a) 100% RAP, (b) 100% RCCA-G/1, (c) 100% RCCA-G/2 (S-1), and (d) 100% RCCA-G/2 (S-1)

Table 2. Experimental program of the study

	Cement content (%)	100% RAP		100% RCCA-G/1		100% RCCA-G/2 (S-1)	
		UCS	Impact echo	UCS	Impact echo	UCS	Impact echo
Number of samples for each test	0	3	3	3	3	3	3
	2	3	3	3	3	3	3
	4	3	3	3	3	3	3
	6	3	3	3	3	3	3

moisture content to ensure adequate compaction and maximum dry density. Maximum size of the aggregates was limited to 32 mm (1.25 in.) throughout the test program. This was done to ascertain compaction homogeneity of the test samples.

5 Methods

5.1 Specimen Preparation

Standard test method Tex-113-E were followed for sample preparation. Samples were prepared in 4 lifts, each compacted by 50 blows to achieve the required compaction level at optimum moisture content. An automated mechanical compactor, which meets the TxDOT specifications, was used for compacting. Samples were 152.4 mm (6 in.) in diameter and 203.2 mm (8 in.) in height. Prepared samples were kept in a moist room for 7 days for curing, in accordance with Tex-120-E, before testing.

5.2 Unconfined Compressive Strength (UCS) Test

ASTM D 2166 standard test procedure was followed for the UCS test of the samples after 7 days of curing. A servo controlled tensile/compression testing machine, with a data acquisition system for measuring the axial deformation, was used for the test. Maximum axial load at which the sample failed, was taken as the ultimate load bearing capacity of the sample. The stress vs. strain curves obtained from the test were used to determine the tangent modulus of elasticity for the samples.

5.3 Impact Echo Test

P-wave velocity through each of the samples were measured following the standard test method designated by ASTM C 1383-04. As the both ends of the samples were accessible, tests were performed by direct transmission method. Hammer impact was made on one end and the geophone was placed on the other. The wave generated by the hammer impact propagates through the sample and is identified by the geophone on the other end. Geophone identifies the arrival of the wave by surface deflection in the time domain. Time domain response of geophone was then transferred to the frequency domain by First Fourier transformation (FFT) technique. Finally, an amplitude spectrum was obtained where the arrival of the wave can easily be identified by a dominant peak in the spectrum. By knowing the height of the sample and the travel time of wave, P-wave velocity was calculated. Figure 3(a) shows a typical time domain and Fig. 3(b) shows a typical frequency domain spectrum found from the impact echo test.

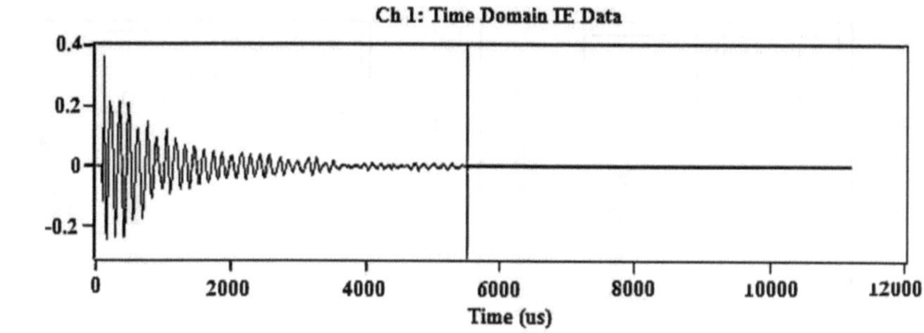

(a)

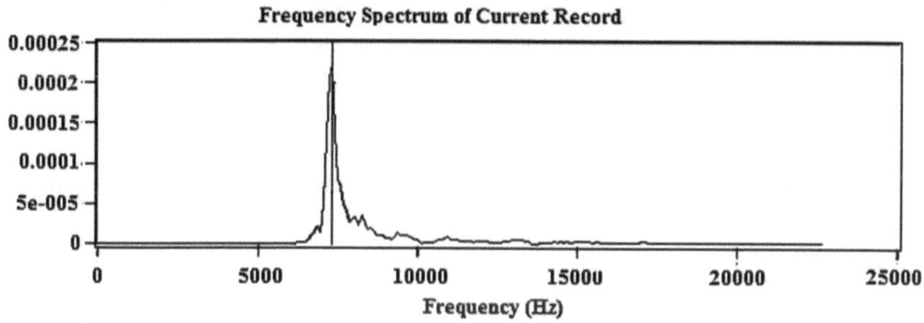

(b)

Fig. 3. (a) Typical time domain spectrum, and (b) typical frequency domain spectrum from impact echo test

6 Results

6.1 Unconfined Compressive Strength Test Results

Compressive strength of recycled materials treated by 4 different cement content has been shown in Fig. 4(a). From Fig. 4(a), compressive strength increases significantly with the increase of cement content. Higher strength was found by RCCA compared to RAP that showed the lowest strength. 2% cement in 100% RAP materials did not increase the strength significantly. This indicates that 2% cement is inadequate to create adhesion between the asphalt coated aggregates. TxDOT specifies guidelines for pavement base construction under Item 276, where minimum strength requirements are class specified. For Class-L, listed minimum 7 days unconfined compressive strength is 300 psi which cannot be achieved by stabilizing RAP. Minimum strength requirements can be attained by RCCA if it is treated by more than 6% cement content. Modulus of elasticity was determined as the offset tangent modulus of the stress-strain curves found from the UCS test. The variation of tangent modulus with different cement dosages has been shown in Fig. 4(b). Little variation of elastic modulus was observed for unbound mixes, whereas; inclusion of cement causes a dramatic increase of moduli values.

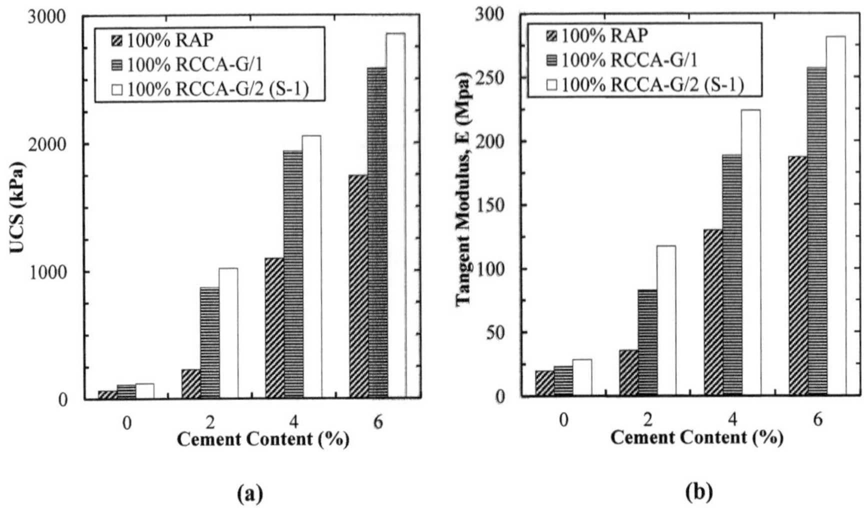

Fig. 4. (a) Variation of UC strength, and (b) tangent modulus with cement content

6.2 Impact Echo Test Results

For all 3 different types of recycled materials, Fig. 5 shows the variation of P-wave velocity with cement content. It can be observed that P-wave velocity increases with the increase of cement content. No significant difference in velocity was detected for the untreated specimens. Cement makes the samples denser by filling voids, which eventually increased the wave velocity. With the addition of 2% cement, the increase of the velocity in the aggregate blends ranged from 63% to 97%, except for 100% RAP materials. For 100% RAP materials 2% cement increased the wave velocity by only 18%. This indicates that 2% cement is inadequate to create adhesion between the asphalt-coated aggregates. Addition of 6% cement caused the highest increase in P-wave velocity. However, the increase in P-wave velocity at 6% cement content was not that high compared to the increases found by 4% cement. This indicates the proximity to optimum cement content.

6.2.1 Dynamic Modulus of Elasticity

Dynamic modulus of elasticity was determined from P-wave velocity, density, and Poisson's ratio using the proposed correlation by Müller-Rochholz (1979). Maximum dry density was considered as the samples were prepared at optimum moisture.

$$V_p = \sqrt{\frac{E(1-\mu)}{(1+\mu)(1-2\mu)\rho}} \tag{1}$$

Where,

E = Dynamic modulus of elasticity;
μ = Poisson's Ratio; and
ρ = Density

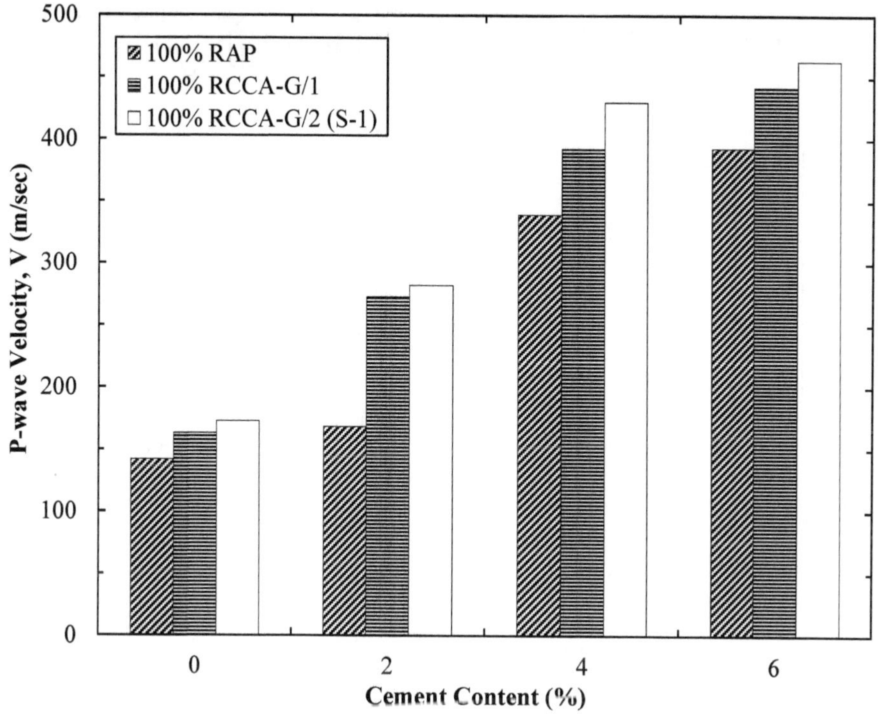

Fig. 5. Variation of P-wave velocity with cement content

As shown in Fig. 6(a), for the approximation of Poisson's ratio of RAP materials, dynamic modulus of elasticity was determined at 0% and 6% cement contents for a wide range values of Poisson's ratio. Poisson's ratio, at which the dynamic modulus of elasticity matched the tangent modulus of elasticity found from the UCS test, was taken as the Poisson's ratio at that specific cement content. For 100% RAP materials, Poisson's ratio was found to be 0.395 at 0% cement content and 0.357 at 6% cement content. Assuming linear variation of Poisson's ratio with cement content, a line joining the Poisson's ratio at 0% and 6% cement content allowed to estimate the Poisson's ratio at 2% and 4% cement. The same procedure was followed to estimate the Poisson's ratio for 100% RCCA materials, which has been shown in Fig. 6(b) and (c).

Figure 6(d) shows the variation of Poisson's ratio of RAP and RCCA with cement contents. Poisson's ratio at 0% cement content was the same (0.395) for all the materials. It was observed that Poisson's ratio tends to decrease with the increase of cement content, though no significant decrease was observed at 2% cement content for 100% RAP materials. At 2% cement, Poisson's ratio was a bit lower than the Poisson's ratio found at 0% cement for 100% RCCA. Minimum Poisson's ratio was found to be 0.342 for RCCA-G/2 (S-1) materials at 6% cement. This is higher than the typical values of cement treated granular materials. The minimum Poisson's ratio may have

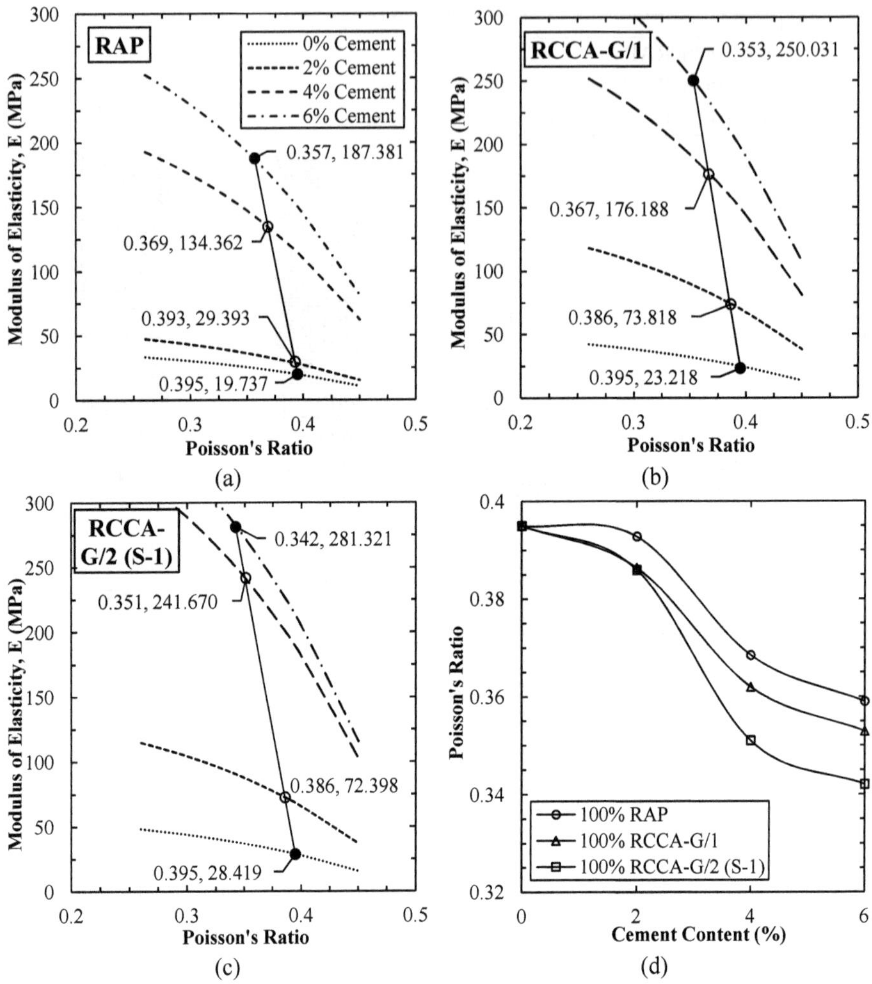

Fig. 6. Approximation of Poisson's ratio (a) 100% RAP, (b) 100% RCCA-G/1, (c) 100% RCCA-G/2 (S-1), and (d) Variation of Poisson's ratio with cement content

happened due to the larger aggregate size and relatively lower cement content to fill the voids in aggregate blends. According to the study conducted by Popovics et al. (1998), the value of Poisson's ratio for concrete varies within the range of 0.2 to 0.33, but for the unbound granular materials Poisson's ratio varies from 0.3 to 0.4. Based on this analysis, Poisson's ratio for all the combinations at 4 different cement contents is presented in Table 3.

The variation of the dynamic modulus of elasticity with cement content for 3 different aggregate types has been shown in Fig. 7. From Fig. 7, it is observed that dynamic modulus of elasticity increases with the increase of cement content. Lowest moduli values were found for RAP materials, which might have occurred due to poor

Table 3. Poisson's ratio for different combinations

Cement (%)	100% RAP	100% RCCA-G/1	100% RCCA-G/2 (S-1)
0	0.395	0.395	0.395
2	0.393	0.386	0.386
4	0.369	0.367	0.351
6	0.357	0.353	0.342

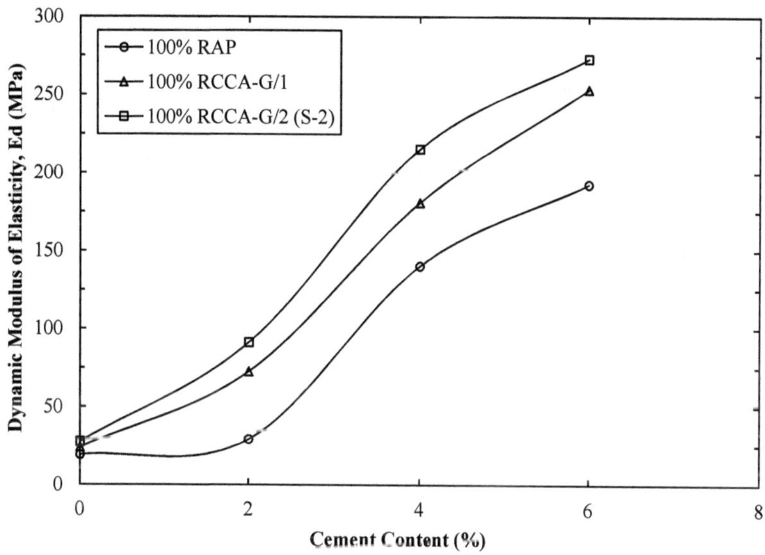

Fig. 7. Changes in dynamic modulus of elasticity with cement content

interlocking between the RAP aggregates. RAP aggregates are coated with asphalt which generates slip surface in the specimen and reduces the strength of transition zone.

7 Analytical Modeling

P-wave velocity, unconfined compressive strength, and tangent modulus increased with the increase of cement content. Minimum values of wave velocity, strength, and stiffness were found at 0% cement content, whereas maximum values were associated with 6% cement. Moreover, at 4% and 6% cement contents, dynamic modulus of elasticity falls within 10% range of the tangent modulus found from the UCS test. At lower cement content, such as 0% and 2% cement, the variation of modulus of elasticity was higher. Inadequate fines to fill the voids might be the reason for lower

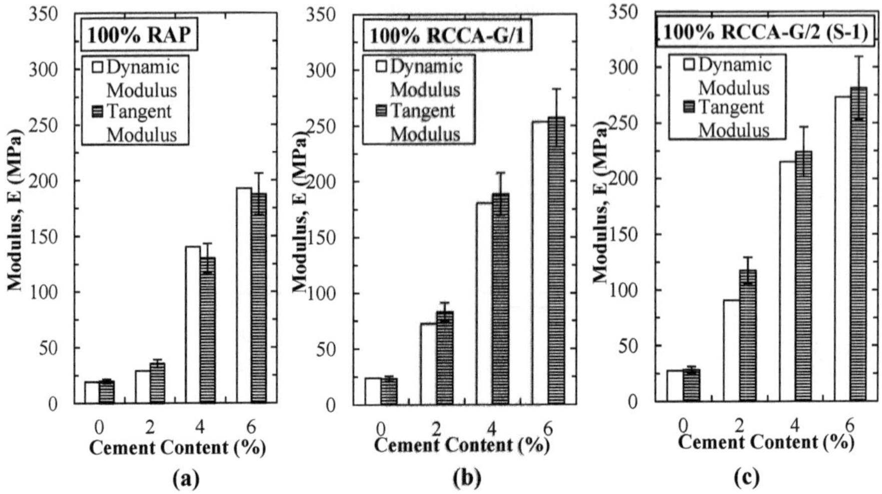

Fig. 8. Comparison of dynamic and tangent modulus of elasticity of (a) 100% RAP, (b) 100% RCCA-G/1, and (c) 100% RCCA-G/2 (S-1)

P-wave velocity at 0% and 2% cement content. This eventually predicted lower values of dynamic modulus. Figure 8 shows the comparison of modulus of elasticity found from P-wave velocity measurement and UCS testing for 3 different aggregate types at four different cement contents.

The variations between dynamic and tangent modulus of elasticity were insignificant. Linear regression analysis between the P-wave velocity and tangent modulus were performed regardless the amount of cement content. Linear regression was associated with higher value of coefficient of determination, but nonlinear trend was observed in the residual plot. Higher order terms, such as quadratic and cubic were implemented, yet the trend kept persisting. This indicates the necessity of variable transformation. Both x and y variables were transformed using log function, which eventually yielded good trend with symmetric distribution of residuals. This indicated the inevitability of nonlinear power regression of these two parameters. The initial theoretical equation used for regression analysis has been expressed by Eq. 2.

$$E = K_1 \, V^{K2} \tag{2}$$

Where, K_1, K_2 = Regression Coefficients

After taking logarithmic function on both sides, the equation takes the following form:

$$\log(E) = \log(K_6) + K_7 \log(V) \tag{3}$$

Regression analysis was performed based on Eq. 3 to find the trend of tangent modulus with P-wave velocity. Figure 9 shows the regression analysis with very closer distribution of data points around the fitted line. Coefficient of determination of this nonlinear power regression is 98% and the standard deviation is also minimum. Both the intercept and the variable coefficient were found to be significant in terms of P-values. This indicates the rejection of null hypothesis, suggesting that the coefficients are significant in terms of statistical judgments. Listed F-values in the ANOVA table for each coefficients were well above zero. This also supports the reliability of these accepted coefficients. Standardized residuals were well distributed within ±2 standard deviation around the mean. Normal probability plot followed the straight line pattern indicating the Gaussian Distribution of data points around the mean. Considering all of this, the model seemed satisfactory in terms of statistical definitions.

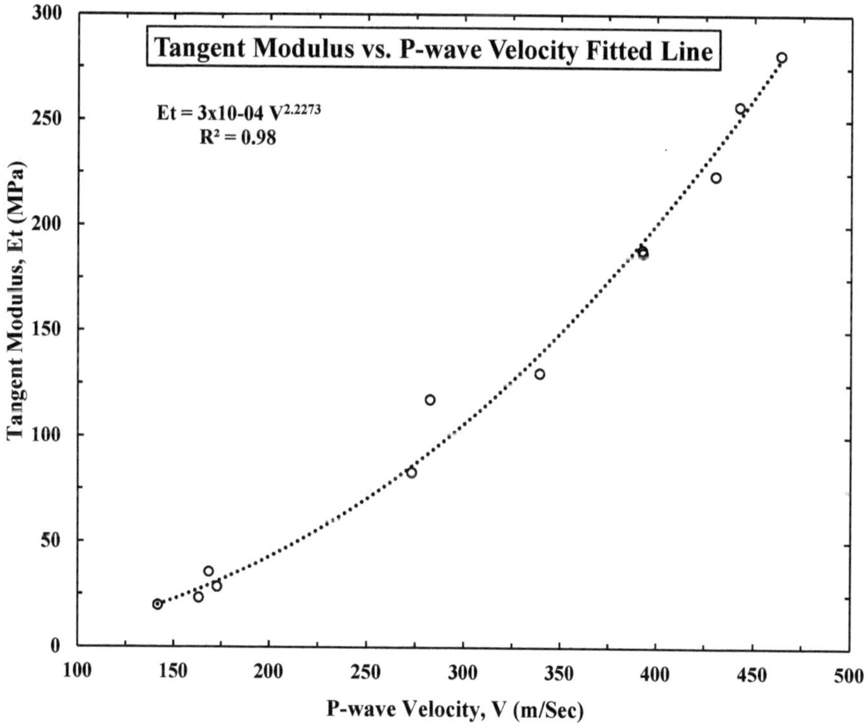

Fig. 9. Non-linear regression between P-wave velocity and tangent modulus

As the moduli values of the aggregate blends were determined by the tangent of the stress-strain curve found from the UCS test, it was anticipated that the association of UC strength and P-wave velocity will also be nonlinear. Nonlinear trend between strength and P-wave velocity of concrete has also been reported by Cho et al. (2011).

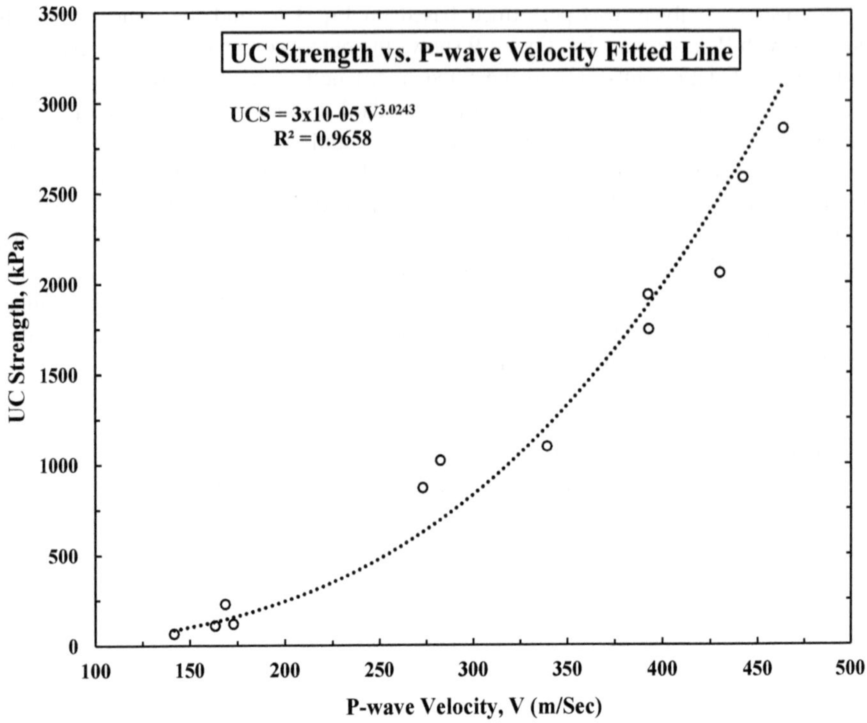

Fig. 10. Non-linear regression between P-wave velocity and UC strength

As such, regardless the amount of cement content, the fitted line of the regression analysis of P-wave velocity and UC strength of 3 different aggregate types has been shown in Fig. 10.

7.1 Model Verification

For the verification of the regression models, Grade-2 type 100% RCCA from another source (source-2) was utilized. Standard procedures as described before were followed for sample preparation, UCS and impact echo test at four different cement contents. The comparison between predicted and actual test values of modulus of elasticity has been shown in Fig. 11(a). Predicted values are within 9–18% range of the actual values, higher variation was observed at higher cement content. Strength comparison between the actual and predicted values has also been shown in Fig. 11(b). Predicted strengths are within the range of 9–11% compared to the actual values.

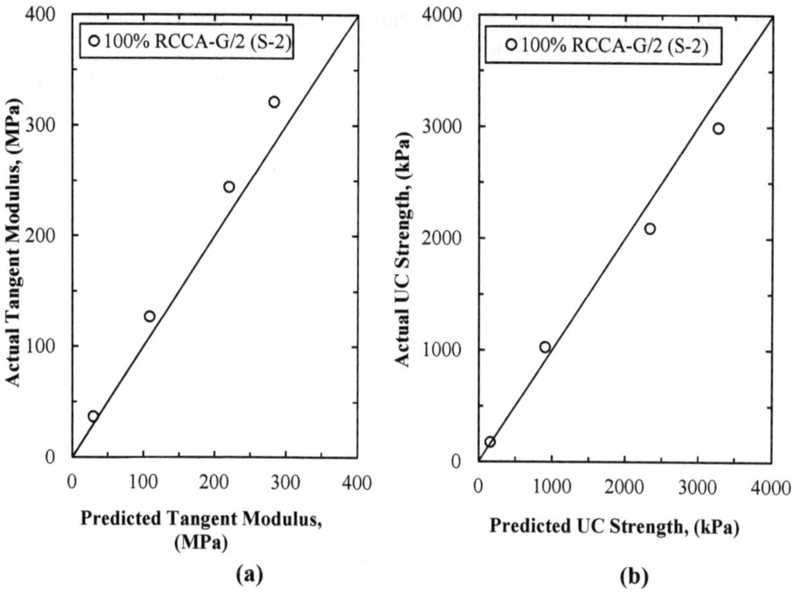

Fig. 11. Comparison between predicted and actual values (a) modulus of elasticity, and (b) unconfined compressive strength

8 Conclusion

Noteworthy advancements in theoretical and experimental domains of impact echo method are now offering a new standard of non-destructive testing of pavement materials. In this study, impact echo method was utilized in characterizing the strength and stiffness properties of three different types of recycled pavement base materials. Based on this study, it was concluded that the impact echo could be a potential tool in characterizing the strength and stiffness properties of cement treated recycled base materials. Some of the salient findings of this research are summarized in the following section:

- Cement inclusion has significant effect on P-wave velocity, unconfined compressive strength, and modulus of elasticity of recycled materials. Velocity, strength, and stiffness responses increase with the increases of cement content. Different strength requirements for base construction can be achieved with recycled materials by utilizing varying percentages of cement.
- Higher strength and elasticity were found for RCCA materials compared to RAP. Addition of 2% cement in 100% RAP materials did not increase the strength and stiffness significantly. This indicates that 2% cement is inadequate to create adhesion between the asphalt coated aggregates.
- In the case of RCCA, higher values of P-wave velocity, unconfined compressive strength, and modulus of elasticity were observed for Grade-2 type materials compared to Grade-1 type. Higher percentage of fines in Grade-2 might be the reason for higher values of velocity, strength, and stiffness. Fines fill the voids,

making the samples denser. This eventually yielded higher values of P-wave velocity, UC strength, and modulus of elasticity.

- Strength and stiffness values of treated and untreated recycled aggregates can reasonably be estimated by impact echo method. Both the regression correlations are likely to hold good agreement with test strength and stiffness values if the P-wave velocity remains within the range of 175 m/s to 475 m/s.

References

Hoyos, L.R., Puppala, A.J., Ordonez, C.A.: Characterization of cement-fiber-treated reclaimed asphalt pavement aggregates: preliminary investigation. J. Mater. Civ. Eng. **23**(7), 977–989 (2011)

Gonzalez, G.P., Moo-Young, H.K.: Transportation applications of recycled concrete aggregate. FHWA State of the Practice National Review (2004)

Ordonez, C.A.: Characterization of Cemented and Fiber-Reinforced RAP Aggregate Materials for Base/Sub-base Applications. The University of Texas at Arlington (2006)

Hansen, K.R., Copeland, A.: Annual asphalt pavement industry survey on recycled materials and warm-mix asphalt usage: 2009–2012 (No. IS-138) (2013)

Faysal, M., Mahedi, M., Aramoon, A., Thian, B., Hossain, M.S., Khan, M.S.: Strength characterization of untreated and cement-treated recycled flex-base materials. In: Geotechnical and Structural Engineering Congress 2016, pp. 1233–1244 (2016a)

Faysal, M., Mahedi, M., Aramoon, A., Thian, B., Hossain, M.S., Khan, M.A., Khan, M.S.: Determination of the structural coefficient of different combinations of cement-treated/untreated recycled base materials. In: Geotechnical and Structural Engineering Congress 2016, pp. 1198–1208 (2016b)

Mahedi, M., Hossain, M.S., Faysal, M., Khan, M.S.: Potential applicability of impact echo method on pavement base materials as a non-destructive testing technique. In: Proceedings, 96th Annual Meeting, Transportation Research Board, Washington, DC (2017)

Euch Khay, S.E., Euch Ben Said, S.E., Loulizi, A., Neji, J.: Laboratory investigation of cement-treated reclaimed asphalt pavement material. J. Mater. Civ. Eng. **27**(6), 04014192 (2014)

Jones, D., Wu, R., Louw, S.: Comparison of full-depth reclamation with foamed asphalt and full-depth reclamation with no stabilizer in accelerated loading test. Transp. Res. Rec. J. Transp. Res. Board **2462**, 126–135 (2014)

Medina, C., Zhu, W., Howind, T., de Rojas, M.I.S., Frías, M.: Influence of mixed recycled aggregate on the physical–mechanical properties of recycled concrete. J. Clean. Prod. **68**, 216–225 (2014)

Wilburn, D.R., Goonan, T.G.: Aggregates from Natural and Recycled Sources; Economic Assessments for Construction Applications; a Materials Flow Study (No. 1176). US Department of the Interior, US Geological Survey (1998)

Graveen, C.: Nondestructive Test Methods to Assess Pavement Quality for Use in a Performance-Related Specification. Doctoral dissertation, Purdue University (2001)

Müller-Rochholz, J.: Determination of the elastic properties of lightweight aggregate by ultrasonic pulse velocity measurement. Int. J. Cem. Compos. Lightweight Concr. **1**(2), 87–90 (1979)

Popovics, J.S., Song, W., Achenbach, J.D., Lee, J.H., Andre, R.F.: One-sided stress wave velocity measurement in concrete. J. Eng. Mech. **124**(12), 1346–1353 (1998)

Cho, Y.S., Kim, J.H., Hong, S.U.: Compressive strength prediction of porous concrete using nondestructive tests. J. Civ. Eng. Archit. **5**(12), 1053 (2011)

Predicting Performance of Flexible Pavement Using Finite Element Method

Anand B. Tapase(✉) and M.S. Ranadive

Department of Civil Engineering, College of Engineering, Pune, Maharashtra, India
tapaseanand@gmail.com, msrtunnel@yahoo.co.in

Abstract. The modern vehicles having very high loads and high tyre inflation pressures are the outcome of improvements in the automobile segment. The combined effect of higher tyre pressure and excessive loading leads to premature failure of roads in the form of potholes, cracking, etc. which has drastically brought down the life of a road. One of the major cause that contributes to road accidents is deteriorated condition of roads in terms of cracking, potholes and uneven surface due to rutting. This research documents the use of finite element analysis by treating tyre-pavement interaction as an axisymmetric two-dimensional problem. For predicting performance of flexible pavement which is subjected to different inflation pressures, loading conditions, variation in thickness and material properties of different layers in the flexible pavement, modeling is done using a developed program in FORTRAN. From the analysis, it is observed that the value of ε_t at bottom of BL increases by 215.33% and 254.48% by an overloading in standard axle load by 2 times and 2.5 times respectively. Also, a noticeable increase of around 252.95% and 336.40% in ε_v at top of subgrade is seen due to overloading the standard axle load by 2 times and 2.5 times respectively. It is noticed that the damage caused to a pavement by an overloaded axle load of around 2 times or 2.5 times the standard axle load is much more than the damage by the standard axle load. From the analysis, it is noticed that the pavement starts deteriorating earlier as the overloaded vehicles are consuming the designed life of the pavement, so from the present work, it is concluded that the pavement should be designed considering the uncontrollable overloading of the vehicles instead of relying on the standard or legal limits.

1 Introduction

The ongoing decade (2011–2020) is termed as the decade of action for road safety by the World Health Organization (WHO). The report titled 'Saving millions of lives' published by WHO highlighted that road traffic crashes take the lives of nearly 1.3 million people globally every year, and injure 20–50 million more. Road accidents are resulting into the cause behind the major loss of lives in developing nations like India. According to the provisional data released by the government of India, the total number of road fatalities stood at 146,000 in 2015. Compared to 2014, which saw 141,526 road accident deaths, the number in 2015 has increased by around 3%. The Road Transport and Safety Bill are brought by the Ministry of Road Transport and Highways (MoRTH) to strengthen road safety as well as improve ease of transport across the country [10].

L. Mohammad (ed.), *Advancement in the Design and Performance of Sustainable Asphalt Pavements*, Sustainable Civil Infrastructures, DOI 10.1007/978-3-319-61908-8_11

One of the major cause that contributes to road accidents is deteriorated condition of roads in terms of cracking, potholes and uneven surface due to rutting [7].

The outcome of improvements in automobile segment is not only fuel efficient vehicles but also the improvements in tyre technology have resulted in very high loads and high tyre inflation pressures even up to 1 MPa. On the other hand, the life of a road is designed on repetition of standard axle load of 81.6 kN wherein overloading of vehicle is not considered [8]. The combined effect of higher tyre pressure and excessive loading leads to premature failure of roads in the form of potholes, cracking, etc. which has drastically brought down the life of road [20].

The frequency of accidents and financial losses has increased enormously, due to damaged roads. Rao et al. [20] estimated that about 80% of trucks plying on Indian roads are overloaded. Thus, on one side transport sector contributes to the economic, industrial, social and cultural development of any country, on other truck overloading is making a huge cavity on it in terms of increased maintenance cost, vehicle operating cost, accidents, travel time and deteriorated environment.

The increasing traffic along with overloaded vehicles, actual field conditions are responsible for early development of distress symptoms like raveling, undulations, rutting, cracking, bleeding and pot holing of bituminous surfacing [15]. In order to reduce the number of road accidents caused due to deteriorated condition of a road, there is an urgent need to prepare a proper blend of design procedure/charts which will accommodate the present day requirements [15]. In this connection, the complex characteristics of the present day systems demand an application of analytical tool, wherein the application FEM holds an ideal assurance.

In continuation with the earlier investigation related to varying the thicknesses, material properties of different layers, effect of temperature variation reported by Tapase et al. [3] and Ranadive et al. [16, 17], this paper describes the usefulness of finite element analysis for exploring the performance of flexible pavement at various combinations of loading and pressure intensities. Useful design charts/procedure for particular material property and thickness combination can be successfully extracted for various trials of loading and pressure intensities such that fatigue and rutting strains developed are within the allowable limit to prevent fatigue cracking and rutting deformation [3]. The pavement life in a number of standard axles can be estimated for various objectives mentioned by incorporating the fatigue and rutting criterions recommended by IRC: 37-2012 [8]. From the past studies and through the recommendations of the Indian Road Congress (IRC) and AASHTO code of practice [2], wheel load, tyre pressure and their combinations as a number of trials are shortlisted for the present study.

2 Loading and Pressure Intensities

In flexible pavement, design damage is conceived in the form of fatigue, permanent deformation of subgrade or rutting and shape loss. The traffic loading due to light vehicles like cars, in general, are ignored since it works out to an insignificant value. There are legal limits for axle loads and gross weights of vehicles but they are neither observed by the transporter nor enforced strictly by the authorities [20].

The design thickness depends on the axle load, tyre pressure, and the wheel configuration, which are taken as standard. However, these values are different in different codes of practices. The dents caused are therefore of different magnitudes. So, the number of repetitions of vehicles of various axle loads is converted into an equivalent standard axle load repetitions (termed Equivalent Single Axle Load or ESAL repetitions in various design guidelines). The standard axle load in India is 81.6 kN and 149.68 kN for single axle and tandem axle respectively [4, 11, 18]. A standard single axle load assumed by other countries is approximately equal to 81.6 kN (8.16 tonnes). Legal axle load is that maximum axle load, any value beyond which is not permitted to move over the road. In India, the legal axle load for the single axle is 102 kN [5, 9, 12]. Standard axle load is that axle load based on which all the calculations related to pavement damage have been standardized [11]. This means, in design charts, the thickness values are read against standard axle load repetitions. Therefore, the various axle load repetitions on in-service road need to be converted to equivalent standard axle load repetitions, before proceeding to use a pavement design chart. Excessive overloading of the axle uses up the design life much faster causing fatigue and deterioration of the road leading to premature failure. The fourth power law proposed by AASHTO code of practice is generally used to predict the rate of road deterioration. As per the fourth power law, 10% overloaded truck can deteriorate pavement up to 40% compare to standard load limited vehicle. The top-down cracking within a year or two due to heavy vehicles is reported in from different parts of India [9, 20].

The modern trucks which usually have more than 0.800 MPa tyre pressure [17] along with overloading are the main cause of pavement damage. As the tyre pressure increases it apparently reduces the tyre-pavement contact area, which may, in turn, result in an increase in the contact stresses and strains and then more damaging effects to the pavement. Since the response of the pavement is greatly affected by loading and tyre pressures, different combinations of axle loads intensities and inflation pressure are investigated here to analyze the considerable impact on pavement damage initiation from fatigue and rutting point of view. Total five trail pressure intensities from 0.56 MPa, 0.575 MPa, 0.700 MPa, 0.84 MPa and 0.98 MPa are checked against standard axle load (81.6 kN) and legal axle load (102 kN) which are allowed on Indian roads also two overloading condition i.e., 2 times (160 kN) and 2.5 times (200 kN) of the standard axle loads are considered for analysis.

3 Pavement Composition and Finite Element Model

Two-dimensional finite element solution technique is adopted wherein the tyre pavement interaction is treated as an axisymmetric solid (Fig. 1). The accuracy of the finite element analysis is affirmed by comparing the compute results with readily available close form solution as reported in Tapase and Ranadive [3]. Maximum damage due to applied load or contact pressure is observed at the axisymmetric axis and it goes on reducing as we move away from the vertical axis, so the critical parameters at this axis are taken for analysis. Figure 1, gives the details of pavement material properties and thickness combination of different component layers used for analysis. The thickness combination and material properties are common in practices which are kept constant throughout the analysis; hence it is an attempt to correlate the present study with actual field conditions [11, 14]. Point P in Fig. 1 is the critical point at the bottom of bituminous layer, as the tensile strain (ε_t), at the bottom of the bituminous layer is considered as conventional critical parameter for pavement design to limit cracking (Fatigue) and point Q is the critical point at the top of subgrade, as the vertical subgrade strain (ε_V), on the top of the subgrade is considered as conventional critical parameter for pavement design to rutting in the non-bituminous layers [8].

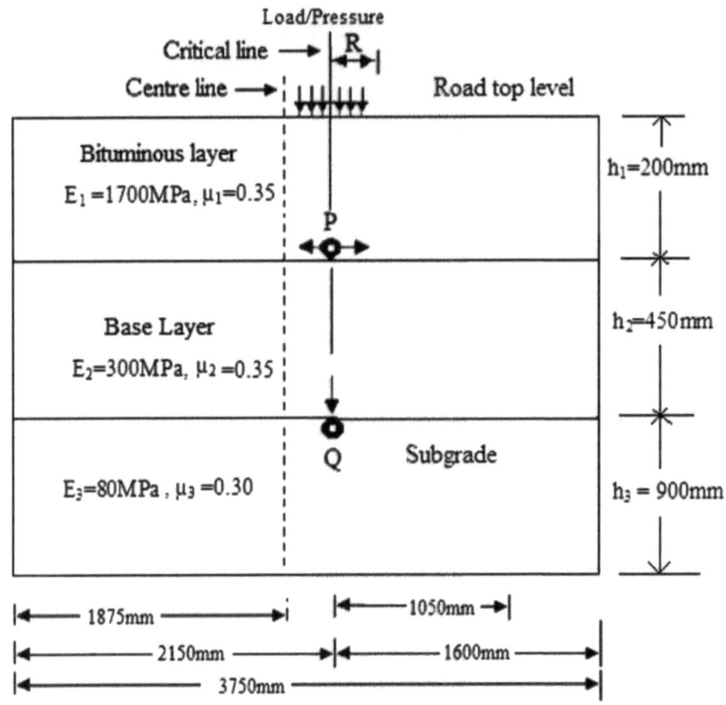

Fig. 1. Details of pavement materials and properties

Dimensions of FE model should be sufficiently large so that constraints imposed at the boundaries have very little influence on the stress distribution in the system. Helwany et al. [6] discretized a three layer pavement system with a right boundary at a distance of about 8 times the loaded radius. Abdhesh et al. [1], placed the right boundary at 110 cm from the outer edge of the loaded area, which is more than 7 times the radius 150 mm of the applied load. Tapase and Ranadive [3] reported the usefulness of a two-dimensional axi-symmetric finite element analysis for exploring the parameter sensitive analysis of flexible pavement wherein, the parameters considered are traffic loading, variation in material properties of different component layers, variation in thickness.

In the present study, the total thickness of the pavement is kept constant, consisting of 200 mm of bituminous layer (comprising of BC and DBM), 450 mm of base layer and 900 mm of subgrade [3, 17] as shown in Fig. 1. The idealization scheme comprises of the bituminous layer having 50 mm thickness of each sub-layer, base layer having sublayer 75 mm thick and bottom most layer as natural subgrade. Depending upon the required type of analysis, a two-dimensional axi-symmetric finite element idealization for the pavement system shown in Fig. 1 being analyzed is developed by means of four noded quadrilateral elements in the present study. The radial extent of the pavement section which represents the actual indefinite lateral extent is assumed at around seven times the circular contact area [1] having a radius equal to trial considered for analysis and vertical bottom rigid boundary at a depth of more than eight times the radius of the circular contact area [1, 11, 13].

The nodes over the base of the subgrade i.e., at depth of more than eight times the radius of the circular contact area, are restrained in both radial (r) and axial (z) directions. The nodes over the axis of symmetry are restrained in radial direction i.e., it automatically takes zero displacements in radial direction. We should note that this condition constitutes an integral part of axi-symmetric deformation analysis. It is assumed that due to the indefinite lateral extent of the pavement section, the nodes over the extreme vertical face of the pavement, i.e., at radial distance equal to seven times radius; do not suffer radial displacements. Hence those nodes are treated as if restrained in the radial (r) direction.

These thicknesses and material properties are generally considered in practice and have relevance with variation in field and climatic conditions [8]; hence it is an attempt to correlate the present study with actual field conditions.

4 Pavement Life in Number of Fatigue and Rutting Cycles

The pavement responses including stresses, strains, and deflection which are evaluated from the structural models are used as inputs for distress models [4, 7]. The distress models include fatigue and rutting criteria. In the present investigation, the calibrated fatigue model and a rutting model is used to estimate the fatigue life in terms of a number of the standard axles and the rutting life in terms of a number of cumulative standard axles respectively, both models for 80% reliability level given in IRC: 37-2012 are employed.

The fatigue cracking of flexible pavements is based on the horizontal tensile strain at the bottom of BL (point P in Fig. 1). As per the IRC:37-2012 guidelines for the design of flexible pavements, Eq. 1 is recommended as fatigue criterion, where normal bituminous mixes with VG30 bitumen are used [8].

$$N_f = 2.21 \times 10^{-04} \times \left[{}^{1}\!/_{\varepsilon_t}\right]^{3.89} \times \left[{}^{1}\!/_{M_R}\right]^{0.854} \tag{1}$$

Where,

N_f = Fatigue life in number of standard axles,

ε_t = Maximum tensile strain at the bottom of the BL, and

M_R = Resilient modulus of the BL

On this criterion, the allowable number of load repetitions (N_f) that causes fatigue cracking is related to the tensile strain (ε_t) at the bottom of BL. If the vertical compressive strain (ε_v) on top of the subgrade (Point Q in Fig. 1) is excessive, permanent deformation occurs on the surface of the pavement structure, and its effect is on the pavement distress due to rutting. As per the guidelines for the design of flexible pavements, Eq. 2 is recommended as rutting criterion for controlling rutting in the subgrade and granular layers.

$$N = 4.1656 \times 10^{-08} \left[1/\varepsilon_V \right]^{4.5337} \quad (2)$$

Where,
N = Number of cumulative standard axles, and
ε_v = Vertical strain in the subgrade

So, the computed strains of critical locations (Fig. 1) are used as input in the fatigue and rutting criteria as recommended in IRC: 37-2012, to estimate the pavement life in terms of a number of fatigue and rutting cycles for hypothetical conditions.

5 Results and Interpretation

Since the response of the pavement is greatly affected by loading and tyre pressures, different combinations of axle loads intensities and inflation pressure are investigated here to analyze the considerable impact on pavement damage initiation from fatigue and rutting point of view. Total five trial pressure intensities are checked against standard axle load (81.6 kN) and legal axle load (102 kN) which are allowed on Indian roads also two overloading conditions i.e., 2 times (160 kN) and 2.5 times (200 kN) of the standard axle loads are considered for analysis. Tables 1 and 2 shows the ε_t at bottom of BL and the ε_v at top of subgrade at various trial pressure and loading conditions respectively. In the table, the results of each pressure intensities at 81.60kN are taken as 100% with which the results for 102 kN, 160 kN and 200 kN are compared. The percentage variation is shown in brackets next to the values of ε_t at bottom of BL. As the pressure intensity is increased, the strain values are also seen to be increased. This can be easily justified as the pressure intensity is increased the contact

Table 1. ε_t at bottom of BL

Pressure	Standard axle	Legal axle	Overloading of standard axle	
	81.6 kN	102 kN	160 kN	200 kN
(MPa)	ε_t at bottom of BL (% increase in ε_t)			
0.56	1.88E−04 (100%)	4.36E−04 (131.91)	5.71E−04 (203.72)	6.32E−04 (236.17)
0.575	1.89E−04 (100%)	4.39E−04 (132.28)	5.75E−04 (204.23)	6.60E−04 (249.21)
0.7	2.00E−04 (100%)	4.69E−04 (134.50)	6.28E−04 (214.00)	6.89E−04 (244.50)
0.84	2.09E−04 (100%)	4.92E−04 (135.41)	6.72E−04 (221.53)	7.52E−04 (259.81)
0.98	2.14E−04 (100%)	5.11E−04 (138.79)	7.13E−04 (233.18)	8.19E−04 (282.71)
Average	2.00E−04 (**100%**)	4.69E−04 (**134.58%**)	6.32E−04 (**215.33%**)	7.10E−04 (**254.48%**)

Table 2. Variation in ε_v and Percentage increase in ε_v

Pressure	Standard axle	Legal axle	Overloading of standard axle	
	81.6 kN	102 kN	160 kN	200 kN
(MPa)	ε_v at bottom of BL (% increase in ε_v)			
0.56	2.38E−04 (100%)	5.76E−04 (142.02%)	8.81E−04 (270.17%)	1.10E−03 (362.18%)
0.575	2.39E−04 (100%)	5.77E−04 (141.42%)	8.81E−04 (268.62%)	1.10E−03 (360.25%)
0.7	2.50E−04 (100%)	5.95E−04 (138.00%)	8.88E−04 (255.20%)	1.10E−03 (340.00%)
0.84	2.64E−04 (100%)	6.16E−04 (133.33%)	8.99E−04 (240.53%)	1.10E−03 (316.67%)
0.98	2.78E−04 (100%)	6.40E−04 (130.22%)	9.18E−04 (230.22%)	1.12E−03 (302.88%)
Average	**1.27E−03 (100%)**	**3.00E−03 (137%)**	**4.47E−03 (252.95%)**	**5.52E−03 (336.40%)**

area goes on reducing which results in the application of load on less contact area ultimately increasing the strain at critical locations. Figure 2 shows a plot of percentage increase in ε_t at bottom of BL due to overloading.

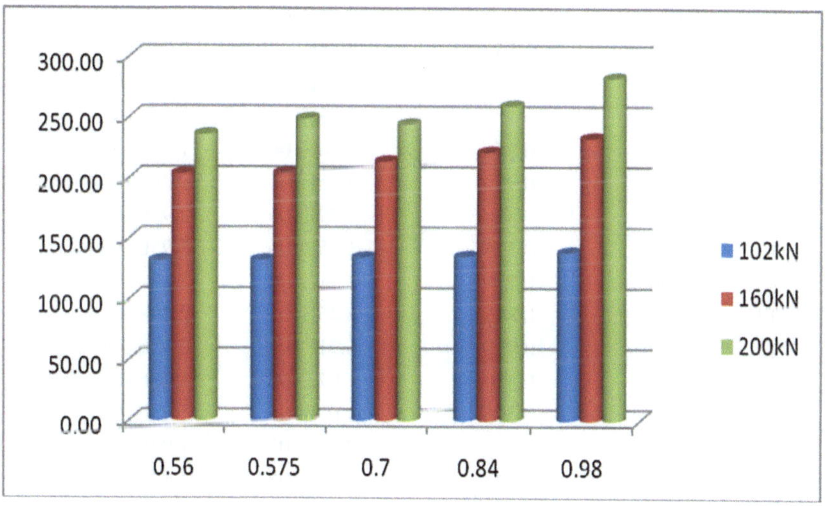

Fig. 2. Percentage increase in ε_t due to overloading

The values of a number of standard axles (Nf) and a number of cumulative standard axles (N) shown in Table 3 are estimating the fatigue life in a number of the standard axles and the rutting life in terms of a number of cumulative standard axles respectively. The value of ε_t and ε_v at critical locations are found to be much higher for pavements subjected to overloading and high tyre pressures compare to standard and legal axle load limits allowed in India. This emphasizes the importance of considering high axle loads and tyre pressures during analysis and design of flexible pavement pavements for predicting the failure of pavement in the form of fatigue (bottom-up cracking) as well as top-down cracking. It is observed that the horizontal tensile strain at bottom of BL increases by 215.33% and 254.48% by an overloading in standard axle load by 2 times and 2.5 times respectively. Also, a noticeable increase of around 252.95% and 336.40% is seen due to overloading the standard axle load by 2 times and 2.5 times respectively. It is observed that the variation in pressure at constant axle load has the least effect on horizontal tensile strain at bottom of BL, as compared to variation in loading at constant pressure. It is noticed that at the higher axle loads the vertical compressive strain is exceeding the allowable limits, which shows a reduction in rutting lives of pavements.

Table 3. Fatigue and rutting lives in terms of number of cumulative standard axles

Case no.	Load intensity (kN)	Uniform pressure (MPa)	Number of standard axles	
			Fatigue life (Nf)	Rutting life (N)
1	81.6	0.56	1.20E+08	1.12E+09
2	81.6	0.575	1.18E+08	1.09E+09
3	81.6	0.7	9.43E+07	8.92E+08
4	81.6	0.84	7.95E+07	6.97E+08
5	81.6	0.98	7.25E+07	5.51E+08
6	102	0.56	4.55E+06	2.03E+07
7	102	0.575	4.43E+06	2.01E+07
8	102	0.7	3.43E+06	1.75E+07
9	102	0.84	2.84E+06	1.50E+07
10	102	0.98	2.45E+06	1.26E+07
11	160	0.56	1.59E+06	2.95E+06
12	160	0.575	1.55E+06	2.95E+06
13	160	0.7	1.11E+06	2.85E+06
14	160	0.84	8.45E+05	2.69E+06
15	160	0.98	6.71E+05	2.45E+06
16	200	0.56	1.07E+06	1.08E+06
17	200	0.575	9.07E+05	1.08E+06
18	200	0.7	7.67E+05	1.08E+06
19	200	0.84	5.46E+05	1.08E+06
20	200	0.98	3.92E+05	9.95E+05

It is noticed that the damage caused to a pavement by an overloaded axle load of around 2 times or 2.5 times the standard axle load is much more than the damage by the standard axle load. Such type of analysis shows that pavement starts deteriorating earlier as the overloaded vehicles are consuming the designed life of the pavement, so it is concluded that the pavement should be designed considering the uncontrollable overloading of the vehicles instead of relying on the standard or legal limits.

6 Conclusions

Based on the work undertaken following are some of the important conclusions which are directly or indirectly applicable in practice.

- It is observed that the value of ε_t at bottom of BL increases by 215.33% and 254.48% by an overloading in standard axle load by 2 times and 2.5times respectively. Also, a noticeable increase of around 252.95% and 336.40% is seen due to overloading the standard axle load by 2 times and 2.5 times respectively.
- It is noticed that the damage caused to a pavement by an overloaded axle load of around 2 times or 2.5 times the standard axle load is much more than the damage by the standard axle load.
- From the analysis, it is noticed that the pavement starts deteriorating earlier as the overloaded vehicles are consuming the designed life of the pavement, so it is concluded that the pavement should be designed considering the uncontrollable overloading of the vehicles instead of relying on the standard or legal limits.

References

1. Sinha, A.K., Chandra, S., Kumar, P.: Finite element analysis of flexible pavement with different subbase materials. Indian Highw. New Delhi **42**(2), 53–63 (2014)
2. American Association of State Highway Officials AASHTO: AASHTO Guide for Design of Pavement Structures. AASHTO, Washington (1993)
3. Tapase, A., Ranadive, M.: Performance evaluation of flexible pavement using finite element method. ASCE GSP 266, Geo-China 2016: Material, Design, Construction, Maintenance and Testing of Pavement, 9–17 (2016). doi:10.1061/9780784480090.002
4. Das, A.: Analysis of Pavement Structures. CRC Press, Taylor and Francis, Boca Raton (2015)
5. Das, A., Pandey, B.B.: M-E design of bituminous road: and Indian perspective. J. Transp. Eng. ASCE (1999). doi:10.1061/(ASCE)0733-947X(1999)125:5(463)
6. Helwany, S., Dyer, J., Leidy, J.: Finite Element Analysis of Flexible Pavement. J. Transp. Eng. ASCE **124**(5), 491–499 (1998). doi:10.1061/(ASCE)0733-947X(1998) 124:5(491)
7. Huang, H.Y.: Pavement Analysis and Design, 2nd edn. Pearson Education, Inc, and Dorling Kindersley Publishing, Inc (2008)
8. IRC: 37-2012: Guidelines for the Design of Flexible Pavements. Indian Roads Congress, New Delhi
9. Kumar, S.S., Sridhar, R., Reddy, K.S., Bose, S.: Analytical investigation on the influence of loading and temperature on top-down cracking in bituminous layers. J. Indian Roads Congr. **69**(1), 71–77 (2008)
10. MORT&H: Guideline for Maintenance of Primary, Secondary, and Urban Roads. Ministry of Road Transport and Highway, Government of India, New Delhi (2013)
11. NCHRP: Evaluation of Mechanistic-Empirical Design Procedure. National Cooperative Highway Research Program, NCHRP Project 1-37A. National Research Council, Washington (2007)

12. Chakroborty, P., Das, A.: Principles of Transportation Engineering. Prentice Hall of India Private Limited, New Delhi (2003)
13. Siddharthan, R., Sebaaly, P.: Heavy off-road vehicle tyre-pavement interaction and response. J. Transp. Eng. ASCE **131**(3), 239–247 (2005)
14. Ranadive, M.S., Katkar, A.B. (2010). Finite element analysis of flexible pavements. Indian Highw. **38**(6)
15. Ranadive, M.S., Tapase, A.B.: Investigation of Behavioral aspects of flexible pavement under various conditions by finite element method. In: Yang, Q., Zhang, J.-M., Zheng, H., Yao, Y. (eds.) Constitutive Modeling of Geomaterials, pp. 765–770. Springer, Berlin (2013). doi:10.1007/978-3-642-32814-5_100
16. Ranadive, M.S., Tapase, A.: Pavement performance evaluation for different combinations of temperature conditions and bituminous mixes. Innov. Infrastruct. Solut. **1**, 40 (2016). doi:10.1007/s41062-016-0040-9
17. Ranadive, M.S., Tapase, A.B.: Parameter sensitive analysis of flexible pavement. Int. J. Pavement Res. Technol. (IJPRT) (2016). doi:10.1016/j.ijprt.2016.12.001
18. Reddy, K.S, Pandey, B.B.: Lateral placement of commercial vehicle on national highways. HRB Bulletin No. 7, Indian Road Congress, New Delhi (2006)
19. Sadeghi, J.M., Fathali, M.: Deterioration analysis of flexible pavements under overweight vehicles. J. Transp. Eng. ASCE **133**(11), 625–633 (2007)
20. Rao, S.K., Srinivasa Rao, V., Murthy, D.V.S., Sadanandam, K.: Analysis of Vehicle Overloading from Axle Load Surveys-Case Studies, pp. 25–31. Indian Highways, New Delhi (2010)

Implementation Initiatives of the Mechanistic-Empirical Pavement Design Guide in Countries with Insufficient Design Input Data – The Case of Lebanon

Ghassan R. Chehab(✉), Rana Hajj Chehade, Lamis Houssami, and Rayane Mrad

Department of Civil and Environmental Engineering, American University of Beirut, Beirut, Lebanon
gc06@aub.edu.lb

Abstract. The new mechanistic-empirical pavement design guide (M-EPDG) delivers a state-of-the-art and practice design procedure that eliminates the AASHTO 1993 empirical design procedure deficiencies. Huge advances with respect to traffic inputs, materials characterization and environmental impacts are integrated in its performance prediction methodology. However, achieving accurate and reliable design and performance prediction results using the M-EPDG requires extensive data collection. This presents researchers and practitioners with major challenges as they plan and work towards full adoption of the M-EPDG. Similar to the case of highway agencies in the U.S., it is imperative for highway agencies outside the U.S., such as those in the MENA region, to initiate a comprehensive and customized implementation plan for adoption of the M-EPDG. The main objective of this study is to present a methodology for facilitating the adoption of the new design guide in such countries. Firstly, the paper presents the current flexible pavement design practices in the countries lacking design input data sufficient for utilizing the M-EPDG. Then, the paper describes the efforts required to gather the necessary input data and presents an approach for utilizing the M-EPDG software. Sensitivity analyses and implementation initiatives are performed for each of the M-EPDG AASHTOWare design input modules, with Lebanon serving as a case study. Performance results are used to determine which of the design parameters require country-specific information, and proposes a plan for acquiring this information and investigates its effectiveness and reliability. Finally, the paper presents recommendations for implementation of the M-EPDG in such countries.

1 Introduction

1.1 Background

For decades, extensive efforts have been exerted to come up with simple yet effective methodologies for pavement design [1]. One of the commonly used methods in the present by countries outside the U.S. and Canada is the AASHTO 1993 design methodology that relies on empirical equations for calculating the thicknesses of the

L. Mohammad (ed.), *Advancement in the Design and Performance of Sustainable Asphalt Pavements*, Sustainable Civil Infrastructures, DOI 10.1007/978-3-319-61908-8_12

different pavement layers [2]. As of late, a mechanistic empirical pavement design procedure (M-EPDG) applied through the use of the AASHTOWare Pavement ME software is being used in lieu of the 1993 empirical method and its derivatives on a trial, research and development, as well as on a permanent basis by various highway agencies in the US [3]. One of the significant changes with the M-EPDG is that the approach to pavement design is effectively reversed. In conventional design methods, various inputs are considered and used to obtain the design requirements for the pavement structure. In mechanistic-empirical design, the design of the pavement structure is initially selected and inputted into the software along with traffic, climate, materials and pavement structure data. Then, the M-EPDG computes the response of the pavement due to the load and environmental stresses. This leads to an estimate of the level of damage the pavement will reach over time, in terms of pavement distresses and deterioration in road quality. The performance indicators used for asphalt pavements include rutting, fatigue (or alligator) cracking, longitudinal cracking, transverse cracking as well as the pavement roughness (IRI).

The M-EPDG relies on a high level of detailed input parameters for materials, climate and traffic. The design inputs are classified according to a hierarchy system [4] where the designer can select the level of data accuracy and sophistication based on the economic impact of the project. The selection is also a function of the state-of-knowledge and availability of the data. The hierarchical Design Levels are divided into three levels, with Level 3 representing the lowest level of the hierarchy with default or user-selected values, Level 2 having a higher level of reliability with design inputs based on laboratory test data and/or default predictive equations and Level 1 representing the highest degree of reliability with design inputs that are generally site specific and determined from laboratory and/or field testing and measurements.

Thus, the major challenge for implementing the M-EPDG is to determine all of the parameters required. Such task requires extensive testing and data collection.

The design methods adopted currently in the Middle East region are based on "performance related specifications" such as the percentage of air voids and the voids in the mineral aggregate. However, these parameters are proven to be insufficient to predict the performance of the asphalt concrete pavement over its service life. Therefore, the need for a well-defined implementation initiatives in the middle east region to shift from their current pavement design practices to M-EPDG as their standard design procedure is thought necessary. However, the level of complexity and the large number of input parameters required by the software poses a huge challenge for the proper implementation in countries with insufficient data.

For the aforementioned reasons, this study is carried out to draw guidelines which facilitates the adoption of M-EPDG, in countries that lack the required information for implementation, using Levels 1, 2 and 3 and Lebanon will be tackled as a case study. The three key objectives for the work are as follows:

1. Evaluate the technical feasibility of using the M-EPDG software for design of pavements in Lebanon,
2. Outline current challenges and constraints in implementing the M-EPDG and offer guidelines for developing an implementation plan in countries outside the U.S and Canada,

3. Identification of the design input variables that significantly impact the pavement performance predicted by the M-EPDG and offer a methodology for facilitating the collection of the desired input variables.

1.2 Literature Review

Numerous research projects and studies have been conducted in the U.S. in verifying the functionality of the M-EPDG software, identification of critical and sensitive input parameters, need for local calibration, and development of plans for the comprehensive implementation for various levels of hierarchal inputs. Literature from recently completed or ongoing projects on M-EPDG implementation was reviewed in details to serve as a base for this study.

Starting with regions inside the U.S., a recent survey found that forty-four states in the U.S. still use empirical design methods [5]. In some states as Illinois for example, design methodologies and programs other than the empirical and ME, such as PerRoad and PerRoadXPress, are used for pavement design by the Illinois Department of Transportation (DOT) and Texas DOT [5]. Nevertheless, most of the U.S. states are moving towards incorporating the mechanistic-empirical methods into their pavement design standards [6]. In September 2013, Wisconsin Department of Transportation has carried out AASHTO M-EPDG regional peer exchange meetings with the state highway agencies (SHA) in AASHTO Region 3 states, to promote the adoption of the M-EPDG in different states, in which the different SHA exchange ideas, concerns, difficulties and experiences regarding the implementation of the M-EPDG. A report is prepared that summarizes the efforts made by the different states for implementing the M-EPDG [6].

The state of Indiana is one of the first states that started implementing the M-EPDG. Currently, Indiana has fully adopted the mechanistic-empirical method and mandated the use of M-EPDG as its standard design methodology for all new state highway and interstate pavement design [5]. Indiana Department of Transportation (INDOT) adopted an implementation plan in which the aim was to target each of the three design levels. The work has included the review of current state-of-knowledge in pavement engineering and management, obtaining local input data which can be inputted directly in the M-EPDG as well as data to be processed to match the requirements of the software, and reporting the most critical input parameters for each of the three design levels. The efforts were focused on reaching the highest design level by taking into consideration the contribution of field and laboratory testing resources and developing a strategic plan to establish local calibration and validation of distress models. Detailed sensitivity analysis was carried out to determine the most critical input parameters and thus drawing a guideline for the initiatives needed to gather sufficient input data that are necessary to obtain accurate performance prediction results. The study included all three levels of analysis as well as the large variety of data related to traffic, climate and pavement materials [7]. The study highlights that it is necessary to collect design input data and establish a data base for inputs that fits local conditions [8]. A knowledge of the input parameters and their significance on the prediction of different pavement distresses is a very crucial factor in the implementation strategy.

Similar to the Indiana implementation strategy, a study of an implementation plan of M-EPDG is proposed for New England States. The study focused on the development of a guideline for data documentation in New Hampshire and Connecticut. In order to develop a strategy and provide recommendations on the use of existing data for input parameters, a series of sensitivity analysis studies were performed and sensitivity ranges of existing parameters were identified [9]. The sensitivity analysis was carried out to provide state highway agencies with information on the required level of data collection. Analysis of variance was utilized to examine the responsiveness of anticipated distresses to various input variables. The input variables were classified according to their order of significance, which can be utilized for suggesting changes to the subsisting pavement management systems [9].

Besides the sensitivity studies of the various input parameters and the evaluation and calibration of the M-EPDG to meet site-specific conditions, the Department of Transportation in the Maricopa County in |Arizona focused on developing a guided methanol on the collection of the needed input parameters. The study presented a guided plan on how to collect and compile a site-specific database of traffic information, materials properties and climatic conditions when such database does not exist [8]. Information and surveys were gathered from different departments in the county and from National resources to meet the objectives of the study.

Moving to regions outside the U.S. and towards the Middle East, in the UAE, Abu Dhabi seeks to follow the M-EPDG as their standard procedure in pavement design, however, the complexity of input data requirements as well as its high cost are two major challenges to overcome [10]. Also, research has been conducted about the implementation of the M-EPDG in Qatar [11]. The highway agencies of other countries and cities do not seem to be diverting away from the 1993 AASHTO Guide anytime soon.

As a wrap up, the findings from the review indicate that pavement distresses are sensitive to various design variables [7, 9] that are listed in Table 1. Additionally, the

Table 1. Critical input variables

Distress type	Critical input variables
Longitudinal (top-down) cracking in the wheelpath	• HMA mix stiffness • Foundation support (base/subgrade resilient modulus) • HMA thickness
Transverse (thermal) cracking	• Binder type • HMA thickness • HMA strength • HMA creep compliance • Coefficient of thermal contraction
Fatigue (bottom-up) cracking	• HMA thickness • HMA mix stiffness • Binder content • Percent air voids
Rutting	• HMA gradation • HMA mix stiffness • HMA thickness • Base/subgrade resilient modulus

review of the papers relevant to the implementation of M-EPDG [12] concluded that the specific truck class distribution factors are required for analysis as it have a significant effect on the predicted pavement performance compared to the default values. The effect of monthly and hourly distribution factors was found to be insignificant.

Based on the above, a methodology for carrying out this study was planned as elaborated in the next section.

2 Research Approach

In achieving the stated objectives, it is important to establish comprehensive familiarity with pavement practices in Lebanon in the aspects of design, construction, and management, as well as knowledge of the availability, collection methodology, and suitability of data needed for the M-EPDG. While several Lebanese government agencies are involved in roadway-related works; however, only two entities are in charge of pavement operations of major highways: the CDR, Council of Development and Reconstruction, for design and construction of new international and primary roadways, and the Ministry of Transportation and Public Works for secondary roadways and routine maintenance and rehabilitation. It is often the case where for some projects the agencies interchange the responsibilities, or jointly oversee the operations. The vast majority of structural design tasks are delegated to consultants and thus for proper M-EPDG implementation, it is essential to involve pavement consultants and other engineering firms responsible for the various pavement operations in Lebanon.

Traffic and materials data were first collected from the aforementioned agencies and firms. Contrarily, reliable climate files were not readily available. Thus, additional effort was needed to gather climate data for Lebanon.

Trial runs were conducted at level 3 analysis using the obtained preliminary data to gain familiarity with the different levels of analysis offered by the software, to assess the reliability of the data and to establish a guideline for the efforts needed to reach higher levels of analysis. This preliminary study showed that the available traffic data was incompatible with the requirement of the software and that the approach used for climate data lacks accuracy. In addition, available materials data served only level 3 analysis. For these reasons, the Structural and Materials Laboratory at the American University of Beirut was consulted and input materials data satisfying levels 1 and 2 analyses were collected for implementation. Furthermore, the creation of raw '.icm' design input climate file was mandatory. Therefore, the records of the meteorological station Rafic Hariri International Airport of Beirut were collected and the missing parameters were calculated based on reliable correlations and equations. Another approach was to search for states in the U.S. having similar climate to the city under study. Jacksonville, Florida was found to be suitable for the case study of Beirut city as a first trial.

Acquiring suitable input traffic data was the most challenging phase of the study; traffic information were collected from traffic count surveys conducted in 2002 for the Lebanon Physical Plan study. Yet, the truck classification adopted in that study was different than that required by the M-EPDG. Therefore, a methodology was developed to find a correspondence between the two different classifications, as well as to

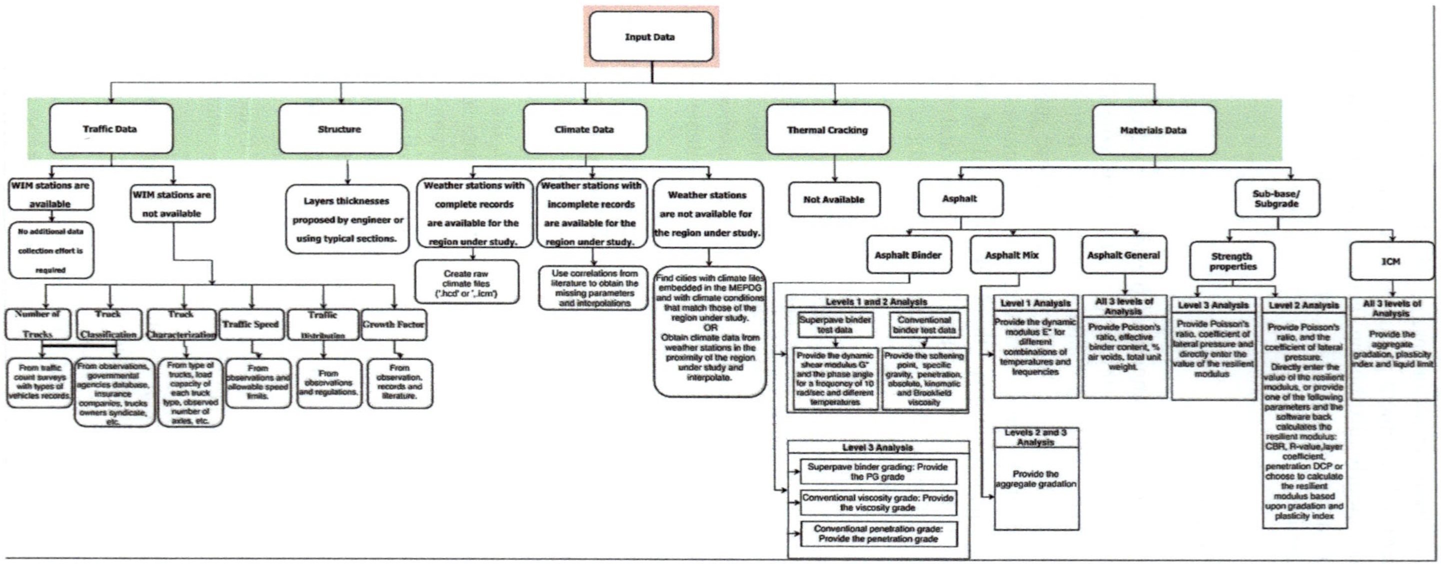

Fig. 1. Proposed methodology for acquiring various input data

calculate the annual average daily truck traffic and other required input data. The proposed methodology is detailed in the following section.

Sensitivity analysis runs - in which design inputs were varied within typical ranges for Lebanon - were performed to test the validity of the suggested data collection methodology for the implementation of the M-EPDG in Lebanon. It should be noted that the methodology proposed herein; even though tested only for Lebanon, can be adopted by other countries lacking sufficient data for the implementation of the M-EPDG, with minor adjustments satisfying the special conditions of each country.

Finally, based on the results of the sensitivity analysis, the critical parameters are chosen and accordingly methodologies for implementation plan are proposed to address the problem of data unavailability and data formats that are incompatible with requirements of the M-EPDG.

A flowchart presented in Fig. 1 summarizes the methodology and steps followed to gather the required input data.

3 Methodology for Collecting Design Input Data

In order to meet the stated objectives of this study, it is important to create input files which would serve as reference files for carrying out the sensitivity analysis. As such, the south coastal highway connecting Beirut to Saida was chosen at a first stage for carrying out the study.

The corresponding climatic, traffic and materials data for the selected roads are collected for the sensitivity analysis. The methodology followed to collect the required input data is elaborated hereafter.

3.1 Traffic Data

The collection of traffic data for pavement design purposes, particularly for the format required by the M-EPDG software, was the most challenging among all other input data. Availability of such data for future implementation of the M-EPDG in Lebanon will definitely require the most effort and investment. The available traffic data originates from surveys conducted for geometric highway design and traffic analysis. Traffic data or vehicle weight are not available for the purpose of pavement design in Lebanon. Pavement engineers often use total traffic count data with estimates of percent trucks, truck distribution, and legal limits for truck loads to calculate ESALs.

For this study, traffic data and information were collected from traffic count surveys conducted in 2002 for the National Physical Master Plan of the Lebanese Territory [13]. The aforementioned study was executed to obtain a database of the national and regional strategic trip making characteristics in Lebanon. Surveys are conducted based on a zone system that presents a subdivision of the 25 Caza areas of Lebanon and a special division of Beirut into 18 regions [13]. Nevertheless, in this paper only selective stations located on major highways, connecting key urban areas and that are subject to a high level of congestion and heavy vehicle loading, are taken as traffic input data for the sensitivity analysis runs performed using the M-EPDG software.

The data stems from short duration traffic counts conducted over the course of one day from 6:00 am to 7:00 pm and in 30 min intervals. Also, the counts are divided according to the vehicle type (car, SUV, Pick-Ups Vans, Mini Buses, Buses, LGV, MGV/HGV, Motor cycle, and other types) and not by the vehicle's Class. The classification of trucks is done according to the European Union Descriptions in which LGV stands for Large Goods Vehicle, MGV stands for Medium Goods Vehicle and HGV stands for Heavy Goods Vehicle [14].

The Software requires a proper characterization of the traffic input data to better design and predict the performance of new and rehabilitated flexible and rigid pavement structures. This characterization encloses site-specific vehicle class distribution and axle load data which are crucial factors for calculating ESALs and traffic loads, monthly adjustment factors, hourly truck traffic distribution, average number of axle groups per vehicle and axle load spectra.

The Ministry of Public Works and Transport in Lebanon does not provide any systematic approach to monitor and classify imported and registered vehicles, trucks in specific, as per guidelines of the Federal Highway Administration in the U.S. It only classifies vehicles according to their weight, of which it distinguishes two types: vehicles that weigh less than 10 Tons and those that weigh more than 10 Tons [15]. This poor grouping and organization does not contribute to the development of pavement design procedures followed by highway agencies and pavement/material consultants in the region. Such characterization was not properly done in the surveys. However, having these surveys as the sole available source of traffic data in Lebanon, it is therefore deemed necessary to carry and apply assumptions to create input files that serve as reference for future use and implementation. These assumptions are based on several factors, one of which is the distribution of the type of load carried by different goods vehicles raging from a small pick-up to HGV. The underlying assumptions for creating the traffic input files when no information is made available for designers to design the corresponding pavement structure are as follows;

- *Vehicle Class Distribution*

Typically, vehicle class distribution is computed using automatic vehicle classification (AVC), Weigh-in-motion (WIM) and vehicle counts. The continuity of data collection and the ability to constantly verify and validate it is the key to having a successful classification.

Also, Classification of vehicles in the U.S and the European Union is based on Gross Vehicle Weight Rating (GVWR), the maximum weight of the vehicle as specified by the manufacturer. The Federal Highway Agency (FHWA), the U.S census Bureau and the U.S. Environmental Protection Agency (EPA) categorize vehicles as medium duty (class 4, 5, 6) with GVWR between 10,001 lbs. and 26,000 lbs., Light heavy duty (class 6) with GVWR between 19,001 lbs. and 26,000 lbs. and Heavy duty (class 7 and 8) with GVWR over 26,001 lbs. [16]. According to the European Union terms, any truck with Gross weight vehicle over 7000 lbs. is referred to as heavy goods vehicle (also large goods vehicle and medium goods vehicle) [14].

Nevertheless, the lack of resources in Lebanon such as toll facilities and management centers together combined with political corruption led to poor if, none existing vehicle classification and gross vehicle weight ranges. Therefore, it is only reasonable to assign classes to vehicles grounded on visual observations. In this paper the following distribution is adopted: buses are classified under class 4 vehicles, LGV's divided into classes 5 and 6, MGV's into classes 7, 8 and 9, and HGV's into classes 10, 11, 12 and 13 (Table 2).

Table 2. Truck traffic growth factors

Vehicle class	Growth rate (%)			Growth function
Class 4	1.46	3.5	7	Compound
Class 5	1.62	3.5	7	Compound
Class 6	0.56	3.5	7	Compound
Class 7	0.56	3.5	7	Compound
Class 8	0.56	3.5	7	Compound
Class 9	0.56	3.5	7	Compound
Class 10	0.56	3.5	7	Compound
Class 11	2.89	3.5	7	Compound
Class 12	2.89	3.5	7	Compound
Class 13	2.89	3.5	7	Compound

Based on visual observations and the vehicle class distribution of trucks according to the Federal Highway Administration, the most common load type carried by pick-ups and Medium Goods Vehicles (MGV) is food, while Heavy Goods Vehicles (HGV) are mostly seen to carry building construction materials and they are the predominant category of trucks in the country [13]. Therefore, based on engineering judgement it is only logical to allocate a higher percentage of the total count of MGV/HGV for HGV's (60% for HGV and 40% for MGV). Table 3 presents an

Table 3. Example of the conversion of the collected traffic data to input data for the south coastal highway for one software run

Vehicle category from vehicle count data	*Buses*	*LGV*	*MGV and HGV*							
Vehicle daily count 2002	300	728	546							
Assigned percentage from vehicle count (%)	100	100	*MGV*				*HGV*			
			40				60			
Assigned vehicle class	*Class 4*	*Class 5*	*Class 6*	*Class 7*	*Class 8*	*Class 9*	*Class 10*	*Class 11*	*Class 12*	*Class 13*
% Class distribution for each category	100	70	30	80	10	10	98	2	0	0
Compound growth factor %	1.46	1.62	0.56	0.56	0.56	0.56	0.56	2.89	2.89	2.89
Vehicle count projected to 2016	367	638	236	189	24	24	347	10	0	0

example of the conversion done to obtain suitable input data for the software from the data collected from the survey. In a later stage, the assigned percentages for the different truck classes within each category are changed to perform the sensitivity analysis. It should be noted that one main objective of this sensitivity analysis is to test the validity of the approach presented herein to convert from the truck classification adopted in the survey (LGV, MGV, HGV) to the classification required for the M-EPDG software.

- *Annual Average Daily Truck Traffic*

Two-way Annual Average daily truck traffic is defined as the total volume of truck traffic passing through a segment of a road in both directions during 24-h period. Since the data is not collected over the entire 24-h period, the measured daily truck traffic is assumed to be representative of that period. It is worthy to note that in Lebanon truck traffic volume is very minimal during late hours of the night (after 8:00 p.m.) and early mornings (before 6:00 a.m.) [13].

- *Truck Growth factor*

Truck traffic growth factors are affected by many superfluous factors, and thus those derived from narrow data collected from very few locations can be biased. A less reliable estimate of traffic growth factors can also be computed from data obtained from short duration counts, since these counts are not as nearly accurate as those available from continuous traffic counts. There is no single correct procedure to follow, it is rather best to develop an estimate based on different data sources collected over a lengthy time period. Also, the design guide software allows end-users to use different growth functions to simulate or predict future truck traffic volumes. In this paper, the compound growth function is used (Table 2). Growth factors are applied for each truck class in order to simulate the annual average daily traffic foreseen in Lebanon in 2026. Due to political events happening in nearby regions such as the ongoing war in Syria – now entering its sixth year- which entailed a state of insecurity, the closure of the Lebanese-Syrian borders was deemed necessary. This has caused almost 60% drop in land transportation from Jordan, Iraq and beyond, especially when it comes to the trade of manufactured goods [17]. The annual average truck traffic volume has decreased greatly and reached a state of plateau ever since the beginning of war. This allowed us to consider a 0% growth factor for trucks for the current period. However, taking into consideration the end of the war in the close future, truck growth factors of 3.5% and 7% were adopted as part of the sensitivity study. This is reflected by the growth in post-war reconstruction periods which are expected to be very high).

Level 3 analysis is commonly used when there is little truck volume information for the roadway in question. This level starts from simple truck volume counts with no segment specific knowledge on the size of the loads those trucks are carrying. An estimate of traffic inputs based on local experience is also considered as level 3.

Having many assumptions, it is safe to note that the inaccuracy of the compiled traffic data used for Level 3 analysis runs using M-EPDG will most definitely drive the reliability of performance predictions. Results from sensitivity analysis are important to assess implications of traffic data quality.

3.2 Climate Data

The climate data for regions outside the U.S. and Canada are not included in the climatic database of the M-EPDG. It is thus important to either create '.hcd' raw data climate files that can be converted by the M-EPDG software to '.icm' climatic files, or directly create '.icm' files for the location desired, using climate data records. Yet, in some cases, there might not be weather stations in the region under study. Therefore, in this study, the possibility of matching, to the extent possible, cities in the U.S. and Canada for which climatic data exist in the M-EPDG database, with similar climatic conditions to those in the region under study is explored. This would solve the problem of having to use the M-EPDG for regions with no climatic data in the database.

For the case of Lebanon, there are seven official weather stations. Out of these stations, few weather stations are currently operational, with the station at Rafic Hariri International Airport in Beirut having the oldest and most comprehensive data on record. In this study, fifteen years of recorded climate data corresponding to Beirut, obtained from the records of the aforementioned station, are considered to create a reference '.icm' design input climate file, for investigating the sensitivity of the pavement mechanical distresses on the climate input data. The number of years required for every study depends on the type of analysis. In explanation, if thermal cracking is of concern, more years shall be considered in the analysis since thermal cracking occurs on the coldest single night of a year, which might occur once every decade or two. In this case, one should make sure that extremely cold events are not missed. However, if the analysis is being carried out in areas with moderate to hot climates, shorter period of data can be used, with a minimum of two years. Nevertheless, the software requires two months of climate data to carry out a run.

The data recorded by meteorological stations typically exclude several parameters that are required for creating the '.icm' file. These include humidity, sunrise and sunset times, solar radiation, percent sunshine and ground water table depth. In order to obtain the required data, correlations available in the literature and almanacs as well as databases are used. The data obtained includes hourly records of temperature, wind speed and direction, sky cover, precipitation and dew point. For the hours with missing records, the data of the preceding and following hour are interpolated. From these records, the required parameters for creating the '.icm' files were calculated in the following manner;

- The humidity is calculated from the dew point and temperature values recorded on hourly basis by the stations, using the August-Roche-Magnus [18] approximation.
- The sunrise and sunset times are calculated using NOAA Solar Calculator, based on the longitude and latitude of the stations under study.

Table 4. Percent sunshine from cloud cover records

Cloud cover	Corresponding % sunshine
Clear	90%
Scattered	75%
Broken	50%
Partial obscuration	25%
Obscured	10%
Overcast	0%

- The solar radiation values are obtained from a study carried out by Dawatec [19] for calculating the solar insulation levels in Lebanon for year 2013. Due to the lack of additional data, the values were assumed to be the same for all the other years.
- The inverse of the sky cover records is translated into percent sunshine in tangent to road. For example, if the cloud cover is 100% overcast it is equivalent to 0% sunshine. The conversions followed are summarized in Table 4.

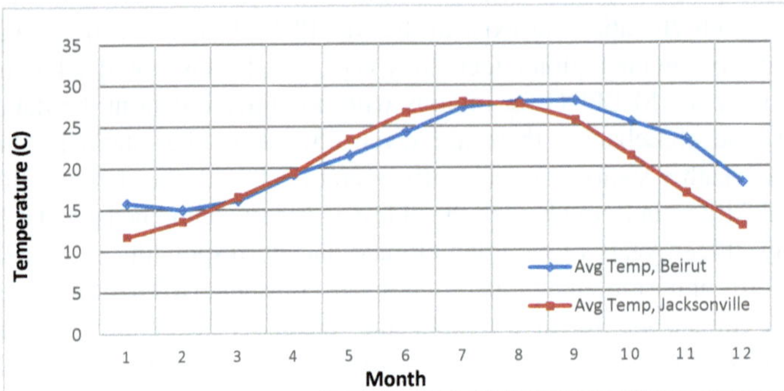

(a)

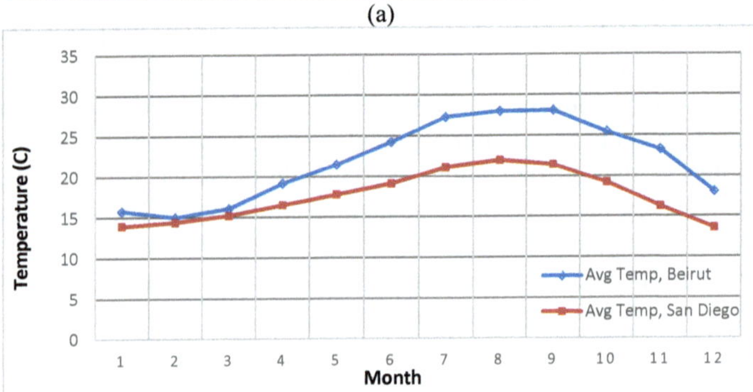

(b)

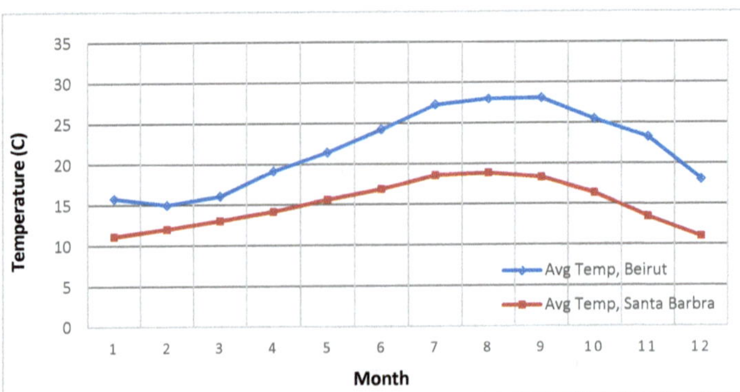

(c)

Fig. 2. Average monthly temperatures for Beirut and selected cities in the US: (a) Beirut vs. Jacksonville, (b) Beirut vs. San Diego, and (c) Beirut vs. Santa Barbra.

- The water table levels for the different seasons are obtained from database available for Engineering Consulting companies.

Based on the above, the reference '.icm' climate file was prepared for Beirut, with the format required by the software.

Following the above, an exercise was followed to select the cities in the U.S., with existing climatic data in the M-EPDG, that have similar climatic conditions as Beirut, Lebanon. It was found out that Jacksonville, San Diego and Santa Barbra mainly resemble Beirut, with respect to average monthly temperatures, as can be seen in Fig. 2.

In here it should be noted that only the monthly average temperatures were considered and not maximum monthly highest and lowest temperatures, since the analysis of thermal cracking distress is not supported in Version 1.0 of the M-EPDG software. Thus, extreme single day temperatures are not of concern in this study. In addition, the percent sunshine and wind speed and precipitation data were omitted from the analysis due to the lack of records of these data. Above and beyond, percent sunshine and wind speed data moderately influence the pavement distresses while precipitation have a negligible influence on flexible pavement performance [20].

3.3 Materials Data

This study tackles flexible pavement generally consisted of three layers:

1. The asphalt concrete layer
2. The sub base
3. The semi-infinite subgrade

For the HMA layer, level 1 analysis is selected which requires detailed input data parameters. The dynamic modulus (E*) of the asphalt mix is to be specified for different combinations of temperatures and loading frequencies. The dynamic shear modulus (G*) and the phase angle (delta) of the asphalt binder are also required for different temperatures and for a constant oscillation rate of 10 rad/s; these two parameters are obtained using the DSR test (Dynamic Shear Rheometer) conducted in the laboratory.

The effective binder content, percent air voids and the total unit weight are required input parameters for all three levels of analysis.

The above mentioned properties of a typical asphalt mix were obtained from the American University of Beirut material and structural laboratory database for mixes used in Lebanon.

These values are taken the same for all the runs performed since our aim is to study the sensitivity of the methodology used and not the parameters themselves.

For countries with no available values of the asphalt mix's dynamic modulus, levels 2 and 3 can be selected. Both require the aggregate gradation of the asphalt mix which can be determined using a simple sieve analysis.

For the asphalt binder, level 2 requires the same input parameters as level 1; one can choose to enter the dynamic shear modulus and the phase angle (delta) of the asphalt binder or to provide the conventional binder test data such as the softening point, absolute viscosity, kinematic viscosity, specific gravity, penetration and Brookfield viscosity.

In the case where there is limited availability of the asphalt binder data, the designer can choose one of the three options provided by level 3, either the PG grade, the viscosity grade or the penetration grade of the binder.

For the granular base and the subgrade, both levels 2 and 3 necessitate the aggregate gradation. Level 3 requires the designer to input directly the resilient modulus while in level 2, the software backcalculates the resilient modulus provided one of the several input parameters: CBR, R-value, layer coefficient, penetration –DCP, the plasticity index and the gradation.

Level 2 analysis is selected for this study and the resilient modulus is calculated based on gradation as following: [21].

Design inputs for unbound layers, subgrade soil, soil classification and CBR values were obtained from the database of design consulting companies and the Lebanese Construction Specifications.

$$CBR = 28.09(D_{60})^{0.358}; \quad (D_{60} \text{ in mm}) \tag{1}$$

$$M_R = 2555(CBR)^{0.64} \tag{2}$$

4 Software Runs

Sensitivity analysis is carried out on the assumed input data that are region specific. These inputs are varied for a range of typical values for Lebanon. In every run, one input variable is varied. Based on the results obtained, the influence of each input variable is determined by analyzing the predicted performance. This can be used to identify the parameters which affect the response prediction significantly and should be monitored carefully.

Knowing that the missing parameters in regions outside the U.S. are mainly traffic data and climate data as explained previously, the validity of the proposed methodology for obtaining the traffic and climate data is evaluated as follows:

- The sensitivity to the approximated percentages of different categories of trucks on the performance of the pavement is studied by varying the percentages of MGV and HGV.
- The sensitivity of the distribution of trucks classes within the same trucks category LGV, MGV and HGV) is considered by trying different combinations of realistic percentages of classes within each category. These combinations are based on visual observations as well as recommendations from different concerned parties such as pavement consultants well aware of traffic behavior in Lebanon.
- The effect of variation of the growth factors for the different classes of trucks due to the political and economic situations of every region is measured by taking into account different typical values of the growth factors.

- The effect of using M-EPDG climate files for cities in U.S. and Canada with similar climate to the region under study is quantified by doing runs using different climate files, in reference to a control climate file prepared for Beirut.

The values of the input parameters used for the sensitivity analysis are summarized in Table 5.

Table 5. Input variables for the sensitivity analysis runs

Parameter under study	Run number	Inputs
Distribution of trucks categories (MGV, HGV respectively)	1 *(Control Run)*	40%, 60%
	2	50%, 50%
	3	60%, 40%
	18	30%, 70%
	19	70%, 30%
Distribution of LGV classes (classes 5,6 respectively)	1 *(Control Run)*	70%, 30%
	4	60%, 40%
	5	50%, 50%
	6	40%, 60%
Distribution of MGV classes (classes 7, 8, 9 respectively)	1 *(Control Run)*	80%, 10%, 10%
	7	60%, 20%, 20%
	8	50%, 25%, 25%
	9	40%, 30%, 30%
Distribution of HGV classes (classes 10, 11, 12, 13 respectively)	1 *(Control Run)*	98%, 2%, 0%, 0%
	10	90%, 10%, 0%, 0%
	11	80%, 10%, 5%, 5%
	12	60%, 20%, 10%, 10%
Growth factor	13 *(Control Run)*	Class specific growth factors *(Table 2)*
	20	0 for all classes
	21	3.5 for all classes
	22	7 for all classes
Climate	25 *(Control Run)*	Created climate file for Beirut
	1	Jacksonville
	23	Santa Barbra
	24	San Diego

5 Results and Sensitivity Analysis

The results of the sensitivity analysis are plotted as shown in Fig. 3, to examine the impact of the input variables of concern, on the performance. The input parameters that affect the different distresses are listed below, in the order of significance.

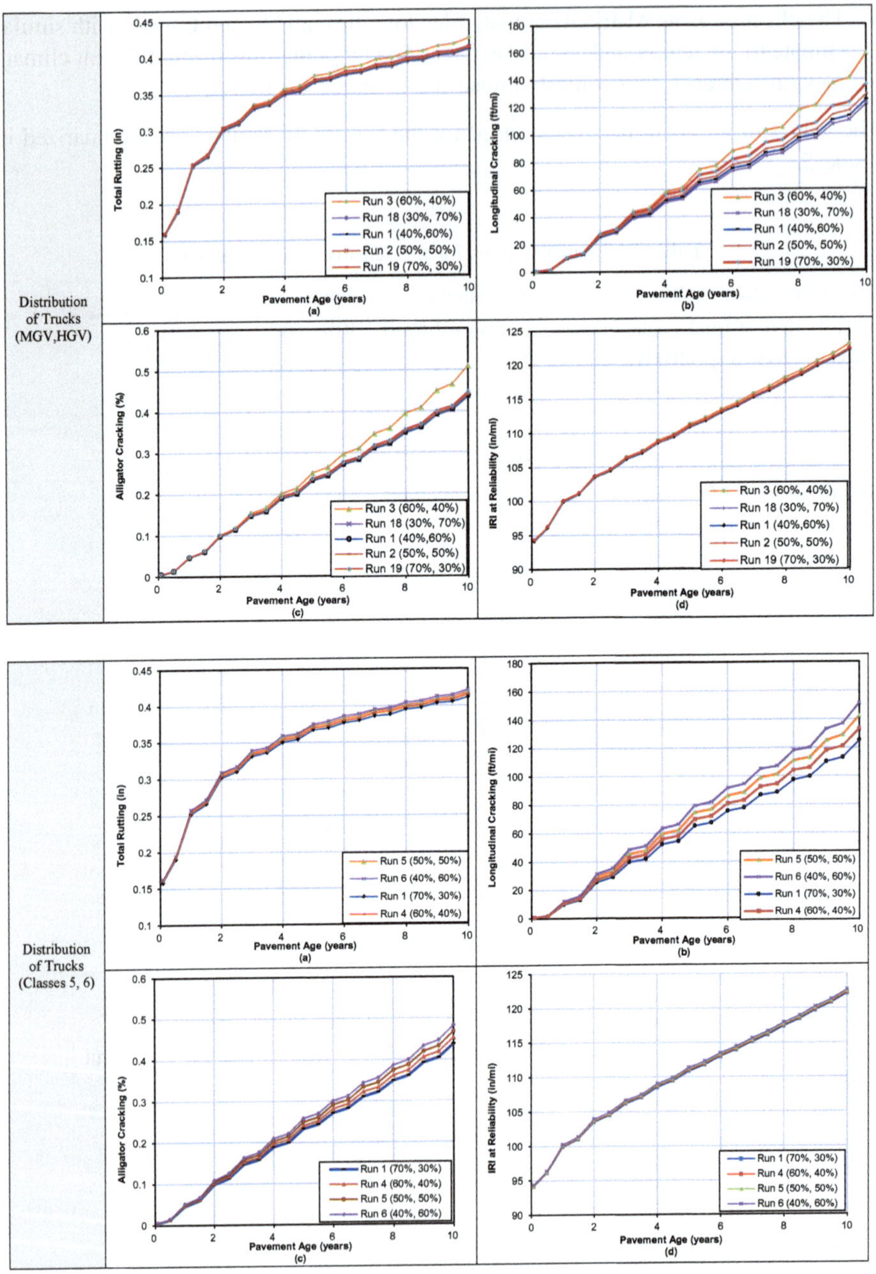

Fig. 3. Distresses plots of sensitivity analysis: (a) total rutting, (b) longitudinal cracking, (c) alligator cracking, (d) IRI at reliability

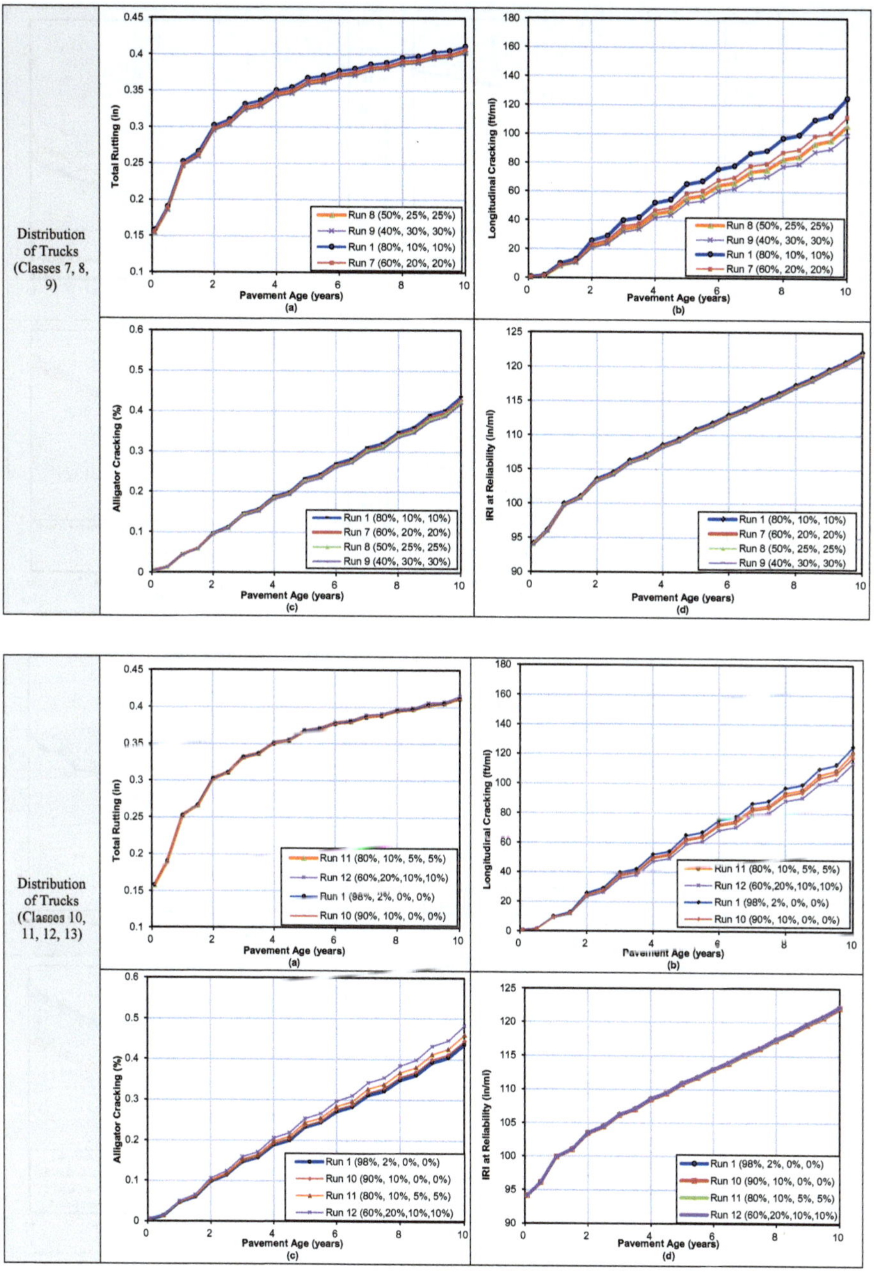

Fig. 3. (continued)

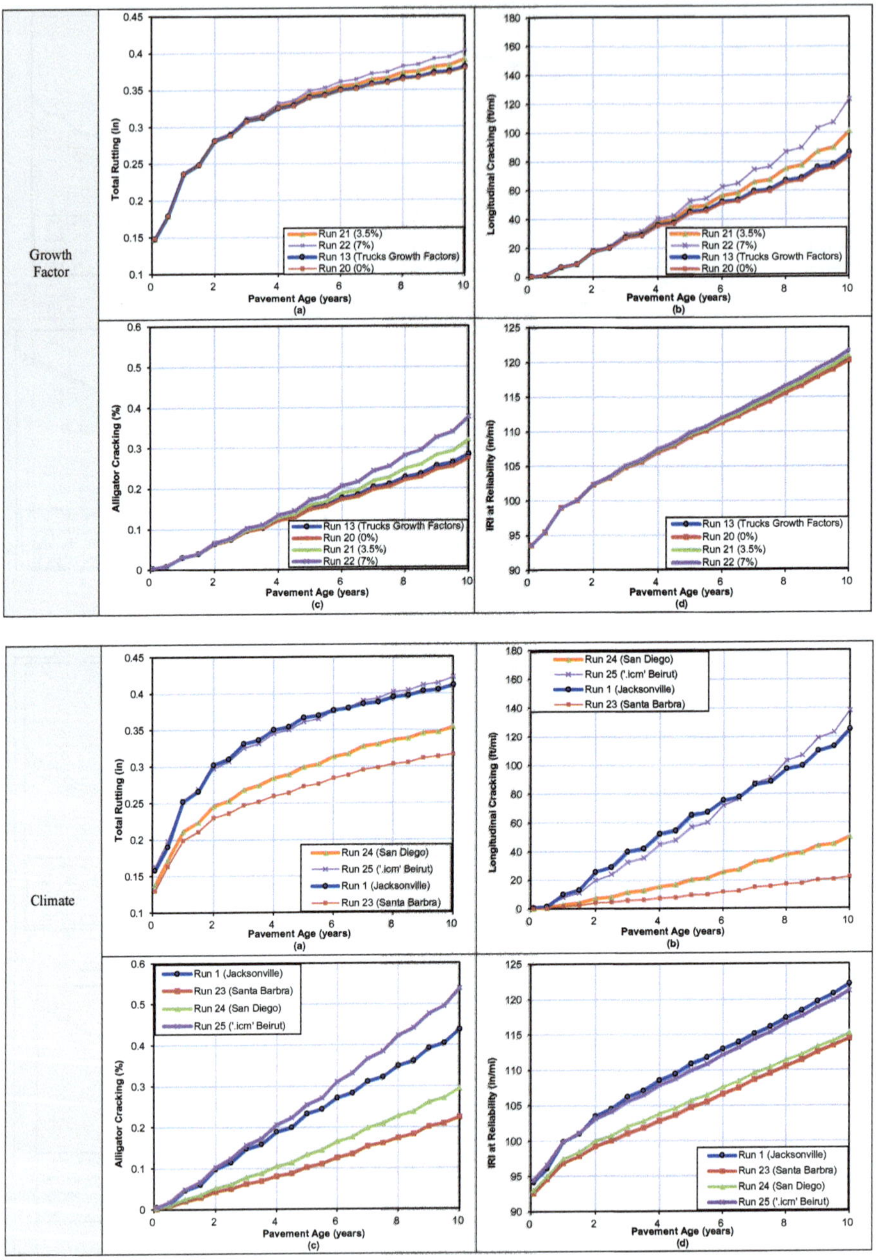

Fig. 3. (continued)

- **Total Rutting**
  Total rutting is significantly sensitive to the climate input data, where AC (asphalt concrete) rutting showed a variation raging between 30–50% when the average

monthly temperature at the hottest months of the year varies on average 6–8 °C. Rutting is also sensitive to the growth factor and equivalently the number of trucks passing (which can also be translated into AADTT) as well as the categories of trucks (MGV and HGV). However, total rutting is less sensitive to the truck traffic class distribution (TTC) within the same truck categories.

- **Top-Down (Longitudinal) Cracking**
  As demonstrated in the graphs, top-down cracking proved to be mostly sensitive to the climate data. Besides, longitudinal cracking is significantly sensitive to the growth factor and hence number of trucks passing. Also, the changes in truck categories and distributions have an appreciable effect on the top-down distress.
- **Bottom-Up (Fatigue) Cracking**
  Bottom-up (Alligator/Fatigue) cracking is significantly affected by the total number of trucks passing or the growth factor, annual average daily truck traffic (AADTT) and climate. Besides, fatigue is extensively sensitive to the climate input data. Finally, bottom-up cracking is less sensitive to trucks classes distributions, yet the change in percentages among truck categories has an appreciable effect.
- **IRI**
  IRI is almost equally sensitive to growth factor, climate and trucks categories distribution, but not much sensitive to the truck traffic class distribution.

5.1 Traffic Inputs

Mild variations on the percentages of trucks passing have a slighter influence on the predicted distresses compared to number of trucks passing or growth factor and the climate. Therefore, reasonable errors in approximating the percentages of trucks classes as well as a mild variability in the trucks distributions is tolerable upon collecting traffic input for a design project using the M-EPDG. Hence, for designing pavement sections in regions with insufficient data, one can rely on the collected traffic counts data as well as rational approximations of truck classes distributions based on observations and vehicles records of Governmental departments and agencies.

On the other hand, the predicted response is highly sensitive to the approximated growth factor. For this reason, an expert study shall be carried out to obtain a reliable growth factor, based on expected political and economic conditions in every region.

5.2 Climatic Inputs

Climate data proved to significantly affect the predicted distresses. As seen from the results, runs 24 and 25 with Santa Barbra and San Diego climate files, all the distresses results seem to be significantly shifted from the actual performance. For that reason, it is optimum to create raw climate files for the region under study. Yet, if no data is available for creating '.icm' files, the option of selecting a climate file embedded in M-EPDG that closely matches the climate at the selected region remains remarkably valid. The selection of the file is critical. The selected file should have almost matching monthly average monthly temperatures during the hottest months of the years, with a tolerable deviation of 6–8 °C. Nevertheless, it should be highlighted that the analysis carried out excludes thermal cracking. As explained previously and knowing that thermal cracking occurs on a single night during extreme cold events, one should make sure that such events are incorporated in the selected climate files to detect thermal cracking. Therefore, the user can manually edit specific hourly temperatures to account for thermal cracking in the analysis.

6 Conclusions

A methodology for implementation of Mechanistic – Empirical Pavement Design Guide in the regions outside the U.S. and Lebanon is presented in this study, while tackling Lebanon as a case study. A sensitivity analysis to the proposed methodology is carried out, to study the validity of the proposed methodology for obtaining the missing input variables, namely climate and traffic data. Regional specific values of the input parameters are selected and varied within tolerable and reasonable ranges.

Based on the results obtained, the variables which affect the predicted performance were presented in their order of significance for every distress. The interpretation of the results is used to finalize the proposed methodology for data collection.

The findings reveal that highway design agencies can rely on available traffic counts surveys and records of vehicles types to create a reliable traffic input file. Yet, the major efforts reside in creating climate files. It is a labor-intensive and hectic procedure to develop weather files and create raw '.hcd' or '.icm' files to be used in the M-EPDG. However, the findings of this study indicate that the need for creating climate files can be eliminated if an M-EPDG embedded climate file with matching temperatures is available.

Acknowledgements. The authors appreciate and thank the University Research Board at the American University of Beirut for funding this study. Thanks also go to our colleagues Ms. Zeina Bsaibes, Ms. Yara Hamdar and Mr. Hussein Kassem who helped in completing this study.

References

1. Monismith, C.L.: Evolution of long-lasting asphalt pavement design methodology: a perspective. In: International Symposium on Design and Construction of Long Lasting Asphalt Pavements. Auburn University, Alabama, USA (2004)
2. Pavement Interactive: 1993 AASHTO Flexible Pavement Structural Design. Pavement Interactive (2008)
3. Pierce, L., Smith, K.: Continuously Reinforced Concrete Pavement: Design Using the AASHTOWare Pavement ME Design Procedure. Federal Highway Agency FHWA-HIF-13-025 (2013)
4. Applied Research Associates Inc. Manual 3: Design of New and Reconstructed Flexible Pavements, Chapter 3. Publication NCHRP 1-37A, National Cooperative Highway Research Program, Transportation Research Board, National Research Council (2004)

5. David, T., Robins, M., Tran, N., Rodezno, C.: Flexible Pavement Design-State of the Practice. Rep. Auburn University. National Center of Asphalt Technology (NCAT), Report 14-04 (2014)
6. Linda, P., Smith, K.: AASHTO MEPDG Regional Peer Exchange Meetings Final Technical Report. Publication no. F H W A - HI F - 15 -0 21. Wisconsin Department of Transportation. N.p., Federal Highway Adminstration (2015)
7. Nantung, T., Chehab, G., Newbolds, S., Galal, K., Li, S., Kim, D.: Implementation initiatives of the mechanistic-empirical pavement design guide in Indiana. J. Transp. Res. Board **1919**(2005), 142–151 (2005)
8. Souliman, M., Mamlouk, M., Zapata, C., Cary, C.: Data collection to support implementation of the mechanistic-empirical pavement design guide for county roads. J. Transp. Res. Board **2225**, 67–77 (2011)
9. Ayyala, D., Chehab, G., Daniel, J.: Sensitivity of M-EPDG Level 2 and 3 Inputs using Statistical Analysis Techniques for New England, International Review of Civil Engineering (I.RE.C.E.) (2010)
10. Khattab, A., El-Badawy, S., Al Hazmi, A., Elmwafi, M.: Evaluation of Witczak E* predictive models for the implementation of AASHTOWare-Pavement ME Design in the Kingdom of Saudi Arabia. Constr. Build. Mater. **64**, 360–369 (2014)
11. Sadek, H., Masad, E., Sirin, O., Al-Khalid, H., Sadeq, M., Little, D.: Implementation of mechanistic-empirical pavement analysis in the State of Qatar. Int. J. Pavement Eng. **15**(6), 495–511 (2014)
12. Tran, N., Hall, K.: Development and Significance of state-wide volume adjustment factors in mechanistic—empirical pavement design guide. Transp. Res. Rec. J. Transp. Res. Board **2037**, 97–105 (2007)
13. Lebanon, Council of Development and Reconstruction (CDR): Schéma D'aménagement Du Territoire Libanais. By DAR and Institut D'Aménagement Et D'Urbanisme De La Région D'Ile-de-France (IAURIF) (2002)
14. European Commission: Heavy Goods Vehicles (2016)
15. El Hajj Chehade, A.: Chairman. Ministry of Interior; Department of Traffic Machinery and Vehicles (R. Hajj Chehade, Interviewer) (2016)
16. Chehab, G.: Associate Professor. American University of Beirut; Department of Civil and Environmental Engineering. (R. N. Mourad, & R. Hajj Chehade, Interviewers) Beirut, Lebanon. "Vehicle Weight Classes & Categories." Vehicle Weight Classes & Categories. U.S. Department of Energy: Alternative Fuels Data Center, n.d. June 2016 (2016)
17. Al-Monitor, S.: How the War on Syria Left Its Mark on Lebanon's Economy (2016)
18. Alduchov, O., Eskridge, R.:. Improved Magnus' form approximation of saturation vapor pressure. National Climatic Data Center, National Oceanic and Atmospheric Administration, Asheville, North Carolina (1995)
19. Dawatec: Solar Irradiance in Lebanon (2013)
20. Schwartz, W., Elkins, G., Li, R., Visintine, B., Forman, B., Rada, G., Groeger, J.: Evaluation of LTPP Climatic Data for Use in Mechanistic-Empirical Pavement Design Guide Calibration and Other Pavement Analysis. Federal Highway Administration FHWA-HRT-15-01: 1-3 (2015)
21. Applied Research Associates Inc.: Guide for Mechanistic-Empirical Design of New and Rehabilitated Pavement Structures, Appendix CC-1: Correlation of CBR Values with Soil Index Properties. Publication NCHRP, National Cooperative Highway Research Program, Transportation Research Board, National Research Council (2001)

Structural Evaluation of Flexible Pavement Using Non-destructive Techniques in Low Volume Road

Vinod Kumar(✉), Sunny Deol, and Rakesh Kumar

Department of Civil Engineering, SVNIT Surat, Surat, Gujarat 395007, India
vinodkumarmtech@gmail.com, gsuunydeol@gmail.com,
krakesh@ced.svnit.ac.in

Abstract. Low volume roads (LVR) are most accessible road network in India, and it contains thin asphalt surface (less than 100 mm). For this type of roads, a light weight deflectometer (LWD) may be useful for structural evaluating the pavement layer moduli and overlay design. In some parts around the globe, still uses conventional Benkelman beam deflectometer (BBD) method for structural evaluation purpose. In this paper, an experimental stretch 550 m length and width 7 m was considered for structural evaluating the pavement using LWD and BBD. Pavement deflection response data collected using a Benkelman Beam deflectometer (BBD) and lightweight deflectometer used to identify structural integrity of pavement. Deflections collected by LWD using 20 kg falling weight with radial geophone spacing 300 mm and 600 mm are used in this study. The proposed methodology for overlay design using LWD, LWDmod, and KENLAYER software's results was discussed. Obtained backcalculated layer moduli results are used to calculate critical stresses and strains generated by standard axle load of 8 tons by using KENLAYER software. The LWDmod overlay thickness analysis ranges from 28.00 to 180 mm, and BBD based calculated overlay thickness ranges 10 to 210 mm were observed. The correlation of deflection values between LWD and BBD was found with R^2 value of 0.741. The practical shortcoming of deflection correlation between BBD and LWD may be overcome by using more data points of LWD in Indian conditions for further research.

Keywords: Structural evaluation · Lightweight deflectometer · Benkelman beam deflectometer · Kenlayer

1 Introduction

Low volume road will contribute more accessible roads than any other roads. India recorded the second largest road network next to the USA. About 80% of roads are characterized as low volume roads. Present road length is over 3.3 million km, spatial road density of about 1 km/sq.km of the area Ministry of Road Transport and Highways (MoRTH 2012). Recently, Government of India estimated that approximately for low volume roads (LVR) the 5-year routine maintenance cost is in the range of 6–13% of construction cost during the base year 2013 (Barodiya and Pateriya 2014). As per Indian

L. Mohammad (ed.), *Advancement in the Design and Performance of Sustainable Asphalt Pavements*, Sustainable Civil Infrastructures, DOI 10.1007/978-3-319-61908-8_13

Road Congress IRC SP 72-2015, the low volume is considered as traffic intensity is less than 2 million standard axle loads. The current method of structural evaluation system largely depends on upon static deflection techniques for Indian low volume roads (Reddy and Veeraragavan 1997). The non-destructive testing techniques are recommended in road construction and evaluation practices for Indian highways to implement the mechanistic–empirically based analysis and design codes such as RC: 37-2012, IRC: 115-2014 and IRC: 117-2015. The LWD is a most acceptable tool in construction as well as a thin pavement structural evaluation tool in Germany and USA. Fleming et al. (2007) used LWD on top of asphalt layer thickness of 127–180 mm and evaluated the stiffness of a layer. The use of LWD on the fully constructed road is also gaining some interest for structural evaluation and overlay design (Fleming et al. 2006). India has adopted conventional static load device BBD for measuring static deformation modulus for overlay thickness design for the past few decades. The resilient behaviors of pavement layer materials are being assessed globally, and in India, National Highways are using Non-destructive field investigation tools such as falling weight deflectometer (FWD). However, uses of portable dynamic deflectometer for structural evaluation on low volume roads studies are very limited in India and across globally. Considering the limitations of static BBD, there is a need for the prerequisite to carry out comparative studies of both regular dependency methods with new approaches to eliminate the ambiguity in the selection of structural evaluation tool for low volume roads. Indian Road Congress has been published the utilizing the sophisticated structural evaluating equipment such as Falling weight deflectometer (FWD) in high volume flexible and rigid pavements for determination of in situ layer strength parameter (IRC: 115-2015 & IRC: 117-2015). Application of FWD on LVR is uneconomical and high cost for each test point. There is no methodology proposed at yet by using LWD as a structural evaluating tool in low volume roads. This paper may find some interest to pavement designers and researchers especially on low volume flexible pavement. The main objective of this article as (a) To comparison of overlay thickness by using static method and dynamic method, (b) Based on the LWDmod backcalculation of existing resilient moduli of individual layers, the remaining life of the pavement will be analyzed by considering critical stress-strain points using KENLAYER software, and finally (c) To determine the deflection relationship between static and impulse method.

2 Proposed Methodology

In this paper, a methodology was proposed for structural evaluation and overlay design determination on thin flexible pavements (less than 100 mm). The use of the light-weight deflectometer is the nondestructive technique for structural evaluation of thin bound and unbound pavement layers. Interpretation of LWD deflection results through pavement analysis algorithm are used to determine the layer properties. Newcomb et al. (1989) states that seasonal variation of material properties plays a vital role into the structural analysis to analyze the performance of the overlay. The assessment of how the material property will change its characteristic strength is hard to determine the performance of overlay. The process of converting pavement deflections into resilient moduli of existing pavement layer is called backcalculation (Abaza 2005; Irwin 2002).

Particular care has to be taken while calculating the layer moduli and overlay thickness. This is due to many significant parameters are required to run the backcalculation program other than the deflection basin and pavement thickness. For example, seed moduli, the presence of bedrock and material nonlinearity significantly affect the calculated layer moduli (Ullidtz and Coetzee 1995). In this study, the backcalculated program provided by Dynatest Company named as LWDmod was used to estimate the overlay thickness and structural evaluation purpose. The methodology followed in this paper is shown in Fig. 1 and backcalculation process is shown in Fig. 2.

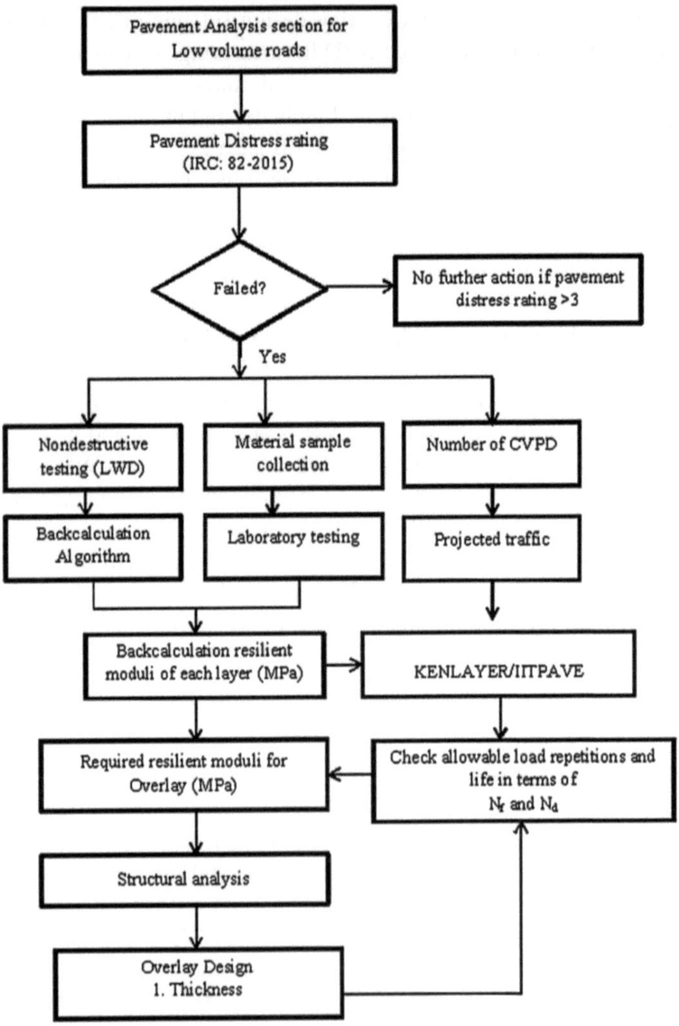

Fig. 1. Pavement overlay design methodology for low-volume road using LWD

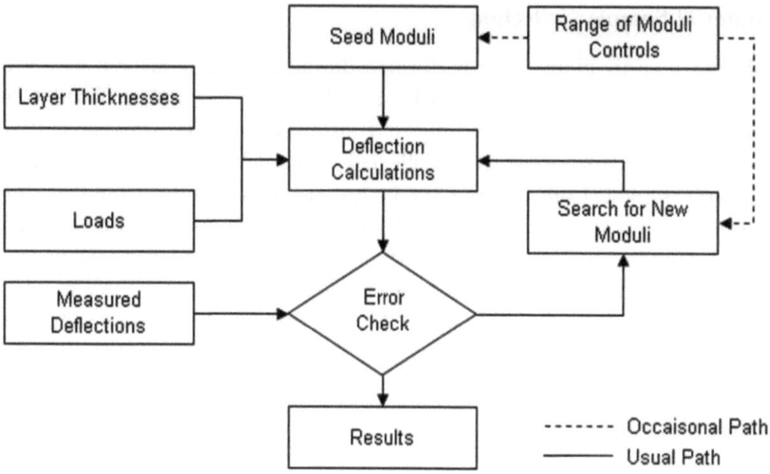

Fig. 2. Backcalculation procedure. *Source*: Internet; Pavement interactive

3 Test Track Stretch Description

An experimental stretch section of length 550 m was selected and was divided into 11 subsections i.e. 50 m each as per IRC: 81-1997. Subsequently, traffic survey was carried out on the test section for the duration of ten months from June 2015 to March 2016. The commercial vehicle per day (CVPD) on the test section was 410 which are less than 450 CVPD, which satisfies the criteria of low volume road as per (IRC SP: 72-2015). The test corridor is a 2-lane undivided corridor with two-way traffic. Visual pavement condition survey was carried out as per IRC: 82-2015 and distress rated for selected low volume road are rated as 1. Nondestructive testing using LWD ASTM (E 2583-07; D 4695-03(2008)) and BBD (IRC: 81-1997) protocol was used to perform the test. Firstly, the static method was conducted later LWD test was conducted and pavement temperature was recorded simultaneously at the time of deflections. The schematic non-destructive test plan was shown in Fig. 3.

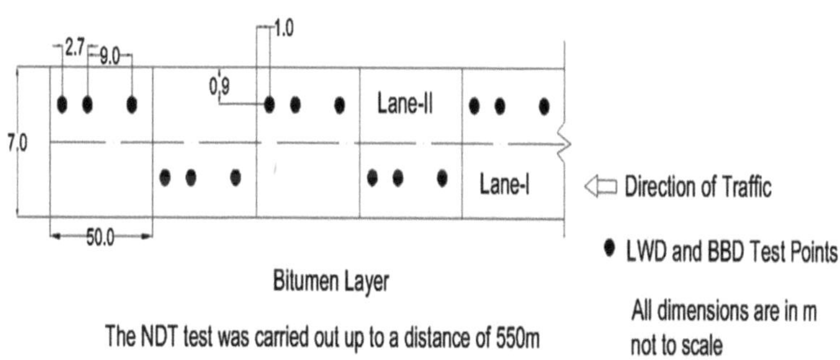

Fig. 3. Schematic test plan

3.1 Material Sample Collection

Detailed field investigations were performed during the pre-monsoon season on the selected stretch using conventional BBD and LWD as per Indian Road Congress (IRC) and American society for testing and materials (ASTM) standards. Pavement responses such as deflections were collected using LWD and BBD followed by pavement core sample from test pits on 11 test sections for identifying in situ crust composition which acts as seed values for backcalculation algorithms to estimate in situ layer moduli. The pit test size is 0.6 m × 0.6 m. The reason for coring the pavement is a small error in pavement layer thickness can lead large effect of the magnitude of the backcalculated modulus value (Molenaar 2009). Further, to investigate the subgrade soil properties in laboratory.

3.2 Laboratory Investigation

Laboratory investigations, including conventional index, volumetric properties along with Conventional California bearing ratio (CBR) test was performed on soil samples collected from test pit excavated at each test section.

3.2.1 Subgrade Soil Properties

Tests pits data were used to assess any variations in soil characterization, which associated properties, were collected to perform laboratory investigations. Laboratory investigations were performed as the Indian standard specifications to determine conventional index and volumetric properties such as liquid limit, plastic limit, average dry density, and average moisture content as shown in Table 1. Table 1 presents the type of soil and a wide range of its associated properties for 11 test samples measured by the laboratory investigations. For each test, three samples were prepared, and an average of three test results was considered as shown in Table 1.

Table 1. Soil properties of test section

Parameters	Subgrade soil
	Low plasticity
Indian soil classification	CL
Liquid limit (%)	30
Plastic limit (%)	19
Plasticity index (%)	11
Free swell index	4
In situ dry density(kN/m^3)	1.854
In situ moisture content (%)	10.47
CBR value (%)	4.6

3.2.2 Pavement Condition Survey

Most of the pavement engineers believe that the required of overlay for strengthening of existing pavement, linear elastic system mainly contributing in calculations and analytical procedures. For calculation point of view, the important and significant parameter related to existing pavement condition must be identified and accurately evaluated to be used in the calculation process. The preliminary step in planning and maintenance operation is an evaluation of the condition of the pavement by visual inspection survey to identify the distresses, roughness and friction. These parameters are necessary for rehabilitation and management. The pavement condition survey was carried out by using Indian protocol IRC: 82-2015 and distress are rated as 1(Poor condition). Total pavement condition stretch was studied 550 m length, and the performance of road was quantified by considering into distress account like cracks, raveling, patching, pothole, and depression. Quantification has been done by pavement severity level assigned weighing points (1 through 3), and observed severity level is 0.8 (which is less than 1) based on weighting and rating condition of pavement is poor.

4 Non-destructive Test

4.1 Static Deflection Test

Conventional evaluation for flexible pavement such as rebound deflection by using BBD, rut depth and crack measurements were recorded periodically three times a year in India (Veeraragavan et al 1991). The BBD deflection studies were carried out as per Canadian Good Roads Association (CGRA) procedure by loading the pavement using a standard dual wheel load of 41 KN as mentioned in the IRC: 81-1997. Static deflection measurement on the experimental stretch is shown in Fig. 4.

Fig. 4. BBD test

4.2 Dynamic Deflection Test

In this study Dynatest, 3031 LWD test was performed on selected pavement test locations by generating impulse load using 20 kg drop mass, from a maximum drop height (457 mm) on top of a circular plate having a 300 mm plate diameter as per ASTM protocols (ASTM E2583-07, 2007). Umashankar et al. (2015) used LWD as quality control compaction tool during the construction of dense bituminous macadam layer 130 mm (DBM) and wet mix macadam layer 250 mm (WMM) on the national highway using 10 kg falling weight from 1 m drop height using 300 mm diameter plate. Steinert et al. (2006) study that 20 kg falling weight on low volume road asphalt pavement thickness 127 mm to 180 mm using LWD and FWD on top of asphalt pavement. Based on Steinert et al. (2006) studies a strong correlation exists between LWD and FWD derived composite modulus of surface and LWD center geophone was used in his analysis. Steinert et al. (2006) state that LWD can be utilized as a structural evaluating tool on restrictions of loads. In this study, an attempt has been made using LWD falling weight 20 kg and 300 mm contact diameter plate. The drop of 20 kg induced an impulse load of 13.2 kN to 16.5 kN was observed on the pavement surface. Produced impulse load is approximately equal to single wheel load.

4.3 LWD Deflection Measurements

Deflections obtained from all transducers were recorded in personal data assistant (PDA). In this study, the mean load impulse time history is varying from 17 ms to

Fig. 5. LWD test

25 ms. However, the only center transducer is selected for the analysis of measured deflections as center transducer generates maximum deflection beneath the load. In this study, LWD test with 20 kg falling weight and maximum falling height 457 mm was performed on top of the thin bituminous layer. In situ LWD experimental set up is shown in Fig. 5. LWD test was repeated at each test location by dropping six multiple drops (deflections) of which three drops were considered as seating drops and remaining were used for backcalculating pavement layer moduli.

5 LWDmod Computer Program

A Personal Digital Assistance (PDA) was used to record the applied pressure deflections that are measured by the sensors from LWD test. Recorded data was transferred into LWDmod software to perform the Backcalculation analysis (Du Tertre et al. 2010). The LWDmod program is a mechanistic-empirical design approach. The concept behind this program is based on the Odemark-Boussinesq theory. It requires input parameters such as layer thickness, poisons ratio, seed value, air temperature, surface temperature, material temperature and non-linearity properties of the subgrade is needed for structural evaluation. The backcalculation analysis was conducted until a root mean square error of 2% or less was achieved (Elbagalati et al. 2016). The moduli values are assigned to various layers and deflections are calculated and compared with measured ones. If the differences are too large, a new set of moduli will be assumed by the computer program, and the deflections are calculated again. This process is known as iterative backcalculation procedure (Molenaar 2009). This process will continue until the good match between calculated and measured deflections. In the case of overlay design existing surface modulus, required surface modulus and modulus of design material is required. LWDmod concept involved in the calculation of layer moduli of different layers is presented in the subsequent sections.

5.1 Estimation of LWD Surfaces Modulus and Resilient Modulus

The dynamic deflections obtained from LWD test on top of bituminous were processed to estimate composite moduli of the surface layer by considering Boussinesq's static linear-elastic half-space theory (Fleming et al. 2007) by using the constitutive equation as shown in Eq. 1.

$$E_0 = f\left(1 - \nu^2\right)\sigma_o.a/d_0 \tag{1}$$

Where, E_o = Composite moduli of surface/subgrade layer (MPa); f = Plate Rigidity factor (2 is a standard value for a flexible plate); ν = Poisson's Ratio, normally 0.35; σ_o = maximum contact stress (kPa); a = Plate radius (m); d_o = Maximum deflection (mm).

The traditional LWD stiffness (E_0) is a composite term that reflects the surface layer and the underlying sub-layers. These findings have reported by various researchers such as (Kavussi et al. 2010; Adam et al. 2009; Fleming et al. 2007). To isolate the elastic moduli of different layers backcalculation of layers was performed using LWDmod computer program. Calculation of elastic moduli value of top layer moduli is based on best fit method. The concept behind this best fit method is it contains main processes. Firstly, it is searching iterative consecutive theoretical maximum deflections that encapsulate that measured maximum deflections. In this study the calculated and measured pavement deflection root mean square (RMS) value observed is less than 3% it is best fit as per Chen et al. (1999). Top layer ΔE_1 to subgrade resilient moduli values are calculated as per iteration process through Eq. 2 to Eq. 4 below as per Molenaar (2009) and LWDmod (2009).

$$\Delta E_1 = \max(\frac{1}{3}E_1(seed), \frac{1}{3}\left[10^{logE1(seed)} + \frac{\mathrm{d}_{max} - d^!_{max}(\mathrm{seed})}{d^!_{max}(\mathrm{seed}+1) - d^!_{\mathrm{max}(seed)}} * (\log E_1(seed{+}1)\right.$$
$$- \log E_1(seed))) - E_1(seed))$$

Where: E_1 (seed) is the initial top layer modulus should enter by the user

$$\mathrm{E}_1(\mathrm{seed}+1) = \mathrm{E}_1(\mathrm{seed}) + \frac{1}{3}\mathrm{E}_1(\mathrm{seed})$$

$$\log E_1 = \log E_1(i) + \frac{\mathrm{d}_{max} - d^!_{max}(\mathrm{i})}{d^!_{max}(\mathrm{i}+1) - d^!_{max(i)}} * (\log E_1(i+1) - \log E_1(i) \qquad (3)$$

Where: E_1 (i) is the top layer elastic modulus that results in $d^!_{max}(\mathrm{i}) \leq \mathrm{d}_{max}$ (or vice versa), E_1 (i + 1) is the top layer elastic moduli that result in $d^!_{max}(\mathrm{i}+1) \geq \mathrm{d}_{max}$, E_1 (i + 1) = E_1 (i) + ΔE_1

$$E_m = C\left(\frac{\sigma_1}{P_a}\right)^n \qquad (4)$$

E_m = Modulus of elasticity of subgrade (MPa).

5.2 Calculation of Required Overlay Thickness

The overlay thickness moduli provided should reduce the subgrade strain to the design level. The moduli of unbound material increases when the confining stress increases. The moduli of cohesive subgrade decrease with increase in stress level (AUSTROADS 1992). After application of required overlay thickness, the stress levels generated by the load in the existing pavement structure decreases. The procedure of determining the overlay thickness is general mechanistic analysis uses MET (Method of Equivalent Thickness) and transformed pavement structure principle. The multilayer pavement structure is transformed to an equivalent single layer system of same moduli. The process of transforming multi-layer pavement structure into a single layer is shown in Fig. 6. Firstly, calculate equivalent layer thickness (t_e) and then transformed pavement thickness h_e. Overlay thickness required on existing pavement is calculated by using Eq. (5).

$$t_e = t_0 \times \frac{(E_{AC})^{\frac{1}{3}}}{(E_{GR} \times E_{SG})^{\frac{1}{6}}} \quad (5)$$

Where, E_{AC} = Required Asphalt overlay modulus (MPa), E_{GR} = Granular base modulus (MPa), E_{SG} = Subgrade modulus (MPa), t_0 = asphalt overlay thickness (mm), h = existing layer thickness (mm).

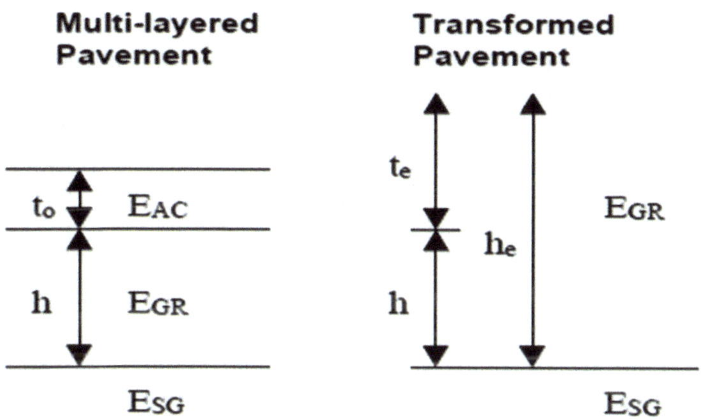

Fig. 6. Transformed pavement principle

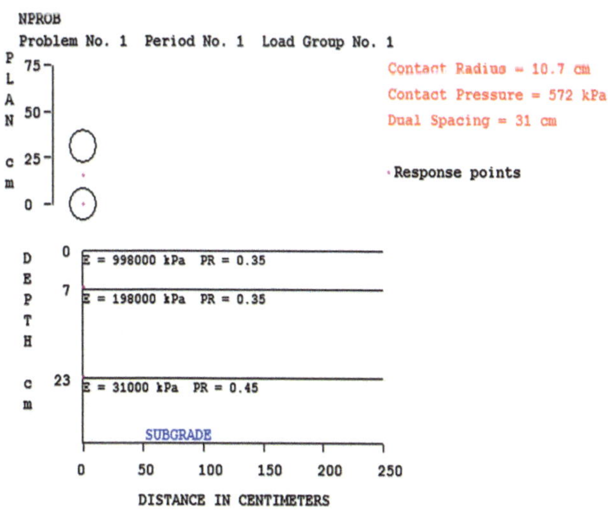

Fig. 7. Kenlayer critical analytical points

5.3 Backcalculated Results Using LWDmod Software

The acronym is Lightweight deflectometer moduli. Backcalculation of layer moduli from LWD deflections is an inverse problem and predicted vertical deflections are matched with measured vertical deflections (Senseny et al. 2013). It uses static analysis for forward modeling of data uses peak deflections for the observed and uses gradient search algorithm as an inverse solver. In this analysis first three drops are considered as seating drops remaining three drops are considered in this analysis. In this computer program seed values must be chosen as an input value and the layer moduli will be calculated automatically. Pavement temperature was recorded as 36 °C, and air temperature is 37 °C. Maximum three layers can be used in this software and back calculated layer moduli results are shown in Table 2 below.

Table 2. KENLAYER analysis of backcalculated layer moduli

Chainage (m)	E_1 (MPa)	E_2 (MPa)	E_m (MPa)	Tensile strain	Compressive strain	N_f (MSA)	N_d (MSA)
0–50	1016	196	30	5.21E−04	1.94E−03	1.86	0.05
50–100	1035	193	39	5.21E−04	1.69E−03	1.81	0.09
100–150	1052	186	31	5.44E−04	1.94E−03	1.51	0.05
150–200	1095	186	33	5.41E−04	1.87E−03	1.46	0.06
200–250	1049	191	31	5.32E−04	1.92E−03	1.65	0.05
250–300	960	190	29	5.37E−04	2.01E−03	1.81	0.04
300–350	1038	186	37	5.39E−04	1.77E−03	1.59	0.07
350–400	1012	202	38	5.01E−04	1.70E−03	2.15	0.09
400–450	1024	197	34	5.15E−04	1.81E−03	1.92	0.06
450–500	1088	194	37	5.18E−04	1.73E−03	1.71	0.08
500–550	998	198	31	5.15E−04	1.90E−03	1.99	0.05
Mean	1033.3	192.6	33.	0.001	0.002	1.77	0.06
Median	1035	193	33	0.001	0.002	1.81	0.06
STD	38.573	5.39	3.55	0	0.0001	0.21	0.018
COV (%)	3.733	2.798	10.57	2.591	5.933	11.86	27.04
Min	960	186	29	0.001	0.002	1.46	0.04
Max	1095	202	39	0.001	0.002	2.16	0.09

E_1 = Resilient modulus of bituminous layer (MPa), E_2 = Resilient modulus of Granular layer (MPa), E_m = Resilient modulus of Subgrade (MPa).

5.4 Comparison Between Static and Dynamic Measurements

Comparisons are made simple to simplify regarding gross evaluation. A correlation was developed and presented in the following equation. The following Eq. 6 was developed by using Fig. 8 data.

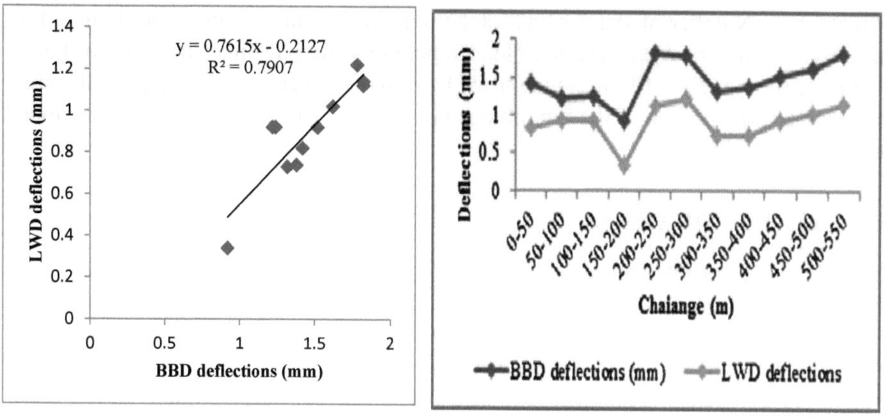

Fig. 8. Comparison of BBD and LWD deflection and regression equation

$$\text{LWD} = 0.7615\text{BB} - 0.21 \tag{6}$$

Considering the account of LWD is simple and faster test than BBD, it may be used to substitute of BBD on low volume roads to evaluate the structural capacity. The deflections obtained by unique tests followed approximately the same pattern. It is important that correlation between LWD and BBD is just for selected stretch location. The correlation for any new section must be established with field test data. Zhou et al. (2010) conducted FWD and BBD test at in-service pavement at a length of 3.2 km and found the correlation between FWD and BBD with a coefficient of determination was 0.9578 and stated that FWD and BBD are strongly correlated.

5.5 Structural Analysis for Calculating Remaining Life

Experimental study stretch was analyzed for critical stress and strain using KENLAYER software (Ziari and Khabiri 2007). After back calculating layer moduli from LWDmod software are fed into forward calculation process using KENLAYER to find critical strain parameters (Lenngren 1991). The critical strain parameters are used to estimate remaining life a pavement (Chua et al. 1994; Vennalaganti et al. 1994). The remaining life of existing flexible pavement can be obtained from field calibrated Eqs. 7 and 8 may be used for Indian conditions (Das and Pandey 1999). The absorbed crust thicknesses of the unique layers are such as bitumen layer thickness 70 mm; granular layer thickness 160 mm was considered in the analysis

$$N_f = (0.1001)(1/E_b)^{1.4747}(1/\in_t)^{3.565} \tag{7}$$

$$N_d = 2.56 \times 10^{-8}(1/\in_z)^{4.5337} \tag{8}$$

Where: N_f = Number of standard axle load repetitions can sustain before fatigue failure, N_d = Number of standard axle load repetitions the pavement can sustain before failing in rutting, E_b = Elastic modulus of bituminous layer, ϵ_t= Critical tensile strain developed at the bottom of the bituminous layer. ϵ_z= Critical vertical strain developed at top of subgrade.

5.6 KENLAYER Analysis

The life of the operational pavement, the critical response points was analyzed using data shown in Table 2. Critical response points considered in this analysis are shown in Fig. 7. Consequently, it is expected that the condition of existing pavement will influence the accuracy of layer moduli backcalculation determination by using Light-weight deflectometer technique. The backcalculated subgrade moduli results show somewhat consistency with laboratory test results (see Table 1) and CBR value was converted into resilient moduli using Eq. 9. As per IRC 37-2012, the resilient moduli of subgrade CBR was found to be 46 MPa.

$$E_{subgrade} = 10 \times \mathrm{CBR} \tag{9}$$

5.7 LWD Overlay Comparison with Conventional Design

In Indian climatic condition, new overlay layer resilient modulus of the bituminous layer at 35 °C has been considered in the guidelines for pavement analysis. The standard temperature 35 °C a well-coordinated pavement performance study in India (IRC: 37-2012). Comparative analysis between BBD and LWD results are shown in Table 3. LWDmod overlay design calculation is based on specifications to the final

Table 3. Comparison of overlay thickness using BBD and LWDmod analysis

S. no	Characteristic deflection (mm)	Traffic load (MSA)	Overlay "mm" as per IRC:81-1997	LWDmod software Overlay (mm)	Assumed required surface modulus (MPa)	Required Resilient moduli (MPa)	Type of mix
1	1.745	0.5	10	28	500	1000	VG10 bitumen
2		1	20	88	550	1700	VG30 bitumen
3		2	70	130	600	3000	VG40 bitumen
4		5	120	180	650	1650	Modified bitumen IRC: SP:53-2010
5		10	140				
6		20	170				
7		100	210				

Note: MSA = million standard axle load; VG = Viscosity Grade.
Source: *IRC: 37-2012*.

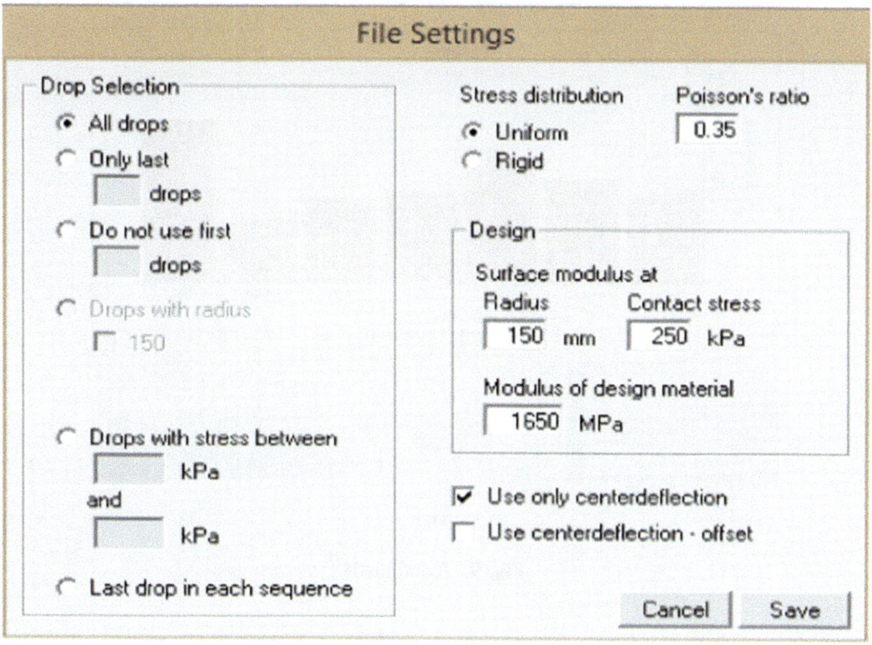

(a)

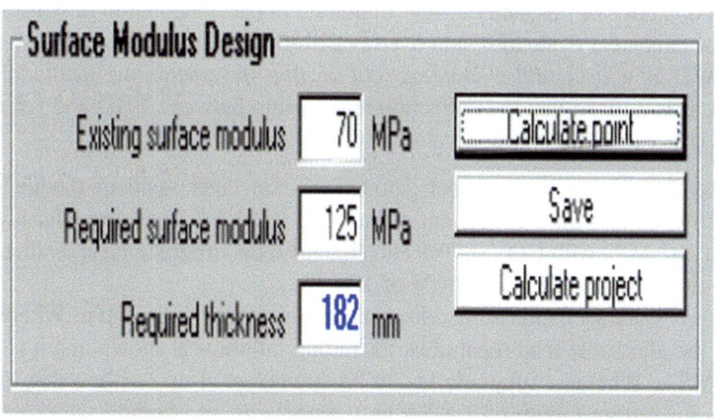

(b)

Fig. 9. Screenshots of LWDmod computer program an example: (a) input file setting for an overlay; (b) required surface modulus; (c) overlay thickness at different chainages

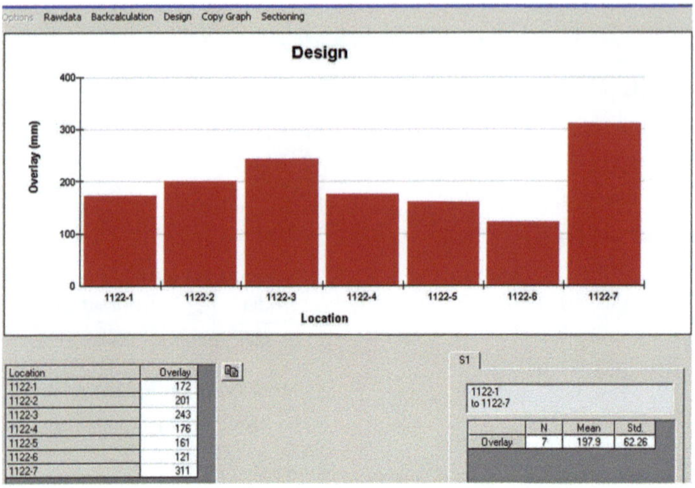

(c)

Fig. 9. (continued)

surface deflection modulus. LWDmod screen shot while performing the overlay analysis is shown in Fig. 9. Observed comparison and conclusions are drawn below.

6 Conclusions

The main objective of this study is very important in backcalculation of pavement layer moduli for pavement evaluation and overlay determination. An attempt has been made to use LWD as a structural evaluating tool on thin pavements the methodology was proposed. It is noticed that the deflection relationship between BBD and LWD follow the same pattern with regression value of 0.79.

- The backcalculated layer moduli shows bitumen layer resilient moduli range of 960–1095 MPa with a coefficient of variance (COV) 3.73%, Granular layer value 186 to 202 MPa with COV 2.79% and similarly, for subgrade layer resilient moduli range of 29 to 39 MPa with COV of 10%.
- Observed backcalculated subgrade mean moduli value is 33 MPa. Whereas, from Eq. 8 the allowable load repetitions for rutting failure was shown that it is less than 0.09 MSA. Whereas, subgrade has to be strengthened up to allowable load repetitions for rutting failure of 2 MSA before applying an overlay.
- Conventional overlay thickness determination by using BBD test shows 10 to 70 mm overlay thickness up to 0.5 to 2 MSA. LWD overlay thickness for VG 10 and VG 30 shows 28 and 88 mm.
- The proposed methodology may be used for structural evaluation on low volume roads. The LWD test procedure is simple to operate, and frequency of test points can be increased by altering the test locations.

- The work presented and analyzed in this paper uses limited data collection from the selected experimental stretch, and more performance data are required to be collected from other part of stretches of low volume road to make criteria more rational.

References

ASTM, E2583-07: Standard test method for measuring deflections with a light weight deflectometer. Annual Book of ASTM Standards, ASTM International, West Conshohocken, PA, 04.03 (2007). doi:10.1520/E2583-07

Adam, C., Adam, D., Kopf, F.: Computational validation of static and dynamic plate load testing. Acta Geotech. **4**(1), 35–55 (2009). doi:10.1007/s11440-008-0081-0

Du Tertre, A., et al.: Combined PFWD and surface wave measurements for evaluation of longitudinal joints in asphalt pavements. In: TRB 2010 Annual Meeting CD-ROM (2010). doi:10.3141/2152-04

Barodiya, P., Pateriya, I.K.: Estimating maintenance cost of rural roads in PMGSY. J. Indian Highw. **8**, 13–27 (2014)

Das, A., Pandey, B.B.: Mechanistic-empirical design of bituminous roads: an Indian perspective. J. Transp. Eng. **125**, 463–471 (1999). doi:10.1061/(ASCE)0733-947X(1999)125:5(463)

Elbagalati, O., et al.: Development of the pavement structural health index based on falling weight deflectometer testing. Int. J. Pavement Eng. **2016**, 1–8 (2016). http://dx.doi.org/10.1080/10298436.2016.1149838

Fleming, P., Frost, M., Lambert, J.: A review of the lightweight deflectometer (LWD) for routine insitu assessment of pavement material stiffness. In: TRB 2007 Annual Meeting CD-ROM. 07-1586, pp. 1–15 (2007). doi:10.3141/2004-09

Irwin, L.H.: Backcalculation: an overview and perspective. In: Pavement Evaluation Conference. Roanoke, VA (2002)

IRC SP 72-2015: Guidelines for the Design of Flexible Pavements for Low Volume Rural Roads, First Revision, New Delhi, India

IRC:37-2012: Guidelines for the Design of Flexible Pavements (Third Revision) New Delhi, India

IRC.115-2014: Guidelines for Structural Evaluation and Strengthening of Flexible Road Pavements Using Falling Weight Deflectometer (FWD) Technique, New Delhi, India

IRC:117-2015: Guidelines for the Structural Evaluation of Rigid Pavement by Falling Weight Deflectometer, New Delhi, India

IRC:81-1997. Strengthening of flexible Road Pavements Using Benkelman Beam Deflection Technique, First revision, New Delhi, India

IRC: 82-2015: First Revision of Code of Practice for Maintenance of Bituminous Road Surfaces, New Delhi, India

Abaza, K.A.: Performance-based models for flexible pavement structural overlay design. J. Transp. Eng. (2005). doi:10.1061/(ASCE)0733-947X(2005)131:2(149)

Kavussi, A., Rafiei, K., Yasrobi, S.: Evaluation of PFWD as potential quality control tool of pavement layers. J. Civ. Eng. Manag. **16**(1) (2010). doi:10.3846/jcem.2010.11

Lenngren, C.A.: Relating deflection data to pavement strain. Transp. Res. Rec. **1293**, 103–111 (1991)

Zhou, L., Wu, Q., Ling, J.: Comparison of FWD and Benkelman beam in evaluation of pavement structure capacity. In: Geo Shanghai International Conference, pp. 405–411 (2010). doi:10.1061/41104(377)51

LWDmod: Dynatest International, Glostrup, Denmark (2009)

Molenaar: Structural design of pavements—part VI. Structural evaluation and strengthening of flexible pavements using deflection measurements and visual condition surveys (2009)

MORTH (Ministry of Road, Transport, and Highways): Basic road statistics of India. Transportation Research Wing, New Delhi (2012)

Newcomb, D.E., Lee, S.W., Mahoney, J.P., Jackson, N.C.: The use of falling weight deflectometer data in monitoring. In: Bush III, A.J., Baladi, G.Y. (eds.) ASTM STP 1026, pp. 470–486. American Society for Testing and Materials, Philadelphia (1989)

Reddy, B.B., Veeraragavan, A.: Structural performance of in-service pavements. J. Transp. Eng. **123**, 156–167 (1997)

Steinert, B.C., Humphrey, D.N., Kestler, M.A.: Portable falling weight deflectometer for tracking seasonal stiffness variations in asphalt surfaced roads. In: TRB 2006 Annual Meeting CD-ROM (2006)

Senseny, C., Krahenbuhl, R., Mooney, M.A.: Genetic algorithm to optimize layer parameters in lightweight deflectometer backcalcualtion. Int. J. Geomech., 473–476 (2013). doi:10.1061/(ASCE)GM.1943-5622.0000222

Ullidtz, P., Coetzee, N.F.: Analytical Procedures in Nondestructive Testing Pavement Evaluation. Transportation Research Record, 1482, Transportation Research Board, Washington, DC, pp. 61–66 (1995)

Umashankar, B., Hariprasad, C., Thejesh Kumar, G.: Compaction quality control of pavement layers using LWD. J. Mater. Civ. Eng. (2015). doi:10.1061/(ASCE)MT.1943-5533.0001379

Veeraragavan, A., Dattatreya, J.K., Prabhudeva, M.: Development of failure criteria for flexible overlays for Indian conditions. IRC J. **52**(1), 183–204 (1991)

Ziari, H., Khabiri, M.M.: Interface condition influence on prediction of flexible pavement life. J. Civ. Eng. Manag. **XIII**, 71–76 (2007). doi:10.1080/13923730.2007.9636421

Application of Artificial Neural Networks for Hot Mix Asphalt Dynamic Modulus (E*) Prediction

Sherif El-Badawy[1(✉)] and Ragaa Abd El-Hakim[2]

[1] Public Works Engineering Department, Faculty of Engineering, Mansoura University, Mansoura, Egypt
sbadawy@mans.edu.eg
[2] Public Works Engineering Department, Faculty of Engineering, Tanta University, Tanta, Egypt
ragaa.abdelhakim@f-eng.tanta.edu.eg

Abstract. The hot mix asphalt (HMA) dynamic modulus (E*) is a fundamental mechanistic property that defines the strain response of asphalt concrete mixtures as a function of loading rate and temperature. It is one of the HMA primary material inputs for the Pavement ME Design. The laboratory testing of dynamic modulus requires expensive advanced testing equipment that is not readily available in the majority of laboratories in the Middle Eastern countries, yet some of these countries are looking for implementing new pavement design methods such as Pavement ME Design. Thus, many research studies have been dedicated to develop predictive models for E*. This paper aims to apply artificial neural networks (ANNs) for E* predictions based on the inputs of the models most widely used today, namely: Witczak NCHRP 1-37A, Witczak NCHRP 1-40D and Hirsch E* predictive models. It also aims at investigating the effect of the different hierarchical binder input levels of the Pavement ME Design on the E* prediction accuracy. A total of 25 mixes from the Kingdom of Saudi Arabia (KSA), and 25 mixes from Idaho state were combined together in one database containing 3720 E* measurements. The database also contains the volumetric properties and aggregate gradations for all mixes as well as the binder complex shear modulus (G_b*), phase angle (δ), and Brookfield viscosity.

The results of this study show that using the same input variables of the three models, the ANNs models generally yielded more accurate E* predictions. Finally, there is a strong evidence of the influence of binder input level on the dynamic modulus E* prediction accuracy of both regression and ANNs.

1 Introduction

For linear viscoelastic materials such as the hot mix asphalts (HMA), the stress-strain relationship under a continuous sinusoidal loading is defined by its complex dynamic modulus (E*) which defines the stiffness characteristics of the HMA as a function of loading rate and temperature. The significance of this material property is that it is one of the primary material inputs in the Pavement ME Design. E* values at different temperatures and frequencies are the main inputs for determining the fundamental linear viscoelastic (LVE) material properties that can be used in advanced HMA and pavement

L. Mohammad (ed.), *Advancement in the Design and Performance of Sustainable Asphalt Pavements*, Sustainable Civil Infrastructures, DOI 10.1007/978-3-319-61908-8_14

models. The dynamic modulus test is used to compute the linear viscoelastic model parameters including the Pronyseries coefficients and the temperature coupling terms for the Pavement Analysis using Nonlinear Damage Approach Program (PANDA) (Rahmani et al. 2013). Despite the demonstrated significance of E*, the test procedure is not simple, time consuming and the equipment is expensive. Thus, there has been continuous efforts to predict dynamic modulus (E*) from HMA material properties. Some of these efforts resulted in widely known and used predictive models and recently, newer techniques other than regression have been used. In the next few paragraphs, the authors will review the most widely known E* models based on these efforts.

1.1 1999 η-Based NCHRP1-37A Witczak Model

Andrei et al. (1999) revised the original Witczak E* predictive equation based on data from 205 mixtures with 2,750 data points. The revised equation is as follows:

$$\log_{10}E* = -1.249937 + 0.02923\rho_{200} - 0.001767(\rho_{200})^2 - 0.002841\rho_4 - 0.058097V_a - 0.82208\frac{V_{beff}}{V_{beff}+V_a} + \frac{3.871977 - 0.0021\rho_4 + 0.003958\rho_{38} - 0.000017(\rho_{38})^2 + 0.00547\rho_{34}}{1+e^{(-0.603313-0.313351\log f-0.393532\log\eta)}} \quad (1)$$

where: $E*$ = HMA dynamic modulus, in 10^5 psi; η = binder viscosity, in 10^6 poise; f = loading frequency, in Hz; V_a = % air voids in the mix, by volume; V_{beff} = % effective binder content, by volume; ρ_{34} = % cumulative retained weight on the 3/4-in. sieve; ρ_{38} = % cumulative retained weight on the 3/8-in.sieve; ρ_4 = % cumulative retained weight on the No. 4 sieve; and ρ_{200} = % passing No. 200 sieve.

This equation is based on a nonlinear regression analysis that incorporates mixture volumetric properties, aggregate gradation, and binder classical viscosity. It is currently one of two options for level 3 inputs in the Pavement ME Design program.

1.2 2007G*, δ-Based NCHRP1-40D Witczak Model

To include the Superpave binder shear modulus (G*) in the predictive model, in 2006, Witczak and his associates reformulated the model (Bari and Witczak 2006; Bari and Witczak 2007). This model was again revised and incorporated for level 3 inputs in the Pavement ME Design (Witczak et al. 2007a, b). The final revised model based on a nonlinear regression analysis using 346 mixtures with 7,400 data points is as follows:

$$\log_{10}E* = 0.02 + 0.758\left(|G_b*|^{-0.0009}\right) \times \begin{pmatrix} 6.8232 - 0.03274\rho_{200} + 0.00431\rho_{200}^2 + 0.0104\rho_4 - 0.00012\rho_4^2 \\ + 0.00678\rho_{38} - 0.00016\rho_{38}^2 - 0.0796\,V_a - 1.1689\left(\frac{V_{beff}}{V_a+V_{beff}}\right) \end{pmatrix} + \frac{1.437 + 0.03313\,V_a + 0.6926\left(\frac{V_{beff}}{V_a+V_{beff}}\right) + 0.00891\,\rho_{38} - 0.00007\,\rho_{38}^2 - 0.0081\rho_{34}}{1+e^{(-4.5868-0.8176\log|G_b*|+3.2738\log\delta)}} \quad (2)$$

Where: E* = HMA dynamic modulus, in psi; $|G_b*|$ = complex shear modulus of binder, psi; δ = phase angle of the binder, degrees; f = loading frequency, in Hz; V_a = % air voids in the mix, by volume; V_{beff} = % effective binder content, by volume; ρ_{34} = % cumulative retained weight on the 3/4-in.sieve; ρ_{38} = % cumulative retained weight on the 3/8-in.sieve; ρ_4 = % cumulative retained weight on the No.4 sieve; and ρ_{200} = % passing No. 200 sieve.

It must be noted that, $G*$ and δ in this model must be determined at loading frequency f_s as follows:.

$$f_s = f_c/2\pi \tag{3}$$

where, f_s = loading frequency of a dynamic loading in "shear" mode (as used in the complex shear modulus test of asphalt binders), Hz; and f_c = loading frequency of a dynamic loading in "compression" mode (as used in the complex dynamic modulus test of AC mixtures), Hz.

Because some of the mixtures in the database used for the model development do not contain G* and δ values for the binders, these values were estimated from the following regression models (Witczak et al. 2007a, b):

$$\eta = \left(\frac{G_b*}{10}\right)\left(\frac{1}{\sin\delta}\right)^{4.8628} \tag{4}$$

$$\log\log\eta = A + VTS\log T_R \tag{5}$$

$$\log(\log\eta_{fs,\mathrm{T}}) = A' + VTS' * logT_R \tag{6}$$

$$A' = 0.9699 * f_s^{-0.0527} * A \tag{7}$$

$$VTS' = 0.9668 * f_s^{-0.0575} * VTS \tag{8}$$

$$\delta_b = 90 - 0.1785 * \log\left(\eta_{fs,T}\right)^{2.3814} * (f_s)^{(0.3507 + 0.0782VTS')} \tag{9}$$

$$|Gb*| = 1.469 * 10^{-9} * \log(\eta_{fs,T})^{12.0056} * f_s^{0.7418}(\sin\delta_b)^{0.6806} \tag{10}$$

where, η = binder viscosity, in cp; T_R = testing temperature, in Rankine; A = regression intercept; and VTS = regression slope of the viscosity-temperature relationship. A' = adjusted "A" for loading frequency; VTS' = adjusted "VTS" for loading frequency; $\eta_{fs,T}$ = binder viscosity as a function of both loading frequency (fs) and temperature(T_R), in cP; δb = binder phase angle, in degree; and $|Gb*|$ = complex binder shear modulus, in psi.

1.3 2003 Pc-Based Hirsch Prediction Model

Christensen Jr., et al. (2003) developed a new E* predictive model for HMA based upon an existing version of the law of mixtures, called the Hirsch model, which combines series and parallel elements of phases. The model is represented in the following equations:

$$E* = Pc\left[4{,}200{,}000\left(1-\frac{VMA}{100}\right)+3|G*|\left(\frac{VFAxVMA}{10{,}000}\right)\right] + (1-Pc)x\left[\frac{1-\frac{VMA}{100}}{4{,}200{,}000}+\frac{VMA}{3xVFAx|G*|}\right]^{-1} \tag{11}$$

where: E* = dynamic modulus of the mixture, in psi; $|G^*| = |G_b^*|$ = shear modulus of the binder, in psi; VMA = % voids in the mineral aggregates; VFA = % voids in mineral aggregates filled with binder; and Pc = contact volume estimated from Eq. (12):

$$Pc = \frac{\left(20+\frac{VFA\times 3|G*|}{VMA}\right)^{0.58}}{650+\left(\frac{VFA\times 3|G*|}{VMA}\right)^{0.58}} \tag{12}$$

This model was developed based on a much smaller database of 206 E* measurements from 18 different HMA mixtures containing eight different binders.

Several studies investigating dynamic modulus prediction have been carried out on other databases from different regions of the United States and the world. Harran and Shalaby (2009) tried to improve the reliability of the E* prediction at high temperatures using parameters that reflect the gradation of aggregates. The analysis was performed on 24 mixtures prepared from various aggregate gradations and types, and several binder grades from Canada and the US. A linear model was developed to adjust the predicted E* at high temperature for fine-graded mixtures. The model was calibrated with data of 11 fine-graded mixtures and showed high goodness of fit.

Far et al. (2009) developed three artificial neural networks (ANNs) models for estimating the E*of HMA layers using data from the Long Term Pavement Performance (LTPP) test sections. These predictive models used ANNs trained with the same set of parameters used in other popular predictive equations: the modified Witczak and Hirsch models. It was found that ANNs structure with three-layer feed-forward back propagation network and sigmoidal transfer function (9-9-1) was the most appropriate. The researchers used an extensive database (14,309 data points) which included Witczak's database besides five other databases.

Ceylan et al. (2007, 2008, 2009a, b) used the Witczak database containing 7400 data points from 346 HMA mixtures to develop ANNs models using the inputs of the widely known predictive equations. The developed ANNs had a feed-forward-error back-propagation architecture with two hidden layers, (4-40-40-1) for Hirsch model inputs and (8-30-30-1) for Witczak 1999 and Witczak 2006 models inputs.

Awed et al. (2011) and El-Badawy et al. (2012) investigated the influence of the binder characterization input level of the Mechanistic Empirical Pavement Design Guide (MEPDG) on the prediction accuracy of the MEPDG E* predictive models for 25 mixtures from Idaho. The evaluated models were Witczack NCHRP 1-37A and Witczack NCHRP1-40D respectively. They concluded that the NCHRP 1-37A E* model along with MEPDG level 3 binder inputs (default A-VTS values based on the binder grade) yielded the most accurate and least biased E* estimates for Idaho mixtures. Nevertheless, the lowest prediction accuracy and highest bias in the E* estimates, especially at the higher temperatures, resulted from the NCHRP 1-40D model with MEPDG level 3 binder characterization.

Jamrah et al. (2014) made an effort to develop analytical improved E* predictive models for HMA used in the state of Michigan. The study showed significant differences between laboratory measured and predicted E* values especially at the high temperatures/low frequencies. Using a laboratory E* data of 64 mixtures sampled from different projects in Michigan, the MATLAB software was used to calibrate the coefficients of the NCHRP 1-40D model. The new coefficients were used in the E* predictive equation and showed improved predictions. In addition, an ANNs model was developed and trained for Michigan mixtures. The structure of the ANNs model was a (9-8-1) feed-forward (back-propagation) network with a tansig transfer function.

Khattab et al. (2014a, b, 2015) investigated the prediction accuracy of the E* models incorporated in the Pavement ME Design which are the NCHRP 1-37A and 1-40D along with the effect of binder input level on each model predictions based on 25 HMA mixtures used in KSA. Results showed that the performance of the investigated models varied by temperature and the binder characterization method. The results also showed that for both models, level 3 binder inputs yielded the highest prediction accuracy and that the NCHRP 1-37A E* model along with level 3 binder inputs yielded the most accurate and least biased E* estimates. They also used laboratory measured E* data to compare Witczak NCHRP 1-40D with the Hirsch E* predictive model. The results showed that both models yielded highly accurate E* values for the investigated mixes. However, both models showed some bias in the predictions. The bias in the predicted E* values was slightly higher for Hirsh model compared to Witczak model.

El-Badawy et al. (2016) compared the prediction accuracy of candidate E* predictive models and ANNs model for the same data from KSA. ANNs trained using the laboratory measured E* data with the same set of parameters used in other popular predictive equations: the NCHRP 1-40D and Hirsch models. The ANNs E*predictions using the NCHRP 1-40D Witczak model set of parameters showed higher prediction accuracy compared to the regression models.

Yousefdoost et al. (2013) tried to establish a dynamic modulus E* database for Australian asphalt mixes. They measured the dynamic moduli of 28 different Australian asphalt mixes over a spectrum of temperatures and loading frequencies. Results of their study indicated that Witczak NCHRP 1-37A, Hirsch, and El-Khateeb Models under-predicted the E* of the Australian mixes while Witczak NCHRP 1-40D model over-predicted it. The level of bias was found to be high and the performance was inconsistent for different temperatures. The overall performance of the models suggested that they are not robust enough for the Australian asphalt mixes.

Ali et al. (2015) investigated the factors affecting the dynamic modulus and phase angle for a wearing and base course mixes and evaluated the NCHRP 1-37A model for international and local asphalt mixes of Pakistan. The results showed that Witczak NCHRP 1-37A model mostly underestimated the dynamic modulus values of selected mixtures.

Georgouli et al. (2015) evaluated the prediction accuracy of Witczak NCHRP 1-37A, Bari & Witczak, NCHRP 1-40D (unmodified) and Hirsch models for 15 asphalt mixtures widely used in Greece. Results showed that Bari & Witczak model over-predicted E*values while Hirsch model under-predicted E*values of the investigated asphalt mixtures. For the database investigated, Witczak NCHRP 1-37A produced closer predictions to the measured values, thus they calibrated the model. The calibrated model was verified and further validated with a high degree of statistical certainty. Later on, (Georgouli et al. 2016) investigated the impact of the predicted E* on the predicted fatigue cracking in the context of M-E Pavement Design. The results showed that fatigue cracking was slightly overestimated when considering the predicted E* values of Witczak NCHRP 1-37A and Hirsch models. However, fatigue cracking was about the same order incase of predicted E* values of the local calibrated model.

2 Problem Statement and Objectives

The covered literature showed mixed results regarding the prediction accuracy and bias of the well know predictive models. However, the few available literature studies utilizing ANNs for the E* predictions showed better accuracy and less bias compared to the regression models. Unlike regression models, ANNs can be constructed without knowing the form of the predictive relationship. However, the primary disadvantage of ANNs is the difficulty in predicting responses when the inputs are outside of the training database (Far et al. 2009). Thus, the primary objective of this study is to compare the prediction accuracy and bias of the well-known E* regression models (NCHRP 1-37A, NCHRP1-40D Modified Witczak, and Hirsh) and ANNs models constructed based on the same inputs of these models using 3720 E* measurements on HMA mixtures from two different geographic regions: Idaho, and KSA. This research aims also to investigate the effect of the binder characterization input level as per the Pavement ME Design on the prediction accuracy of the investigated models as well as the ANNs.

3 Research Methodology

The data management and analysis was performed using Artificial Neural Network modeling analysis by the NeuroSolution5 Software. E* values were also predicted using the Witczak NCHRP 1-37A η-based, Witczak NCHRP 1-40DG_b*-based and Hirsch E* predictive models (Eqs. 1, 2, and 11), at the three hierarchical binder input levels of the Pavement ME Design. For levels 1 and 2 binder characterization inputs, laboratory measurements based on the binder grade are required. For conventional binder grades, traditional asphalt testing such as penetration, ring and ball softening point, absolute and kinematic viscosities, and Brookfield viscosity testing on Rolling Thin Film Oven (RFTO) aged samples are required (this is called level 1-a in this research). On the other hand, for the Superpave performance grade binders, laboratory measured binder shear modulus (G*) and phase angle (δ) values at various temperatures and one angular loading frequency of 10 rad/sec (1.59 Hz) using the dynamic shear rheometer test (DSR) on RTFO aged samples are required (this is called level1-b in this research).Finally, for level 3 binder inputs, default values for the A-VTS parameters, based on the Superpave performance grade, conventional penetration grade or conventional viscosity grade systems are required (ARA, Inc. 2004).

For the ANNs development, in this research, the first set of binder data was based on viscosity direct parameters such as Brookfield viscosity, BRV (η), G_b* and δ. The second set of binder data was based on viscosity-temperature regression coefficients (A and VTS) estimated from the relevant binder data based on the Pavement ME three input levels for the binder characterization using Eqs. (4–8). All the inputs used for the three prediction models with two different data sets are illustrated in Table 1. These inputs are used at every binder input level as per the Pavement ME Design, level 1-a (binder data based on BRV results), level 1-b (binder data based on DSR results) and level 3 (default values). A comparison was made between ANNs and NCHRP 1-37A, NCHRP 1-40D, and Hirsch E* predictive models. The research methodology flowchart is illustrated in Fig. 1.

Table 1. Inputs and outputs used in ANNs models for the investigated regression models

Variables	Model					
	NCHRP 1-37 A		NCHRP 1-40D		Hirsch Model	
	η	A,VTS	G*, δ	A,VTS	G*, δ	A,VTS
Input variable						
Temperature (°F)		x		x		x
Frequency(Hz)	x	x		x		x
Viscosity η (cP)	x					
A		x	x	x		x
VTS		x	x	x		x
G* (psi)			x		x	
δ (°)			x		x	
VMA (%)					x	x
VFA (%)					x	x
%Pc (contact volume)					x	x
% Aggregate passing No. 200 sieve	x	x	x	x		
% Aggregate retained on No. 4 sieve	x	x	x	x		
% Aggregate retained on No. 3/8 in. sieve	x	x	x	x		
% Aggregate retained on No. ¾ in. sieve	x	x	x	x		
% Air voids (by V_{mix})	x	x	x	x		
% Effective asphalt content (by V_{mix})	x	x	x	x		
Output variable						
Dynamic modulus E* (psi)	x	x	x	x	x	x
Dynamic modulus E* (MPa)	x	x	x	x	x	x
ANN architecture	8-36-36-1	10-36-36-1	8-36-36-1	10-36-36-1	5-36-36-1	7-36-36-1

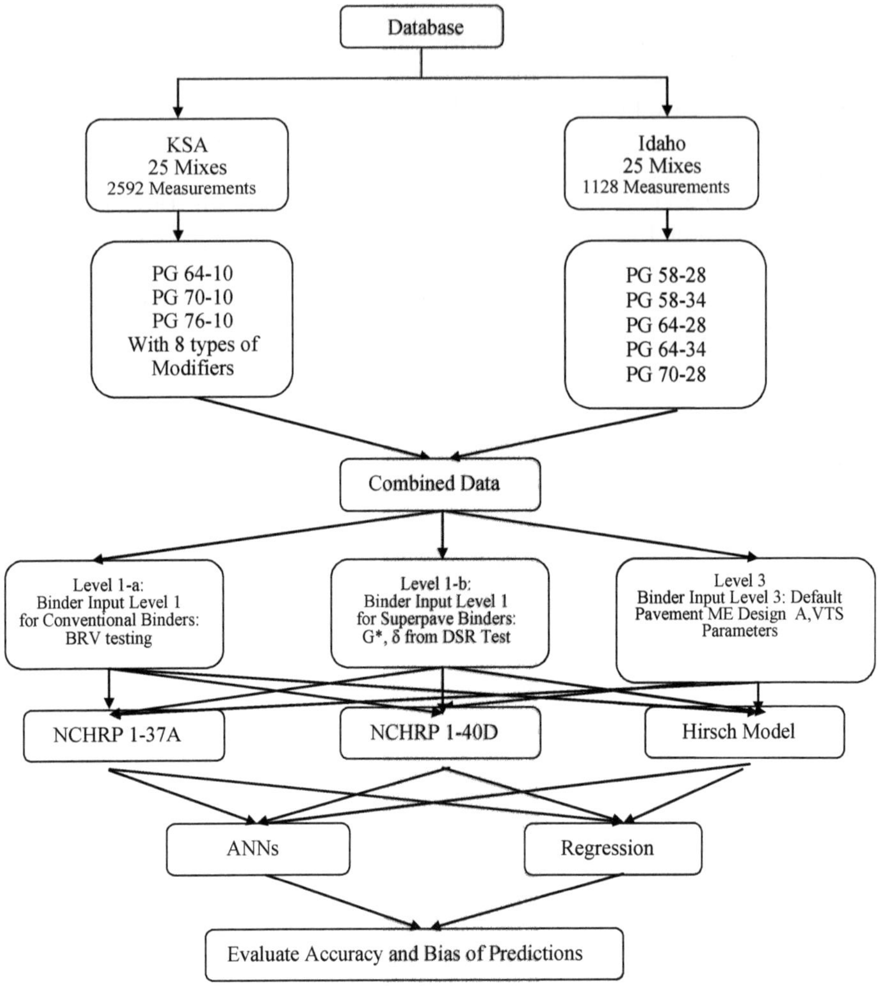

Fig. 1. Research analysis methodology flowchart

4 Database

In this research, the database of 25 HMA mixes from KSA was combined with the data of 25 mixes form the state of Idaho. These mixes contain different aggregate gradations and binder performance grades that cover the different climatic regions in both KSA and Idaho. The KSA mixtures contain binders that are either modified with crumb rubber or different polymer types. The total E* measurements in the combined database are 3720. For more details about the properties of these mixes, readers are referred to (Bayomy et al. 2011; Khatab 2015).

4.1 E* Predicted Using the Investigated Models

First, for level 1-a binder data, the mix binder Brookfield viscosity values measured at various temperatures were fitted with the relationship given by Eq. (4) and the A-VTS parameters were determined. Second, for level 1-b, Eqs. (4)–(10) were used to compute G* and δ values for each mix binder at the same temperature and frequency of E*. Finally, for level 3 binder data, the Pavement ME Design default A-VTS parameters were determined based on the binder PG grade. Then, E* values were predicted using the three investigated models at the same temperature and frequency of the laboratory measured E* values while the binder inputs were varied for each model based on the level of binder inputs just explained. measured and predicted values of E* from each model and binder characterization input were then compared and evaluated.

4.2 ANNs Model Development

In this study, dynamic ANNs were used to develop a reliable and accurate dynamic modulus E* predictions. The data inputs and outputs were normalized to values between 0 and 1 and were randomly divided into two sets: training and testing. Out of the 3720 data points, 80% were used for training and the remaining 20% were used for testing. The trained ANNs models were statistically evaluated using all 3720 data points to evaluate the overall predictive accuracy of the E* ANNs models for different binder input levels, either using directly the binder viscosity (η), or G* and δ as per the Pavement ME Design, or using the A-VTS parameters based on the binder testing data as shown in Table 1.

In order to obtain the best results, several ANNs architectures were investigated and tried. Various network combinations were initially tried and a large number of trial runs were performed. Different number of hidden layer neurons and number of epochs were examined. Different algorithms were also attempted. The best results were obtained when the Multilayer perceptrons feed forward back propagation algorithm was employed with a SigmoidAxon transfer function and a Momentum learning rule. It consists of variable number of inputs according to the model and variable set used (see Table 1), 2 hidden layers with 36 neurons each and 1 output layer (inputs-36-36-1) at 10,000 epochs. The different inputs with the adopted algorithm architectures are illustrated in Table 1 for all models and set of variables.

5 Analysis and Discussion of Results

A comparison of the goodness of fit statistics between measured and predicted values of E* of the ANNs models and the three investigated regression models at the three binder inputs is shown in Table 2 for the logarithmic values of E* and in Table 3 for the arithmetic values of E*. The goodness of fit statistics parameters were computed with reference to the line of equality using Eqs. (13)–(16) (Pellinen 2001).

$$s_y == \sqrt{\frac{1}{n-1}\sum_{i=1}^{n}\left(E^*_{mi} - \bar{E}^*_m\right)^2} \tag{13}$$

$$e_i = \sum_{i=1}^{n}\left(E^*_{pi} - E^*_{mi}\right) \tag{14}$$

$$\mathrm{S_e} = \sqrt{\frac{\sum_{i=1}^{n} e_i^2}{n-p}} \tag{15}$$

$$R^2 = 1 - \frac{n-p}{n-1}\left(\frac{S_e}{S_y}\right)^2 \tag{16}$$

where; n = number of data points; p = number of model parameters; E^*_{mi} = measured dynamic modulus; $\overline{E}^*_m$ = mean value of measured dynamic modulus; E^*_{pi} = predicted dynamic modulus; S_y = standard deviation of the measured E* values about the mean measured; e = error between the predicted and measured E* values; S_e = standard error (i.e., standard deviation of error); R^2 = coefficient of determination.

Both S_e/S_y and R^2 are measures of the model accuracy (degree of scatter with reference to the line of equality). R^2 is the coefficient of determination (higher R^2 means higher accuracy). The S_e/S_y indicates the relative improvement in accuracy. The equality line is a line with a unit slope and zero intercept. Thus, one way to assess the prediction bias is the closeness of the slope and intercept of the linear relationship between predicted and measured E*to 1 and 0, respectively.

It is apparent from Tables 2 and 3 that ANNs models exhibited higher accuracy compared Witczak and Hirsch regression models in terms of R^2 and S_e/S_y. A comparison of the goodness of fit statistics of the studied cases clearly shows a dramatic decrease in the prediction accuracy of the regression models in the arithmetic space compared to the logarithmic space for both Witczak models and Hirsh model. On the other hand, the ANNs predictions in both spaces were very close. In depth discussion of the results of each model is presented in the subsequent sections.

5.1 ANNs versus 1999 η-Based NCHRP 1-37A Witczak Model

The results indicate that the ANNs based on the NCHRP1-37A Witczak inputs generated accurate E* values at all binder input levels compared to the regression model. This is evident by higher R^2 and lower S_e/S_y values. The most accurate prediction of dynamic modulus was achieved at binder input level 3 for both η-based set of variables with neural architecture 8-36-36-1 (Fig. 2) and A-VTS set of variables with neural architecture 10-36-36-1 (Fig. 3). The goodness of fit results for this level of input is shaded in Tables 2 and 3. A comparison could be made between the accuracy of the NCHRP 1-37A Witczak predictive model and the developed ANNs in both logarithmic and arithmetic spaces if we compare Fig. 4 with Figs. 2 and 3, for binder input level 3.

The comparison shows lower scatter around the line of equality and lower bias (slope of the line representing the relationship between predicted and measured values of E* is closer to 1). This finding is in agreement with Khattab et al. (2015).

Table 2. Goodness of fit statistics and bias of the ANNs models compared to the investigated regression models in logarithmic space

Binder input level	NCHRP 1-37A								
	η as input 8-36-36-1			A-VTS 10-36-36-1			Regression		
	Equation	R^2	S_e/S_y	Equation	R^2	S_e/S_y	Equation	R^2	S_e/S_y
Level 1-a	Y = 0.846 x + 0.59	0.88	0.35	Y = 1.015 x – 0.087	0.84	0.41	Y = 0.75 x + 0.75	0.80	0.45
Level 1-b	Y = 0.813 x + 0.71	0.90	0.31	Y = 0.902 x + 0.365	0.92	0.29	Y = 0.79 x + 0.64	0.85	0.39
Level 3	**Y = 0.998 x −0.014**	**0.90**	**0.32**	**Y = 0.911 x + 0.327**	**0.91**	0.30	**Y = 0.85 x + 0.44**	**0.86**	**0.37**
	NCHRP 1-40D								
	G*, δ 8-36-36-1			A,VTS 10-36-36-1			Regression		
	Equation	R^2	S_e/S_y	Equation	R^2	S_e/S_y	Equation	R^2	S_e/S_y
Level 1-a	Y = 0.762 x + 0.919	0.88	0.35	Y = 1.237 x – 0.97	0.76	0.50	Y = 0.78 x + 0.98	0.74	0.52
Level 1-b	**Y = 0.851 x + 0.56**	**0.90**	0.31	**Y = 0.826 x + 0.67**	0.90	**0.31**	Y = 0.74 x + 0.89	0.81	0.44
Level 3	Y = 0.813 x +0.718	0.90	0.31	Y = 0.736 x + 1.024	0.87	0.37	**Y = 0.79 x + 0.74**	**0.84**	0.41
	HirschModel								
	G*, δ, 5-36-36-1			A,VTS, 7-36-36-1			Regression		
	Equation	R^2	S_e/S_y	Equation	R^2	S_e/S_y	Equation	R^2	S_e/S_y
Level 1-a	Y = 0.753 x + 0.953	0.88	0.35	Y = 0.758 x + 0.93	0.87	0.35	Y = 0.81 x + 0.588	0.81	0.44
Level 1-b	**Y = 0.810 x + 0.723**	**0.90**	0.32	**Y = 0.774 x + 0.87**	**0.89**	0.34	Y = 0.76 x + 0.73	0.83	0.41
Level 3	Y = 0.702 x + 1.158	0.86	0.38	Y = 0.751 x + 0.953	0.88	0.35	**Y = 0.80 x + 0.62**	**0.87**	**0.35**

5.2 ANNs Versus 2007 G*, δ-Based NCHRP1-40D Witczak Model

Regarding the NCHRP 1-40D model, using the adopted ANNs function with neural architecture 8-36-36-1 with G*, δ set of variables, and neural architecture 10-36-36-1 with the A-VTS set of variables, resulted in higher prediction accuracy compared to Witczak regression equations for all levels of inputs. Interestingly, the regression equation exhibited its highest accuracy with binder input level 3, (not level 1-b as one would expect) while ANNs although resulted in higher accuracy for all input levels, achieved the highest accuracy when binder inputs of level 1-b were used. This can be seen by the lower scatter about line of equality, the higher R^2, and the lower S_e/S_y. The reason for the lower accuracy of level 1-b compared to level 3 for the Witczak 1-40D model predictions is due to the fact that, G*and δ values used for the original model development were based on regression models not actual measured values in the laboratory. The slope of the non-constrained regression line was closer to 1 in both logarithmic and arithmetic spaces and the intercept was closer to zero in logarithmic space. These results are presented in Tables 3, and Figs. 5, 6 and 7. The findings of the current study are consistent with those of Khattab et al. (2015) who reported that level 1-b exhibited the highest accuracy for ANNs, while regression exhibited the highest accuracy at level 3 for NCHRP 1-40D Witczak Model.

Table 3. Goodness of fit statistics and bias of the ANNs models compared to the investigated regression models in arithmetic space

Binder Input Level	NCHRP 1-37 A								
	η as input 8-36-36-1			A,VTS 10-36-36-1			Regression		
	Equation	R^2	S_e/S_y	Equation	R^2	S_e/S_y	Equation	R^2	S_e/S_y
Level 1-a	Y = 0.896 x + 795.2	0.91	0.31	Y = 0.896 x + 744.9	0.90	0.31	Y = 0.62 x + 421.5	0.58	0.65
Level 1-b	Y = 0.903 x + 711.3	0.91	0.30	Y = 0.908 x + 663.9	0.91	0.30	Y = 0.67 x + 550.9	0.68	0.57
Level 3	**Y = 0.908 x + 695.4**	**0.91**	**0.30**	**Y = 0.904 x + 725.6**	**0.91**	0.30	**Y = 0.77 x + 490.8**	**0.72**	**0.53**
	NCHRP 1-40D								
	G*, δ 8-36-36-1			A,VTS 10-36-36-1			Regression		
	Equation	R^2	S_e/S_y	Equation	R^2	S_e/S_y	Equation	R^2	S_e/S_y
Level 1-a	Y = 0.889 x + 862.5	0.91	0.31	Y = 0.896 x +698.4	0.91	0.31	Y = 0.89 x + 3403.8	0.56	0.67
Level 1-b	**Y = 0.893 x + 762.1**	**0.91**	**0.31**	**Y = 0.892 x + 850.6**	**0.91**	**0.31**	Y = 0.79 x + 588.1	0.69	0.56
Level 3	Y = 0.894 x + 836.7	0.91	0.30	Y = 0.879 x + 945.9	0.90	0.32	**Y = 0.87 x + 739.6**	**0.81**	**0.44**
	Hirsch Model								
	G*, δ, 5-36-36-1			A,VTS, 7-36-36-1			Regression		
	Equation	R^2	S_e/S_y	Equation	R^2	S_e/S_y	Equation	R^2	S_e/S_y
Level 1-a	Y = 0.881 x + 905.6	0.90	0.31	Y = 0.879 x + 906.8	0.90	0.32	Y = 0.52 x + 1567.5	0.54	0.68
Level 1-b	**Y = 0.891 x + 784.4**	**0.90**	**0.31**	**Y = 0.894 x + 868.9**	**0.90**	**0.31**	Y = 0.49 x + 1161.8	0.52	0.69
Level 3	Y = 0.809 x + 1457	0.83	0.41	Y = 0.884 x + 819.7	0.90	0.31	**Y = 0.53 x + 1374.5**	**0.61**	**0.63**

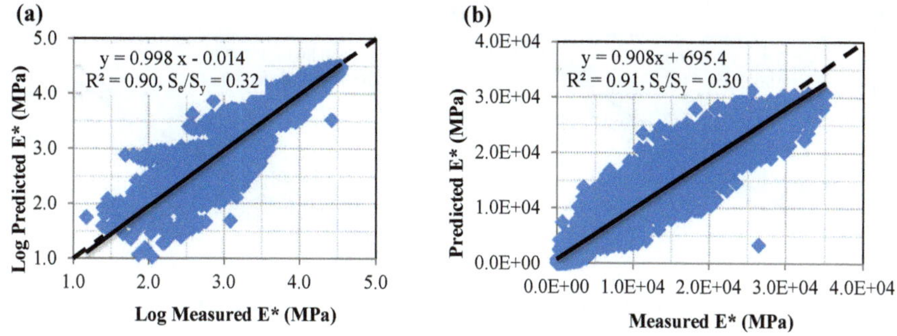

Fig. 2. Comparison of measured versus predicted E* using level 3 binder inputs (η-Based 8-36-36-1 ANNs architecture, NCHRP 1-37A inputs). (a) Logarithmic space, (b) Arithmetic space

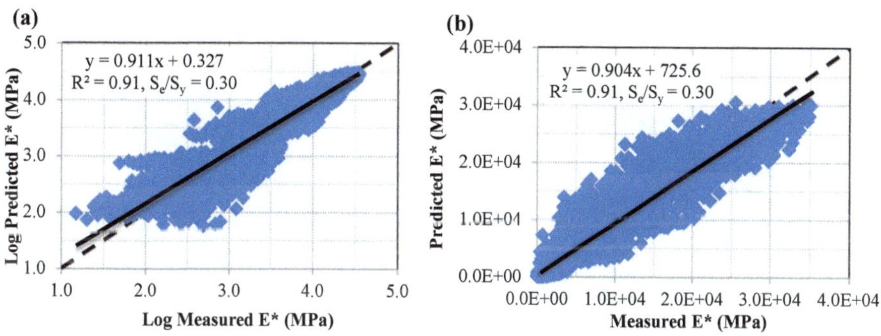

Fig. 3. Comparison of measured versus predicted E*using level 3 binder inputs (A-VTS based 10-36-36-1 ANNs architecture, NCHRP 1-37A inputs). (a) Logarithmic space, (b) Arithmetic space

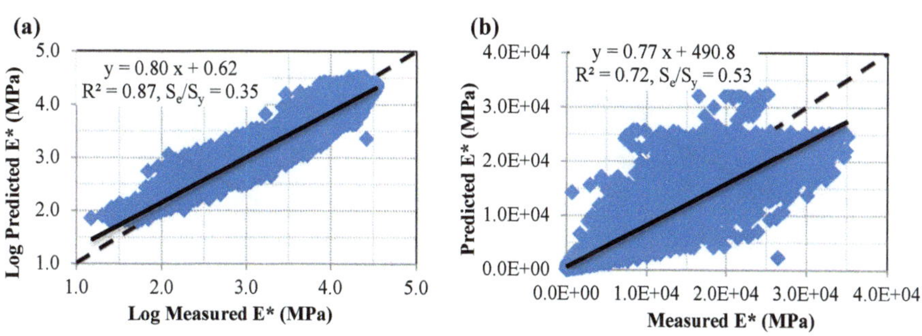

Fig. 4. Comparison of measured versus predicted E*using level 3 binder inputs for NCHRP 1-37A Witczak equation. (a) Logarithmic space, (b) Arithmetic space

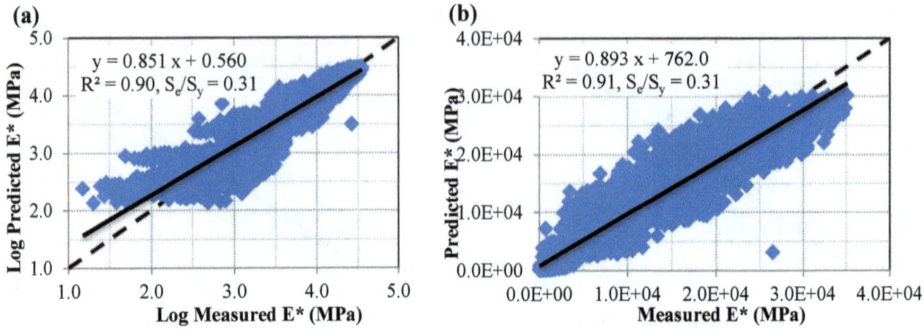

Fig. 5. Comparison of measured versus predicted E*using level 1-b binder inputs (G*, δ-based 8-36-36-1 ANNs architecture, NCHRP 1-40D inputs). (a) Logarithmic space, (b) Arithmetic space

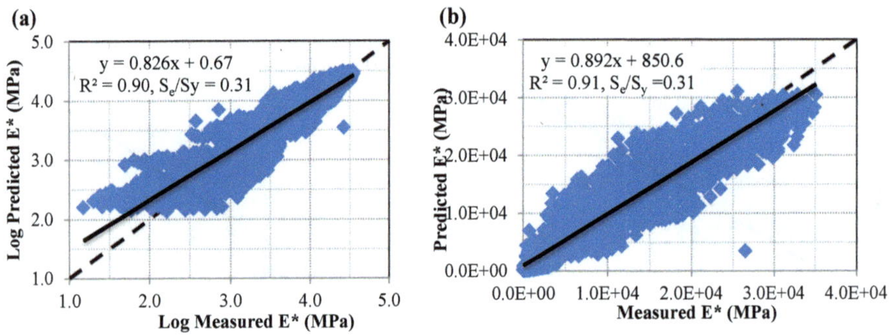

Fig. 6. Comparison of measured versus predicted E*using level 1-b binder inputs (A-VTS based 10-36-36-1 ANNs architecture, NCHRP 1-40D inputs). (a) Logarithmic space, (b) Arithmetic space

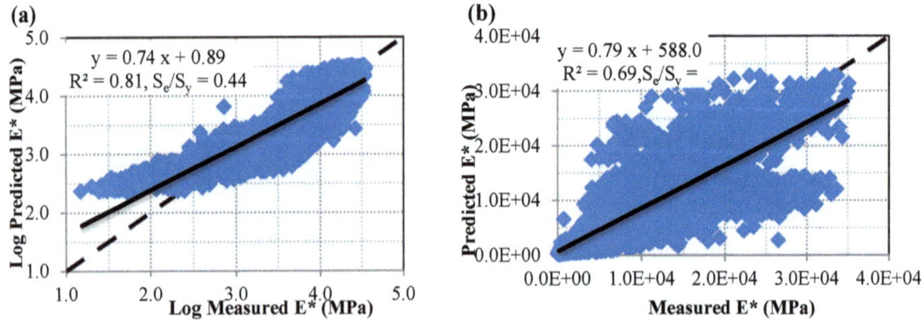

Fig. 7. Comparison of measured versus predicted E*using level 1-b binder inputs for NCHRP 1-40D Witczak equation. (a) Logarithmic space, (b) Arithmetic space

5.3 ANNs versus 2003 Pc-Based Hirsch Prediction Model

Figures 8 and 9 represent a comparison between measured E* and ANNs predictions with both neural architectures; 5-36-36-1 for the G*, δ set of variables, and 7-36-36-1 for the A-VTS. Figure 10 compares between the prediction accuracy and bias of the Hirsh model based on level 1-b binder inputs in both logarithmic and arithmetic spaces. It is interesting to note that of the three models investigated in this study, Hirsch model exhibited the most biased predictions of dynamic modulus. This is obvious from the intercept and the slope of the non-constrained regression lines shown in Tables 2 and 3. In addition, when evaluating the prediction accuracy, one can surmise from the values of R^2 and S_e/S_y presented in Tables 2, 3 that this model was the least accurate among the studied models. Moreover, the observed difference in accuracy and bias between ANNs and regression for this model was significant. Nevertheless, ANNs predictions based on Hirsh model inputs exhibited accuracy and bias similar to the NCHRP 1-37A and NCHRP 1-40D in terms of R^2 and S_e/S_y as well as line slope and intercept. While regression produced the highest accuracy at input level 3, ANNs produced the highest accuracy at input level 1-b.

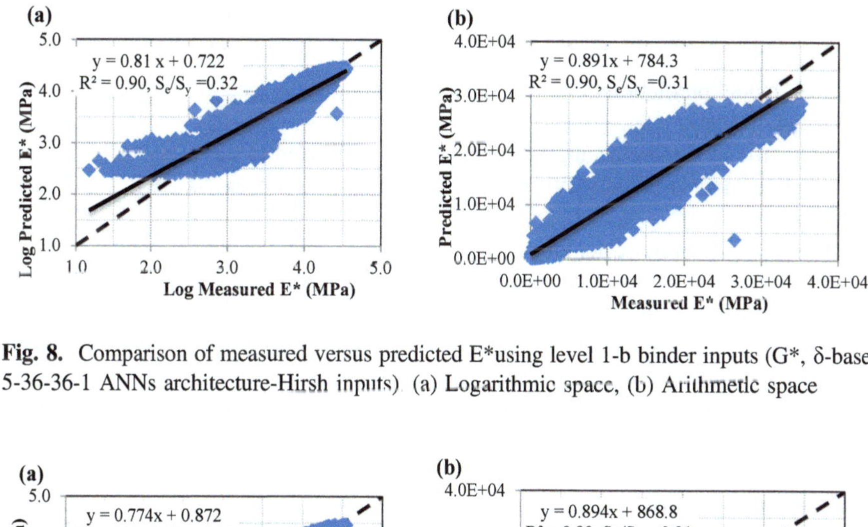

Fig. 8. Comparison of measured versus predicted E*using level 1-b binder inputs (G*, δ-based 5-36-36-1 ANNs architecture-Hirsh inputs). (a) Logarithmic space, (b) Arithmetic space

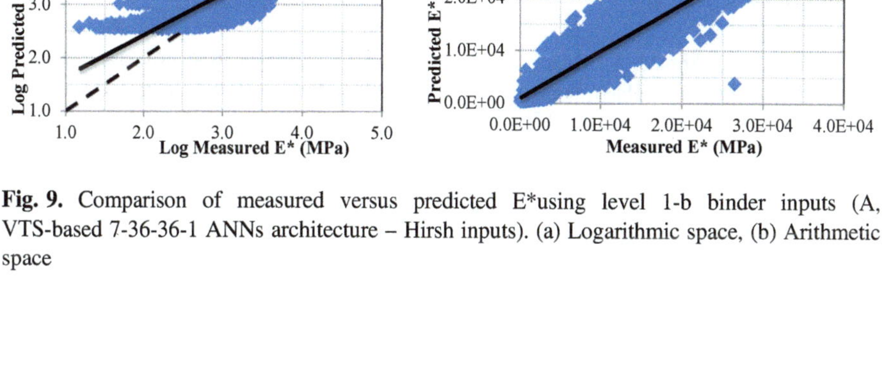

Fig. 9. Comparison of measured versus predicted E*using level 1-b binder inputs (A, VTS-based 7-36-36-1 ANNs architecture – Hirsh inputs). (a) Logarithmic space, (b) Arithmetic space

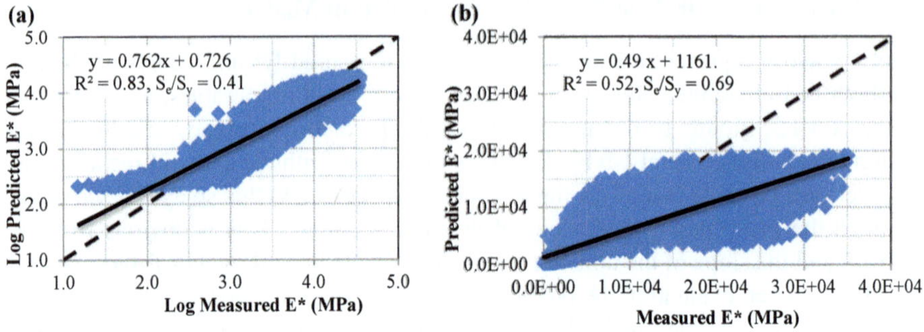

Fig. 10. Comparison of measured versus predicted E*using level 1-b binder inputs for Hirsh model. (a) Logarithmic space, (b) Arithmetic space

6 Conclusions

The results of this research demonstrated the power of the ANNs in E* predictions as compared to regression models. The results also showed the effect of the binder characterization method on the E* predicted from the studied models as well as the ANNs models. Following are the main conclusions of this research work:

- Hirsch model predictions were the most biased and least accurate compared to both Witczak models.
- ANNs produced more accurate E* predictions compared to the most well know E* predictive models namely Witczak NCHRP 1-37A, Witczak NCHRP 1-40D, and Hirsch.
- While in logarithmic scale, the predicted E* values from the studied models were in close agreement to the measured ones, in the arithmetic scale (which is the most important), the accuracy of the models predictions decreased dramatically.
- For the ANNs models, the accuracy of the predictions was comparable in both arithmetic and logarithmic scales.
- Among the investigated binder input levels, the most accurate predictions occurred at level 3 binder inputs. This is true for all three regression models.
- For the ANNs models, among the investigated binder input levels, the most accurate predictions occurred at level 1-b binder inputs.

References

Ali, Y., Irfan, M., Ahmed, S., Khanzada, S., Mahmood, T.: Investigation of factors affecting dynamic modulus and phase angle of various asphalt concrete mixtures. Mater. Struct. **2016** (49), 857–868 (2015). doi:10.1617/s11527-015-0544-3

Andrei, D., Witczak, M.W., Mirza, M.W.: Development of revised predictive model for the dynamic (complex) modulus of asphalt mixtures. In: Development of the 2002 Guide for the Design of New and Rehabilitated Pavement Structures, NCHRP 1-37A, College Park, MD, Interim team Technical Report (1999)

ARA, Inc., ERES Consultants Division: Guide for Mechanistic-Empirical Design of New and Rehabilitated Pavement Structures. NCHRP 1-37A Final Report, Transportation Research Board, National Research Council, Washington, DC (2004)

Awed, A.M., El-Badawy, S.M., Bayomy, F.M., Santi, M.: Influence of the MEPDG binder characterization input level on the predicted dynamic modulus for Idaho asphalt concrete mixtures. In: 90th Annual Meeting of the Transportation Research Board. Washington, DC (2011)

Bari, J., Witczak, M.W.: Development of a new revised version of the Witczak E* predictive models for hot mix asphalt mixtures. J. Assoc. Asphalt Paving Technol. **75**, 381–417 (2006)

Bari, J., Witczak, M.W.: New predictive models for the viscosity and complex shear modulus of asphalt binders for use with the mechanistic-empirical pavement design guide. J. Transp. Res. Rec. **2001**, 9–19 (2007)

Bayomy, F., El-Badawy, S., Awed, A.: Implementation of the MEPDG for flexible pavements in Idaho. ITD Project RP 194, NIATT Project KLK557, National Institute for Advanced Transportation Technology, University of Idaho, Moscow, Idaho, December (2011)

Ceylan, H., Gopalakrishnan, K., Kim, S.: Looking to the future: the next-generation hot mixasphalt dynamic modulus prediction models. Int. J. Pavement Eng. **10**(5), 341–352 (2009). doi:10.1080/10298430802342690

Ceylan, H., Gopalakrishnan, K., Kim, S.: Advanced approaches to hot-mix asphalt dynamic modulus prediction. Can. J. Civ. Eng. **35**, 699–707 (2008). doi:10.1139/L08-016

Ceylan, H., Schwartz, C.W., Kim, S., Gopalakrishnan, K.: Accuracy of predictive models for dynamic modulus of hot-mix asphalt. J. Mater. Civ. Eng. **21**(6). ©ASCE. ISSN: 0899-1561/2009/6-286–293/$25.00. doi:10.1061/_ASCE_0899-1561_2009_21:6_286_ (2009)

Christensen Jr., D.W., Pellinen, T.K., Bonaquist, R.F.: Hirsch model for estimating the modulus of asphalt concrete. In: Asphalt paving technology, pp. 97–121. Journal of the Association of Asphalt Paving Technologists, Lexington, KY (2003)

El-Badawy S.M., Bayomy F.M., Awed, A.M.: Performance of MEPDG dynamic modulus predictive models for asphalt concrete mixtures: local calibration for Idaho. J. Mater. Civ. Eng. **24**(11). © SCE, ISSN 0899-1561/2012/11-1412-1421/$25.00. doi:10.1061/(ASCE)MT.1943-5533.0000518 (2012)

El-Badawy, S., Khattab, A., Al Hazmi, A.: Using artificial neural networks (ANNs) for hot mix asphalt E* predictions. Geo-China **2016**, 83–91 (2016). doi:10.1061/9780784480076.010

Far, M.S.S., Underwood, B., Ranjithan, S., Kim, R.Y.: Transportation Research Record: Journal of the Transportation Research Board, No. 2127, Transportation Research Board of the National Academies, Washington, D.C., pp. 173–186 (2009). doi:10.3141/2127-20

Georgouli, K., Loizos, A., Plati, C.: Calibration of dynamic modulus predictive model. Constr. Build. Mater. J. (2015). doi:10.1016/j.conbuildmat.2015.10.163Elsevier

Georgouli, K., Loizos, A., Plati, C.: Assessment of dynamic modulus prediction models in fatigue cracking estimation. Mater. Struct. J. **2016**(49), 5007–5019 (2016). doi:10.1617/s11527-016-0840-6

Harran, G., Shalaby, A.: Improving the prediction of the dynamic modulus of fine-graded asphalt concrete mixtures at high temperatures Canadian. J. Civ. Eng. (2009). doi:10.1139/L08-123

Jamrah A., Kutay, M.E., Ozturk, H.I.: Characterization of asphalt materials common to Michigan in support of the implementation of the mechanistic-empirical pavement design guide. In: 93rd Transportation Research Board Annual Meeting, 12–16 Jan 2014. Washington, DC (2014)

Khatab, A.: Dynamic Modulus Predictive Models for Superpave Asphalt Concrete Mixtures. M.Sc. Thesis, Public Works Engineering. Mansoura University, Egypt (2015)

Khattab, A.M., El-Badawy, S.M., Al Hazmi, A., Elmwafi, M.: Comparing Witczak NCHRP 1-40D with Hirsh E* predictive models for Kingdom of Saudi Arabia asphalt mixtures. In: 3rd Middle East Society of Asphalt Technologists (MESAT) Conference American University in Dubai, UAE, 6–8 Apr 2015 (2014)

Khattab, A.M., El-Badawy, S.M., Al, Hazmi A., Elmwafi, M.: Evaluation of Witczak E* predictive models for the implementation of AASHTOWare-pavement ME design in the Kingdom of Saudi Arabia. Constr. Build. Mater. J. (2014). doi:10.1016/j.conbuildmat.2014.04.0660950-0618/_2014

Pellinen, T.K.: Investigation of the Use of Dynamic Modulus as an Indicator of Hot-Mix Asphalt Performance. Ph.D. Dissertation. Tempe, Arizona: Arizona State University (2001)

Rahmani, E., Darabi, M., Abu Al-Rub, R., Kassem, E., Masad, E., Little, D.: Effect of confinement pressure on the nonlinear-viscoelastic response of asphalt concrete at high temperatures. Constr. Build. Mater. **V**(47), PP779–PP788 (2013)

Witczak, M., El-Basyouny, M., El-Badawy, S.: Incorporation of the new (2005) E* predictive model in the MEPDG. In: NCHRP1-40D Final Report (2007)

Witczak, M.W., El-Basyouny, M., El-Badawy, S.: Incorporation of the New 2005 E* predictive model in the MEPDG. In: NCHRP 1-40D Final Report, Arizona State University, Tempe, AZ (2007)

Yousefdoost, S., Binh, V., Rickards, I., Armstrong, P., Sullivan, B.: Evaluation of dynamic modulus predictive models for typical Australian asphalt mixes. In: 15th AAPA International Flexible Pavements Conference, 22-25 September, Royal International Conference Centre, Brisbane (2013)

Potentials for Using Mechanically Activated Concrete Powder in Stabilized Granular Pavement Mixtures

Jan Valentin[1(✉)], George Karráa[2], Jan Suda[1], Jakub Šedina[1], Pavel Tesárek[1], and Zdeněk Prošek[1]

[1] Faculty of Civil Engineering, Czech Technical University, Prague, Czech Republic
jan.valentin@fsv.cvut

[2] Czech University of Life Sciences, Prague, Czech Republic
gkarraa@gmail.com

Abstract. Stabilization of granular materials is a common practice in transport infrastructure structures. It is used mainly for improving the roadbed or for achieving higher performance of base layers. The traditional approach is to use cement, lime hydrated, hydraulic road binder or standard fly-ash in mixing with soil or granular material (gravel) to get a hydraulically bond mixture which then shows improved bearing capacity and better resistance to water immersion or frost impacts. However, most of the binders (especially cement) show higher carbon footprint and represent therefore increased socio-economic cost of the structure. On the other hand, yearly many old structures are demolished and waste is created often containing huge volume of concrete material, which can be crushed and reused as a granular material. At the same time this material can be treated by high-speed milling (disintegration) process and it is possible to achieve a partial reactivation of the cement components in the concrete. This paper presents some of the achievements in soil stabilized materials and cold recycling mixtures where activated micromilled concrete powder/fine ground recycled concrete (FGRC) was used as an active filler or a binder substitute. The effect of the micromilled concrete powder with respect to its content or combination with other binders or additives is described further. Regular tests like Proctor Standard, CBR or compressive strength, indirect tensile strength or stiffness determination are done. The effect to act as an alternative binder is studied not only it terms of used content, but also with respect to its impact during test specimen curing. Finally, some recommendations are given in this paper with respect to the practical application of such material.

1 Introduction

The paper focuses on the issue of alternative modification of selected granular material types with the option of mechanical or even mechanical-chemical activation thereof. For this specific process of grinding, the consequence is not only the particle size reduction. The material reactivity may be renewed to a certain degree in the sense of typical characteristics for hydraulic binders as well. The principles of high-energy/high-speed

L. Mohammad (ed.), *Advancement in the Design and Performance of Sustainable Asphalt Pavements*, Sustainable Civil Infrastructures, DOI 10.1007/978-3-319-61908-8_15

grinding are used for this purpose. CTU Prague has examined the high-speed grinding for over 5 years in cooperation with several industrial partners who focus on this type of grinding specifically. In relation to the current trend of identifying and enforcing power-saving technologies, it should be emphasised that high-energy grinding does not necessarily mean a process where a huge quantity of energy is applied. The term merely underlines the fact that the grinding itself is associated with high-energy presence. The energy might result – and results – from the grinding process itself; the energy is not self-perpetuating but it plays an important role, which facilitates the aforementioned reactiveness.

The focus on reclaimed concrete is logical with respect to the fact that the material contains hydrated cement and on a global scale, there are huge quantities of both concrete waste and material obtained by reclaiming it. Even the crushing to various fractions yields particles <4 mm, which include higher proportions of hydrated cement (compared to coarse concrete particles) and offer rather easier grinding as well. Such process can already be called mechanical-chemical activation as the grinding is repetitively shown not only to reduce the particle sizes but also, probably due to the high instant energy, some mineralogical structures are modified and the resulting material, containing the hydrated cement, is reactivated to a certain degree. The best findings in both cases have been obtained at CTU in Prague in the field of lime, cement or lime-cement mortars or mineral adhesives as well as mixtures for masonry components. Particularly the options of exploiting finely ground activated recycled concrete as an alternative binder in road construction have also been tested repeatedly.

1.1 High-Speed Grinding

As has been suggested by the introduction, high-speed grinding is one of the types of high-energy grinding which is characterised by a high quantity of energy transformed per unit of pulverized material during the grinding process. The high-energy or high-speed grinding is not defined exactly anywhere in any literature sources (Richter 2002; Najborodenko 2003). All of the basic properties, which are partly discussed below, are the same as known for the traditional grinding process: that is, refining granularity, increasing specific surface, opening the particle structure etc. In contrast to the traditional grinding, high-speed grinding involves certain phenomena (effects) which have not been observed in standard grinding. These effects are characteristic by consuming part of the spent energy, which, in standard grinding, converts into heat with no other use. With inorganic materials, such effects include for example:

- mechanical-chemical (mechanical) activation of the particles,
- forming higher proportions of micron particles and nanoparticles,
- in some cases, increased effectiveness of exploitation of the energy consumed to create new particle surfaces.

Various types of grinding have been used since ancient times (Fuerstenau and Han 2003; Šebor 1983; Richter 2002; Lynch and Rowland 2005). Grinding was initially used for food processing or pharmaceuticals. Subsequently, it was applied to process mineral materials (mainly to extract metals). Later, grinding became one of the basic technological operations utilised in most industrial manufacturing processes. Intense development of grinding technology was brought on along by the development of other industrial technologies over 200 years ago. Practically no industry can effectively work without grinding, from mineral processing, power industry and metallurgy to chemistry, food industry, papermaking and construction material manufacturing (Bárta 1961; Zbirovský and Michálek 1983; Hlaváç 1988; Hanykýr and Kutzendörfer 2002). It is a fact that most common industrial products could not be produced without some type of grinding.

Grinding is commonly understood as the process of refining particle size while increasing specific surface of the material, but also as the process of opening the particle structure (Snow et al. 2009; Austin and Trass 1997). Grinding yields products of such parameters, which are needed by subsequent processing. The particles need to be opened to provide access or separation to the individual material components (e.g. the ore and mining waste). Refinement of the particle size and enlargement (or modification) of their surface are necessary to achieve the required chemical or physical properties of the material processed. In some cases, grinding also involves homogenisation of heterogeneous mixtures of various materials, like some cement materials, powder metallurgy mixtures, ceramics etc. The grinding of organic and biological materials has a slightly different view of the material coming from inorganic material processing. The classic and maybe earliest historically supported example of grinding is the crushing of cereals to prepare flour. The use of grinding to pulverize and pulp the cellular walls in wood mass during paper and cellulose production, to open cells and extract tannins, enzymes and other organic substances. Grinding is also used to process fossil and other solid fuels for powder and fluid combustion in heat-producing thermal units (Richter 2002).

The grinding process distinguishes whether it occurs between the grinding elements and the material to be refined, or between a layer of material particles positioned between the grinding elements, upon contact between the particle and the grinding elements, contact between two particles or between the medium (environment) and a particle. The grinding mechanisms are presented in the following Table 1.

Table 1. Basic principles of grinding

1	2	3	4	5	6	7
Compression	**Shear**	**Compression pulse**	**Free hit**	**Stroke/Flexion**	**Chopping**	**Cutting**

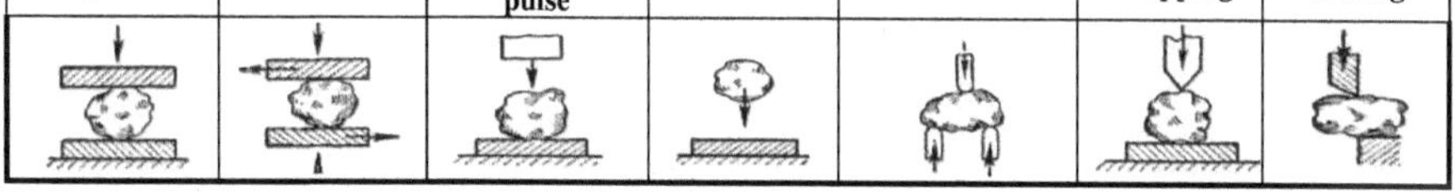

The majority of classic mills (ball mills, ring mills, rod mills, disc mills, cylinder mills, vibration mills etc.) combines mechanisms 1, 2 and 3 to a certain level; impact is made on particles or layers of particles between the grinding elements. From the energy point of view, mechanisms 1 to 3 usually have low efficiency because, prior to pulverizing the particles of the material processed, the adhesion between the grinding element and the particles or particle layer has to be overrun. Then, the particles must pass the plastic deformation stage because the contact between the grinding elements and the material particles occurs under relatively low speed, which can be compensated by the internal structure (elasticity, twinning etc.). During those processes, the major part of the spent energy is transformed into heat and is not efficiently exploited within particle size reduction (or specific surface enlargement) in the material processed. Another disadvantage of these mechanisms is the relatively small contact between the surface of grinding elements and the particles of the material processed. Therefore, in proportion to productivity, mills employing such mechanisms are quite large and very heavy but less efficient. An example of a typical high-speed mill is shown by Fig. 1.

Fig. 1. High-speed disintegrator used within the experiments described by this paper (left outspread annulus of rotors for placing of milling elements)

1.2 Use of FGRC in Stabilized Mixtures in Pavement Structures

The material specifications for bound layers in pavement structures and manufacturing processes for the composites tested and presented by this paper apply to the following mixes:

- BCSM-BE: Bitumen-Cement Stabilized Material – Bitumen Emulsion according to Technical Specification of Czech ministry of Transportation - TP 208,
- BCSM-FB: Bitumen-Cement Stabilized Material – Foamed Bitumen according to TP 208,
- CSM: Cement Stabilized Material according to TP 208,
- CBGM: Cement Bound Granular Mixtures according to European technical standard ČSN EN 14227-1.

The preparation of bitumen stabilized mixtures (BSM) and BCSMs used screened reclaimed asphalt material, fraction 0/22 mm from the Středokluky mixing plant. The homogeneity of the material was very poor, affecting the results of experimental measurements considerably and rendering any consistent conclusions and comparisons quite difficult. The reclaimed material composition even differed within individual control batches while the differences between such batches were rather significant. It was due to the lack of homogeneity that the grading of reclaimed material was determined repeatedly as indicated in Fig. 2. The bituminous binder content in the reclaimed asphalt material fell in the interval of 5.1–6.0 M%. Within the framework of capturing the difference between the parameters tested which were due to lack of homogeneity in the input material and the idea of what values these could amount to, reference mixtures were prepared at the beginning (mixtures marked “old”) and in the end (mixtures marked “new”) of the research of application of mechanically activated recycled concrete (FGRC).

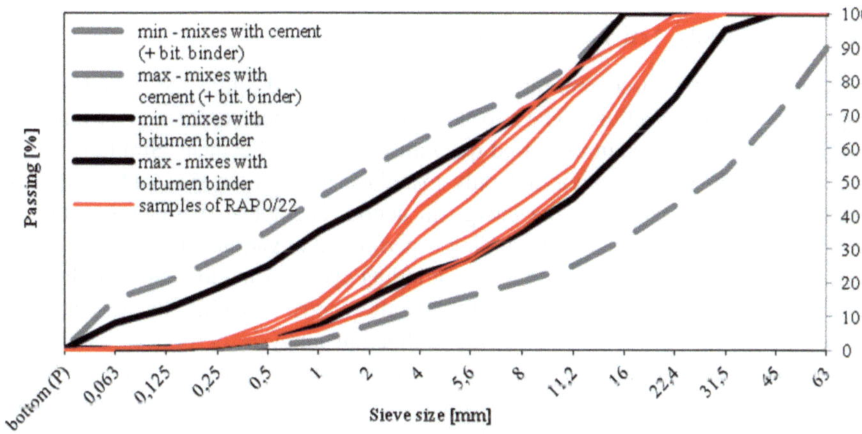

Fig. 2. Used reclaimed asphalt and its grading fluctuation during the production of particular experimental mixtures

Mix BCSM-BE used slow-breaking cationic bituminous emulsion C60B7 according to EN 13808:2013, which is the standard used in cold recycling technology. The BCSM-FB mix contained foamed bitumen. The basic characteristics of the bitumen used in foamed bitumen preparation and the optimum quantity of water added to the binder are indicated in Table 2. Foamed bitumen optimisation was undertaken according to the procedures described in Wirtgen Manual (2012).

Table 2. Foamed bitumen characteristics including the content of added foaming water

Bitumen	Penetration at 25 °C [0.1 mm]	Softening point (R&B) [°C]	Optimum water content in the foam [M% of bitumen]
70/100	83	46.8	3.8

Foamed bitumen was injected into the mix at 170 °C using a Wirtgen WLB10S laboratory unit with forced mixing in a Wirtgen WLM 30 twin-shaft compulsory mixer.

The preparation of cement-bound granular mixtures (CBGM) used grit 0/63 mm and gravel 0/16 mm in a 4:1 proportion as filler; the granularity and proportion were chosen to suit the granularity of the CBGM according to ČSN EN 14227-1.

In the case of bounded recycled mixtures (cement-stabilised macadam), grit 0/63 mm in combination with reclaimed asphalt material 0/22 mm was used to a 1:1 proportion compliant with the requirements for cold recycled mixtures according to TP 208.

Standard Portland slag cement CEM II/B – M 32.5 R (or cement CEM II/A-S 42.5 R) according to European standard EN 197-1 was used as hydraulic binder.

1.3 Mechanically Activated Recycled Concrete

Recycled concrete is used presently as a type of recycled aggregate, which can be supplied in various fractions. The broader potential of finer particles (≤ 4 mm) has been known as they contain increased quantities of hydrated cement. Similarly to e.g. fluidized fly ash, grinding aims not only to reduce the particle size and to increase specific surface but based on the energy in action, attempt at partial restoration of material reactiveness. High-speed grinding is performed in various modes, which differ by grinding intensity as well as number of grinding cycles or modifications to the rotor annulus type. A number of tests are performed afterwards, including the sieve analysis (see examples in Figs. 3 and 4), X-ray diffraction analysis or optical or electron microscope analysis. In the last 2 to 3 years, attention has been paid at CTU in Prague to concrete collected from the reconstruction of motorways D1 or D2, concrete from old railway sleepers or concrete from the runways of the Praha-Ruzyně airport.

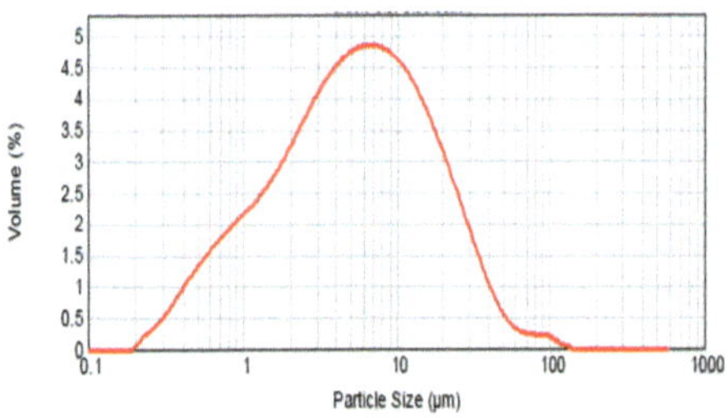

Size (μm)	Vol Under %
0.200	0.00
0.500	3.73
1.000	11.66
2.000	23.51
3.000	33.13
4.000	41.21
5.000	47.98
6.000	53.70
7.000	58.59
8.000	62.81
10.000	69.72
15.000	81.20

Size (μm)	Vol Under %
20.000	87.90
25.000	92.03
30.000	94.65
40.000	97.39
45.000	98.07
50.000	98.52
60.000	99.02
63.000	99.11
71.000	99.32
75.000	99.40
90.000	99.66
100.000	99.80

Size (μm)	Vol Under %
125.000	99.98
150.000	100.00
180.000	100.00
200.000	100.00
250.000	100.00
300.000	100.00
355.000	100.00
400.000	100.00
500.000	100.00
600.000	100.00
750.000	100.00
1000.000	100.00

Fig. 3. Grading curve of fine grained (mechanically activated) recycled concrete – railway cross-tie

Prior to specific applications in the construction industry, "cement paste" was always prepared and basic characteristics thereof, which are usually tested for standard cement when used in concrete and other composites, were determined. As an example we list the findings from applying activated, pulverized recycled concrete from highway D1. A total of five mixtures were designed for cement paste testing: mix A is used as a reference and contains solely cement; mixes B to E contain 33, 50, 67 and 100 M% of finely ground recycled concrete (FGRC). The microstructure was only analysed for

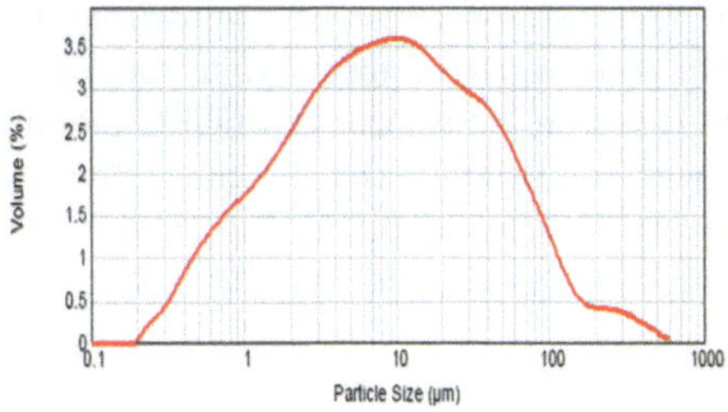

Size (µm)	Vol Under %	Size (µm)	Vol Under %	Size (µm)	Vol Under %
0.200	0.00	20.000	68.71	125.000	96.39
0.500	3.13	25.000	73.29	150.000	97.22
1.000	9.70	30.000	76.87	180.000	97.80
2.000	19.08	40.000	82.27	200.000	98.08
3.000	26.32	45.000	84.35	250.000	98.66
4.000	32.20	50.000	86.13	300.000	99.10
5.000	37.05	60.000	88.96	355.000	99.46
6.000	41.14	63.000	89.65	400.000	99.65
7.000	44.66	71.000	91.24	500.000	99.91
8.000	47.76	75.000	91.91	600.000	100.00
10.000	52.98	90.000	93.89	750.000	100.00
15.000	62.40	100.000	94.84	1000.000	100.00

Fig. 4. Grading curve of fine grained (mechanically activated) recycled concrete – airport Prague – Ruzyne

samples A, B, C and D as mix E did not show sufficient quality for a sample to be prepared. The mixtures consisted of cement paste of Portland cement CEM I 42.5 R with varying quantities of FGRC and without added plasticiser. With respect to the differences in cement and FGRC behaviour upon combination with water, the mixes had different water content (w/c ratio) which ranged from 0.350 (mix A) to 0.417 (mix E). The unifying parameter for the mixtures was their workability defined by the cement paste consistency. The aforementioned solution ensures similar homogeneity for all versions; if mixtures contained FGRC, it reduced the size and frequency of technological pores in the set composite. For the microscopic analysis, samples were prepared with the mixtures and stored in water at (21 ± 2) °C for 28 days. Cylindrical test specimens with 30 mm diameter and 50 mm height were prepared. Having cured for 28 days, the samples were sliced by a diamond saw to approximate slice thickness of 5 mm; their surface was subsequently filed and polished (using an emulsion with nano-diamond of 0.25 µm size in the last step). Test specimens were cured in water at (21 ± 2) °C for 28 days and then prepared for mechanical property testing. The samples were prepared as beams with 40 × 40 × 160 mm dimensions.

Laser particle size distribution showed that 34% of the particles were larger than 20 µm and only 10% were larger than 45 µm. The largest particle had 130 µm and a median particle had the dimension of 12 µm. The specific surface was determined according to the Blaine method with values between 412 and 435 m^2/kg. Material density was determined as the range of 2.653 to 2.670 g/cm^3. The Portland cement used has 45% of the particles over 20 µm and 12% particles over 45 µm. The largest particle does not exceed 90 µm and the median particle is 18 µm. The specific surface of CEM I 42.5 R is 351 m^2/kg. Having compared the particle analysis and specific surface it can be noted that FGRC was pulverized to cement level and, therefore, the material could act as an active microfiller the specific surface of which is even slightly higher than that of cement. Moreover, old, non-hydrated parts of cement particles contained in the recycled concrete could be uncovered by the applied activation and allowed to hydrate. On the other hand, FGRC particles are not small enough to coat the cement particles, thus preventing water from entering and hydrating the cement.

Comparing the images taken by the electron microscope (Fig. 5) allows distinguishing the individual phases of the composite materials. Samples with 50% FGRC, when compared to sample A, clearly show a new phase of dark grey colour which, in this case, represents the aggregate from the recycled concrete. The comparison also shows an obvious reduction in the quantity of non-hydrated cement (white parts) with increasing quantity of FGRC. Even though, based on the cement and FGRC proportions between samples A and C, the quantity of non-hydrated cement should drop by approx. 50%; there was a reduction by almost 75% in sample C that supports the assumption on changes to the hydration process, which was also verified by the calorimetry test.

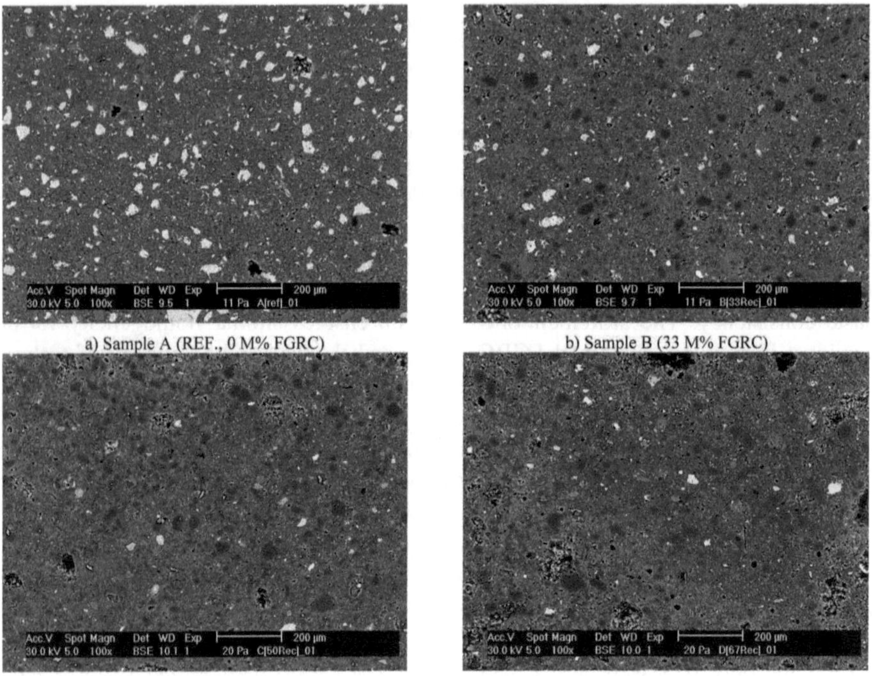

a) Sample A (REF., 0 M% FGRC)

b) Sample B (33 M% FGRC)

c) Sample C (50 M% FGRC)

d) Sample D (67 M% FGRC)

Fig. 5. Microscopic pictures of the sample microstructure (electron microscope with 100× zoom)

Bending tensile strength tests were only conducted with samples A to D. The results presented in Fig. 7 show an obvious drop in bending tensile stress depending on cement substitution. When 33% cement was replaced, there was a roughly 22% drop if compared to the reference sample. If substitutions over 33% are made the drop is even more distinctive. This corresponds very well with the experimental results gained for shrinkage potential, as shown in Fig. 6. Compressive strength testing for the same type of samples is described in Fig. 8. The results show just a slight drop (roughly 2%) in compressive strength when 33% cement is replaced by FGRC. Any further increase in cement replacement induces a distinctive, directly proportionate, drop in strength. A slight reduction in compressive strength confirms the assumption on FGRC acting as a micro-filler and binder replacement (Table 3).

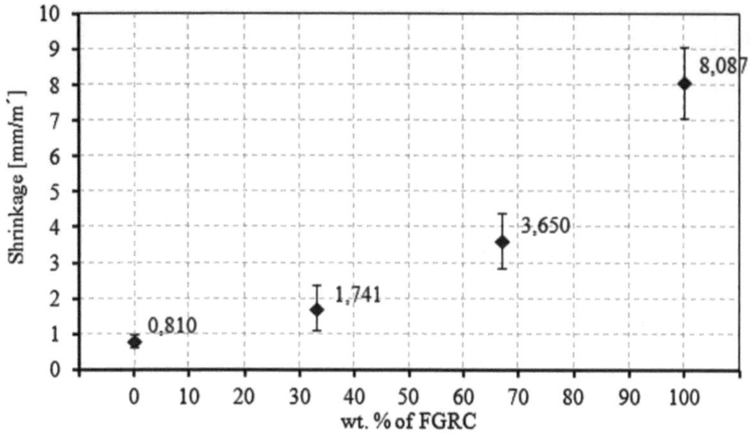

Fig. 6. Level of shrinkage after 28 days curing – dependence of FGRC content in the cement paste

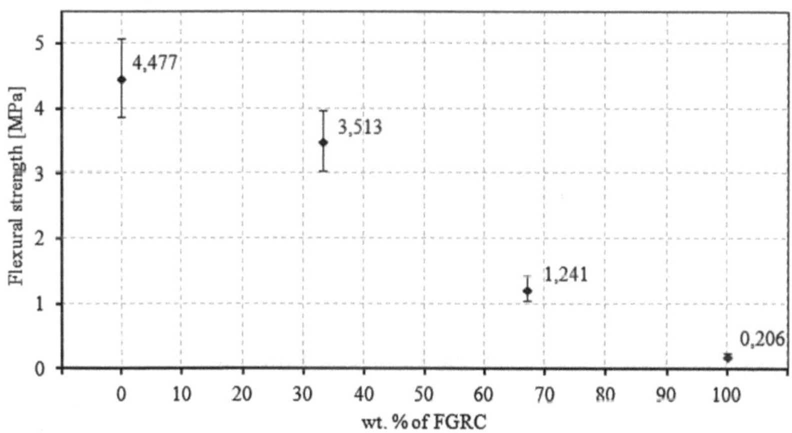

Fig. 7. Flexural strength (bending tension) after 28 days curing – dependence of FGRC content in the cement paste

1.4 Laboratory Mix Design for Stabilized Mixtures

The optimum water content of the tested stabilized mixtures was determined according to ČSN EN 13286-2. The individual mixtures were designed to comply with the real-life bonded or recycling mix designs while being inter-comparable from the point of view of input component economy. The designed composition reflects the expected function of FGRC in the mix or the function of hydraulic binder substitute or mineral additive to the fine component in the total aggregates. The composition of the mixed tested is indicated in the Table 4.

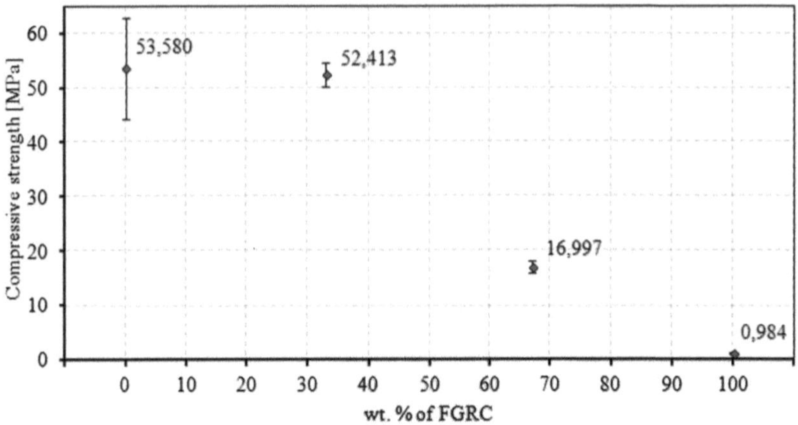

Fig. 8. Compression strength after 28 days curing – dependence of FGRC content in the cement paste

Table 3. Composition of tested mixtures

Mixture	Cement (CEM I 42.5 R) (g)	FGRC (g)	Water (g)	Consistency (mm)
A (ref., 0 M% FGRC)	1000	–	350	130
B (33 M% FGRC)	670	330	383	130
C (50 M% FGRC)	500	500	400	130
D (67 M% FGRC)	330	670	417	130
E (100 M% FGRC)	0	1000	450	130

1.5 Test Specimen Preparation and Used Test Protocols

The technical specifications for stabilised cold recycled mixtures (BSM, BCSM) according to TP 208 stipulate test specimen preparation in cylindrical moulds of 150.0 ± 1.0 mm diameter and 200–300 mm height. During processing, the mix was compressed in the moulds by 5.0 MPa. The test specimen was left in the mould at (20 ± 2) °C for (24 ± 6) hours and, subsequently, kept at 90% to 100% relative humidity and (20 ± 2) °C for another two days. Then, the specimens were kept at 40% to 70% humidity and (20 ± 2) °C. The evaluated mixtures were left under the aforementioned curing conditions for 7, 14 and 28 days. From the perspective of water susceptibility, a set of specimens was put in a water bath for 7 days following the seven days of curing at 40% to 70% humidity and (20 ± 2) °C. In the course of curing, the specimens were cut into two parts to make the specimen height 60 ± 5.0 mm.

The assessment of BSM and BCSM mixtures included the determination of indirect tensile strength, applying a process according to TP 208 where the test specimens were air tempered for 4 h to (15 ± 1) °C and then tested with constant loading rate of (50 ± 1) mm/min until specimen failure. Another characteristic observed was the stiffness modulus according to ČSN EN 12697-26, tested by a non-destructive test

Table 4. Material composition of tested mixtures

Aggregates	Bitumen emulsion (E) M%	Foamed bitumen (F) M%	Cement (C) M%	Pulverized recycled concrete (FGRC) M%	Pulverized concrete and cement 1:1 (FGRC-C) M%
Granular material 0/63: gravel 0/16 (4:1)			6Cref		
			3C	6FGRC	
			1,5C	9FGRC	
Granular material 0/63: RAP 0/22 (1:1)			3Cref		
			1C	4FGRC	
RAP 0/22	3.5E(old)ref				
	3.5E(new)ref				
	3.5E(old)ref		3C		
	3.5E(new)ref		3C		
	3.5E			3FGRC	
	3.5E			5FGRC	
		4,5F(old)ref			
		4,5F(new) ref			
		2Fref			
		4,5F		3FGRC	
		4,5F		5FGRC	
		2,5F		3FGRC	
	(*)	2,5F		9,6FGRC	
		2F		3FGRC	
		2F		6FGRC	
		2F			3FGRC-C
		4,5F(old)ref	3C		
		4,5F(new) ref	3C		
		2Fref	1C		
	(**)	2Fref	3C		
		4,5F	1,5C	1,5FGRC	
		4F	1C	2FGRC	
		4F	1C	6FGRC	
		2F	1C	2FGRC	
		2F	1C	4FGRC	
		2F	1C	6FGRC	
		1F	1C	4FGRC	

*Reclaimed Asphalt Product (RAP) 0/11
**Cement CEM II/A-S 42,5 R
ref - control mixture

applying a repeated indirect tensile stress using NAT apparatus (Nottingham Asphalt Tester). In this case, the determination always took place at 15 °C. Another parameter describing mix durability and stability is moisture susceptibility tested according to TP 208, which is commonly expressed by the indirect tensile strength ratio (ITSR) or indirect tensile modulus ratio (ITMR). This parameter expresses the relation of indirect tensile strength (stiffness modulus) of test specimens after 7 days dry curing to the indirect tensile strength after seven days at 40% to 70% humidity at (20 ± 2) °C and seven days submersion in water.

Cement-bonded mixtures (CBGM) were prepared according to ČSN EN 14227-1, which stipulates test specimen compaction by Proctor compaction hammer while the compacting energy corresponds with the modified Proctor test. CSM mixes were prepared in the same way as BCSM and BSM, where the process is accepted by the technical conditions TP 208. Cylindrical test specimens of those mixes were left in the moulds at (20 ± 2) °C for (24 ± 6) h; subsequently the specimens were kept at 90% to 100% humidity and (20 ± 2) °C for another 27 days.

The parameters monitored for mixtures CSM and CBGM were compression strength after 28 days of curing. The next parameter was the reduction of compression strength after thirteen freeze cycles which characterises resistance to water and frost as stipulated by ČSN EN 14227-1.

For cold recycled mixtures according to TP 208, the requirement for mechanical properties is stipulated from two perspectives. There is the minimum compression strength requirement ($R_{c28} = C_{3/4}$) together with the water and frost resistance characterised by the reduction coefficient ($0.85R_{c28}$) or, contrastingly, the determination of minimum indirect tensile strength (R_{IT7} = 0.3–0.7 MPa) and the moisture susceptibility characterised by the reduction coefficient (0.75 R_{IT7}).

1.6 Test Results and Discussion

Based on the strength parameters below, the possibility of using FGRC as partial substitute for cement binder in CSM and CBGM can be declared unviable (see Fig. 9);

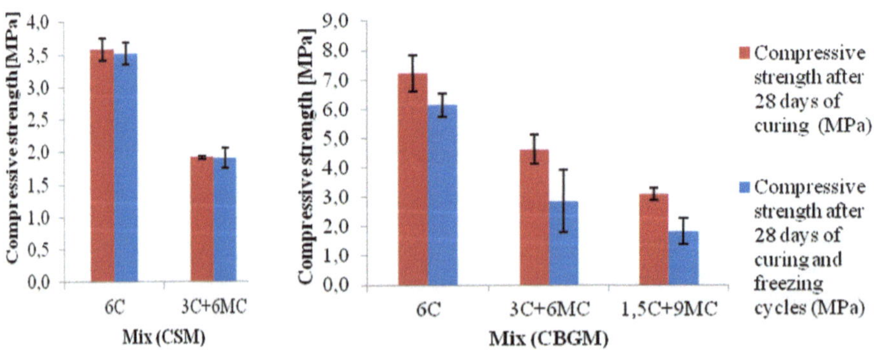

Fig. 9. Strength characteristics of **(a)** Cement Stabilized Material (CSM) and **(b)** Cement Bound Granular Mixtures (CBGM)

compression strength decreases rapidly with increasing values of cement substitution by activated micromilled concrete. The parameter of mix stability from the point of view of resistance to water and frost remains more or less constant with respect to the ratio of strength achieved and reduced cement content. The potential of FGRC use in such mixtures thus might lie in the form of an additional filler component in the aggregate, and low-level substitution of cement to ensure the required mix stability; although the overall strength will be lower, the potential of stress relate cracks (shrinkage) decreases as well.

The basic volumetric parameters of the cold recycled mixtures were determined first. The voids contents measured are indicated in Fig. 10 for all assessed mix options. This shows primarily the impact of lack of homogeneity in the reclaimed material on the individual mixtures during production.

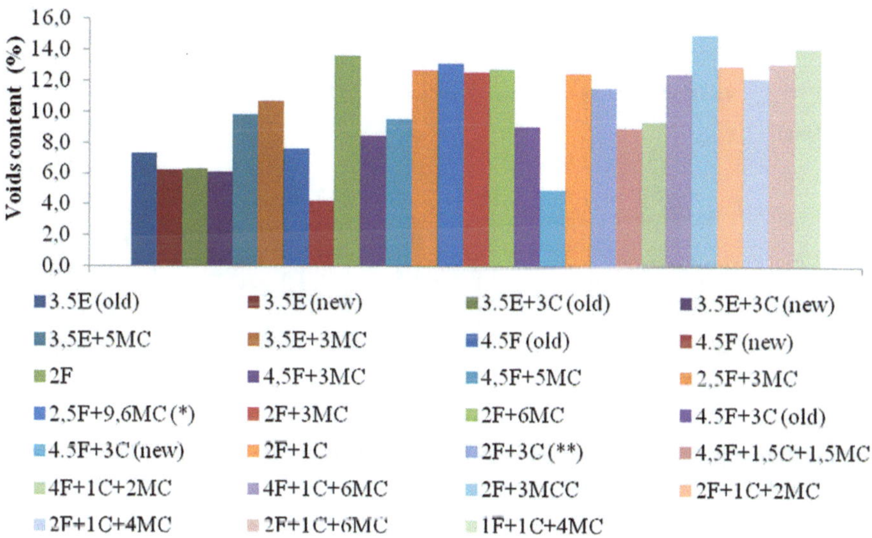

Fig. 10. Voids content of BSM, BCSM

The resulting parameters assessed for cold recycled mixtures using primarily bitumen emulsion and combinations thereof with cement (BSM-BE) including the variants using FGRC are shown in Fig. 11. If the cold recycled mixtures with FGRC are compared to the reference mix with no cement, only a small increase in indirect tensile strength and stiffness is noted. However, a positive trend appears in the field of durability where the stabilized mixtures with FGRC demonstrate a lower level of moisture susceptibility. Contrastingly, if mixtures with FGRC are compared to the reference mix, which only used cement out of hydraulic binders, there is an obvious disproportion from the point of view of all parameters assessed. Any realistic application of FGRC is only viable if quality parameters of cold recycled mixtures with bitumen emulsion improve, which particularly means the durability of such mixtures.

From the perspective of the results achieved, activated fine ground concrete may serve as an improving additive in the form of a filler component. If FGRC acts as an alternative filler, the effect of the material is inconclusive based on the results presented. It is also obvious that the quality of the bitumen emulsion as such is crucial (cf. mix 3.5E "old" and "new").

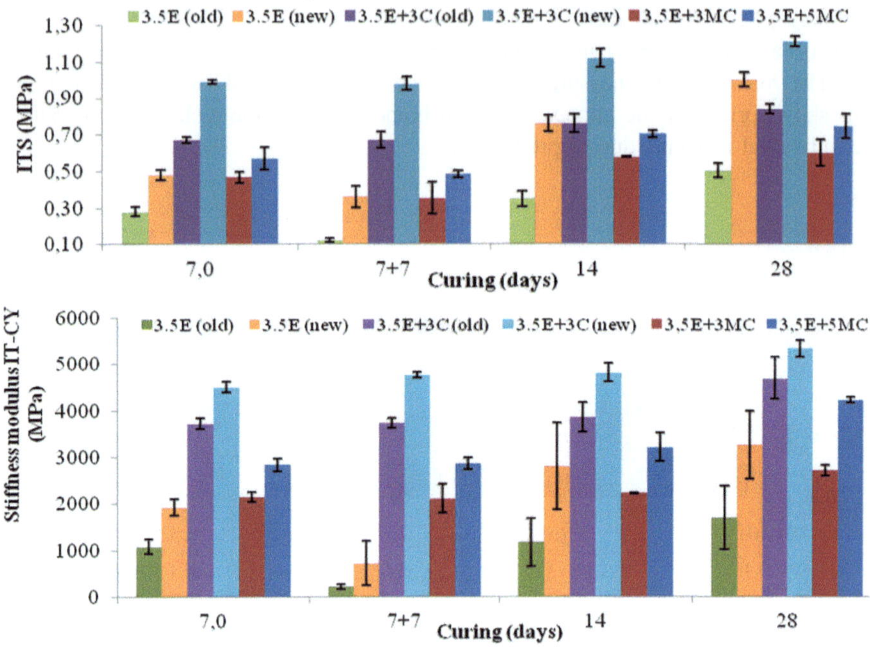

Fig. 11. Bitumen Stabilized Material with Bitumen Emulsion (BSM-BE); **(a)** Indirect tensile strength and **(b)** Stiffness modulus

The main part of the assessment of FGRC applicability in cold recycled mixtures, focused on BSM-FB and BCSM-FB. In the case of these mixtures, the specific principle of the binding effect of foamed bitumen can be employed where the gradual collapsing of individual bitumen foam bubbles this material attempts to adhere to the fine particles of cold and, usually, partly moist aggregate very tightly, creating tiny bitumen binder droplets. These droplets are due to compacting and water distribution activated and combined to form a thin bitumen film coating the fine particles and forming a compact mortar, which facilitates the subsequent adhesion and coating of larger particles of the aggregate.

For the sake of clarity in this paper, the results have been arranged into two categories. The first category quotes mixtures with no applied hydraulic binder, which only contain a combination of bitumen binder and FGRC. The second category includes mixtures containing cement in combinations with bituminous binder and FGRC.

In general, BSM-FB mixtures with FGRC demonstrate, in relation to the reference mixtures, higher values of indirect tensile strength or stiffness only if the FGRC and the foam bitumen contents are equal (FGRC% $\sim$ FB%). In the opposite case, the strength and deformation parameters decline. Any increase or decrease of strength or stiffness largely depends on the composition of the aggregate skeleton, which must be taken into account. However, the fact that mixtures with FGRC demonstrate distinctively lower sensitivity to water curing when compared to reference mixtures can be considered interesting, although such curing is usually crucial for bitumen-based mixtures. From the point of view of the optimal design of the aforementioned cold recycled mix type, it can be defined that the FGRC content should not exceed the foam bitumen content in the mix. The results of strength and deformation parameters for BSM-FB mix are indicated in Fig. 12.

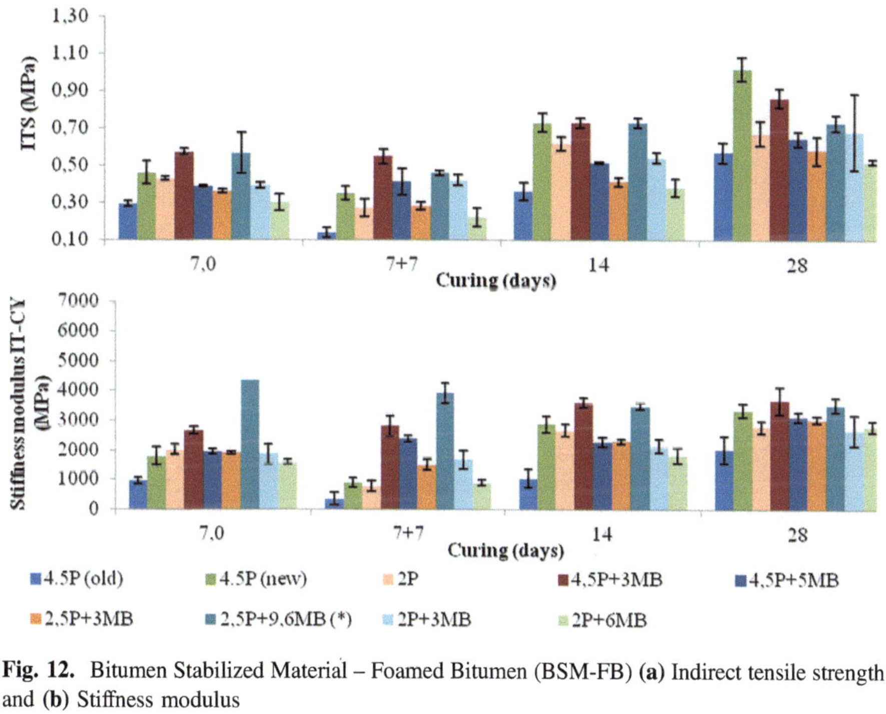

Fig. 12. Bitumen Stabilized Material – Foamed Bitumen (BSM-FB) **(a)** Indirect tensile strength and **(b)** Stiffness modulus

In the case of BCSM-FB cold recycled mixtures, foam bitumen combined with cement was added. In the mixtures with FGRC, cement was added primarily to improve the economic effectiveness of the resulting mix, i.e. mostly 1 M% of the mix. The results here are much more interesting; in relation to the reference mixtures, the reduced quantity of the bituminous binder added and increased quantity of the activated finely ground concrete is associated with a drop in indirect tensile strength; however, there is no significant decrease in the stiffness modulus. Moisture susceptibility of all mixtures is satisfactory. If focusing on mixtures where the sum of cement and FGRC contents does not exceed the quantity of the added bituminous binder, the mixtures equal the parameters of the reference mixtures while certain savings are made in the material costs. In the case of applying the combined binder, FGRC + C with high-speed milling and activation of cement and recycled concrete (ratio of 1:1) in one-step, the obtained results provide no change in the trends within the parameters tested. The results of strength and deformation characteristics of BCSM-FB cold recycled mixtures with the aforementioned binder combinations are indicated in Fig. 13.

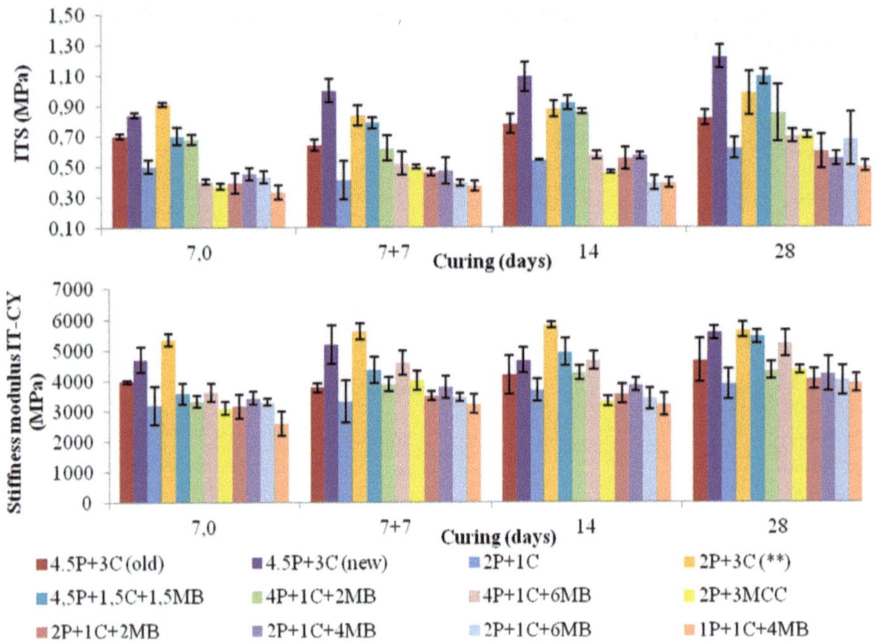

Fig. 13. Bitumen-Cement Stabilized Material – Foamed Bitumen (BCSM-FB) **(a)** Indirect tensile strength and **(b)** Stiffness modulus

2 Conclusions

The presented rather extensive laboratory research study intended to assess the potential of mechanically activated, finely ground concrete in cold recycled mixtures or those close to the traditional hydraulically bonded granular mixtures. The properties of the individual mix alternatives were assessed based on the traditional strength tests or the assessment of durability of the stabilized mix from the perspectives of water immersion or water and frost effects. The results gathered during the experiments suggest that the use of FGRC (tiny fractions of recycled concrete, modified and partially activated by high-speed milling), when applied alone as substitute of the traditional cement, or in combination with cement, does not result in any improvement of the strength characteristics of hydraulically bonded mixtures. Quite the contrary, the results verify the need for cement and the very limited binding potential of FGRC. The only significant benefit could be the reduced risk of shrinkage micro-cracks. Moreover, in the case of FGRC, it was shown that the replacement of 3 M% cement by 6 M% activated finely ground concrete does not affect mix resistance to water and frost effects at all.

Assessing the potential of FGRC in cold recycled mixtures seems much more interesting. A distinctly larger group of various combinations of this type of mix was created and the combined effects with bitumen emulsion or foamed bitumen were compared. If a bitumen emulsion and FGRC system is used, a minor improvement of the strength characteristics or stiffness is observed in comparison to the reference mix; however, the durability of the cold recycled mix improves. When compared to the bitumen emulsion and cement combination, the strength characteristics of the version with FGRC are much lower. If we focus on the combination of this material and foamed bitumen, it is obvious that there is a distinctive improvement of stability in the form of lower water susceptibility (increased value of ITSR). In addition, the stiffness values increase to a certain degree. Similarly, combinations using smaller quantities of foamed bitumen and cement while the binders are substituted by FGRC constitute interesting options.

If the costs of processing activated finely ground concrete which are not overly high due to the high-speed milling technology, and the added potential of recycling waste material are considered while taking into account the very low power demands, such processed material is viable for using in the bonded granular mixtures of pavement base and subbase layers. Technically, the use of FGRC can be supported. However, the selection of a specific application should always be based on good knowledge of the input materials with respect to the findings obtained. We should nevertheless bear in mind the fact that although high-speed milling could achieve material reactiveness, FGRC can never be compared to cement completely and expected to demonstrate similar strength characteristics. In comparison to the option with no hydraulic binder at all, a certain improvement in the strength properties can be expected; however, improved overall stability of the bonded mix is much more essential.

Acknowledgments. This paper was elaborated within the research project No. TA04031256 in the program Alfa of the Technology Agency of the Czech Republic.

References

Lynch, A.J., Rowland, C.A.: The History of Grinding, Society for Minning, Metalurgy, and Exploration, Inc. (SME), 8307 Shaffer Parkway, Littleton, Colorado, USA 80127, ISBN 0-87335-238-6 (2005)

Fuerstenau, M.C., Han, K.N.: Principles of Mineral Processing, Society for Minning, Metalurgy, and Exploration, Inc. (SME), 8307 Shaffer Parkway, Littleton, Colorado, USA 80127, ISBN 0-87335-167-3 (2003)

Richter, M.: Průmyslové technologie—úvod, Skripta, Fakulta životního prostředí Univerzity J.E. Purkyně, Ústí n.L. (2002) (in Czech)

Šebor, G.: Těžba a úprava nerostných surovin, Skripta, ČVUT v Praze, Ediční středisko ČVUT, Praha (1983) (in Czech)

Zbirovský, M., Michálek, J.: Základy technologie, SNTL Praha (1983) (in Czech)

Bárta, R.: Chemie a technologie cementu, Nakladatelství ČSAV, Praha (1961) (in Czech)

Найбороденко, Ю.С., Касацкий, Н.Г., Сергеева, Е.Г., Лепакова, О.К.: Влияние механической активации на высокотемпературный и фазообразование низкокалорийных интерметаллических соединений, Химия в интересах устойчивого развития 10, 199–204, Томск, Россия (2002) (in Russian). http://www.sbras.ru/PSB/phsb/papers/CSD2002-2-27.pdf

Hanykýř, V., Kutzendörfer, J.: Technologie keramiky, Silis, Praha (2002) (in Czech)

Hlaváč, J.: Technologie silikátů, SNTL, Praha (1988) (in Czech)

Austin L.G., Trass, O.: Size reduction of solids crushing and grinding equipment. Chapter 12 in “Handbook of Powder Science and Technology”, 2nd edn, pp. 586–634. Chapman & Hall (1997)

Серго, Е.Е.: Дробление, измельчение и грохочение полезных ископаемых, Справочник по обогащению руд, Подготовительные процессы, под ред. В. А. Олевского (и др.), 2 изд., Москва (1982) (in Russian)

Snow, R.H., Allen, T., Ennis, B.J., Litster, J.D.: Size Reduction and Size Enlargement (2009). http://chc.edu.vn/forum/uploads/file//2009-08-16_092142_Chap20.pdf

Pavement Modification Using Enzymatic Lime

Greeshma Nizy Eujine(✉), S. Chandrakaran, and N. Sankar

Department of Civil Engineering, National Institute of Technology, Calicut, India
nizy_123@yahoo.com, {chandra, sankar}@nitc.ac.in

Abstract. Soil samples containing different clay fractions were subjected to lime stabilization, enzyme stabilization and enzymatic lime stabilizations in the laboratory. Unconfined Compressive Strength Tests were conducted to determine the optimum dosages of stabilizing agents. California Bearing Ratio Tests were then performed on samples stabilized with optimum dosages of lime, enzyme, and enzymatic lime respectively. These stabilized samples were tested at different periods of curing and showed that enzymatic lime stabilized soils delivered comparatively higher strengths. Based on the results of CBR tests, the modified thickness of flexible pavement is also discussed.

1 Introduction

Highway construction often requires ex-situ soil or modified in-situ soil to satisfy the strength criteria. The conventional additives used to modify the in-situ soil include lime, cement and fly ash. Lime has been used a stabilizer since the 4th century. But its effect on the environmental is damaging and thus its use is discouraged nowadays. Its caustic effect is more predominant than it being cheap and easily available (Little 1995; Najmiah and Razap 2007). A possible solution is to reduce the amount of lime addition during stabilization by substituting with other agents. The current study attempts to partially replace lime with an enzyme. Since enzymes are biodegradable, they are environment-friendly (Shankar et al. 2009; Shukla et al. 2003), and the risk of pollution is avoided. Moreover, enzymes have been proven to increase the durability of soil subgrades. While numerous research works are and have been done to understand the behavior of soil-lime mixtures in the presence of other salts/chemicals, no literature has been found on enzymatic lime soil stabilization. However, a number of case studies have been reported (Vedula et al. 2002) stating that enzymes by itself improve soil properties. The scope of the paper includes observation of the changes in CBR values of cured 'lime and enzyme' treated soil systems and the possible savings in pavement design and construction.

2 Mechanism of Stabilization

The utilization of lime in soil modification is a traditional means in a variety of construction applications since the time of Romans and has never entirely disappeared. Lime modifies the soil by cation exchange mechanism rather than the cementing action

L. Mohammad (ed.), *Advancement in the Design and Performance of Sustainable Asphalt Pavements*, Sustainable Civil Infrastructures, DOI 10.1007/978-3-319-61908-8_16

brought by pozzolanic reaction (Sherwood 1993). Certain authors suggest that lime forms a binder with silicate clay particles (Rogers and Glendinning 1997; Khalid et al. 2014) alters the clay surface mineralogy, producing a reduction in plasticity and moisture holding capacity, and an improvement in soil stability. But there are disadvantages to lime stabilization which include lime carbonation and sulfate salt reactions which may lead to the disintegration of bonds on aging. To account for the negative impacts, a number salts and chemicals have been added with lime to soil and tested. Among these already tried and proved agents include cement, fly ash, rice husk, etc.

In this paper, an effort is made to study the effect of the enzyme in lime stabilization. Enzymes are hydrophilic organic catalysts that stimulate very certain chemical reactions if conditions are favorable to the reaction (Choudalakis and Gotsis 2009). Since they do not actively participate in the reaction, very low concentrations suffice. Enzyme additives attach with large organic molecules that are attracted to the clay mineral's net negative surface charge. Once compacted with an enzyme, the soil loses its capacity to reabsorb water and thereby increasing its mechanical benefits (Tingle et al. 2007). Since lime and enzyme use the same mechanism of cation exchange to improve soil properties, the idea to add both lime and an enzyme together - or in other words 'enzymatic lime' - in the soil and to investigate the alterations in soil CBR seemed feasible.

3 Materials Used and Methodology

The soil specimens utilized in the work are prepared using a kaolinitic native soil procured from a place 10 km south of Calicut. Lime used in the study was purchased from local market. The bio-enzyme used is an extract from sugar molasses, which was also available in the local market.

The natural soil was air dried for a week, pulverized manually, sieved through 4.75 mm sieve and preserved in large open containers in an enclosed room. This was considered as one soil specimen. Two sets of additional soil specimen were prepared by wet sieving natural soil in 75-micron sieves to extract fines, and adding it to non-sieved native soil specimen to increase the percentage of clay fines – LC1 and LC2. The three different soil specimens and enzyme have been characterized, the properties of which are given from Tables 1 and 2.

Remolded samples of all soil specimens at OMC and MDD were prepared at different dosages of lime, using the procedure specified in Indian Standard (IS 2720 (Part 10): 1991). The samples were cured in labeled plastic bags and transferred to desiccators to prevent moisture loss. The percentage of lime in the test sample which showed the highest average strength improvement in 1, 2, 3 and 4 weeks of curing during unconfined compressive strength tests was considered as the optimum lime content of the soil. Similarly, optimum enzyme and enzymatic lime dosages were also determined.

For studies on CBR strengths, the different soil specimens were prepared with lime, enzyme, and enzymatic lime, at respective optimum stabilizer percentages and OMC using procedure given in (IS 2720 (Part 16): 1987). This cast and cured test specimens

Table 1. Engineering properties of test soils

S.no.	Property	NS (20% clay)	LC1 (40% clay)	LC2 (60% clay)
1	Specific gravity	2.6	2.5	2.5
2	Soil classification	CH		
3	Liquid limit	79	98	130
4	Plastic limit	48	56	72
5	Shrinkage limit	27	34	38
6	Max. dry density (kN/m^3)	17.5	16.2	15.4
7	Optimum moisture content (%)	32	44	46
8	Unconfined compressive strength of 425 microns sieved samples	60	128	65
9	Unconfined compressive strength of unsieved samples	50	70	45
10	CBR (dry)	3.6	6	8

Table 2. Engineering properties of enzymes

Property	Value
Boiling point	212 F
pH	2.8–3.5
Vapor pressure (mmHg)	As water
Melting point	Liquid
Vapor density (air = 1):	1
Solubility in water	Infinite
Evaporation rate	As water
Specific gravity	
(H2O = 1)	1.00–1.10
Appearance and odor	Lt. gold liquid, characteristic odor
Evaporation rate	As water

were subjected to California Bearing Ratio (CBR) tests to evaluate the effect of stabilizers on CBR property of the soil. This paper discusses in detail the variation in CBR property of treated soils.

4 Results on UCS

The unconfined compressive strength test results on natural soil showed that an addition of 3% lime was optimal for strength improvement. Work done by certain other authors (Bell 1996; Kassim 2000; Eren and Filiz 2009) agree on similar percentages of optimum lime content. Likewise, optimum enzyme dosage was obtained as 80 ml/m^3 (Isaac et al. 2003). The optimum enzyme lime mixture was obtained as 70 ml/m^3 of

Enzyme and 1.75% of lime. The results on natural soil are represented in Figs. 1, 2 and 3. Similar results were found on the treatment of control specimens with the stabilizers. Hence the same optimum dosages were used in the natural soil as well as control specimens for CBR tests.

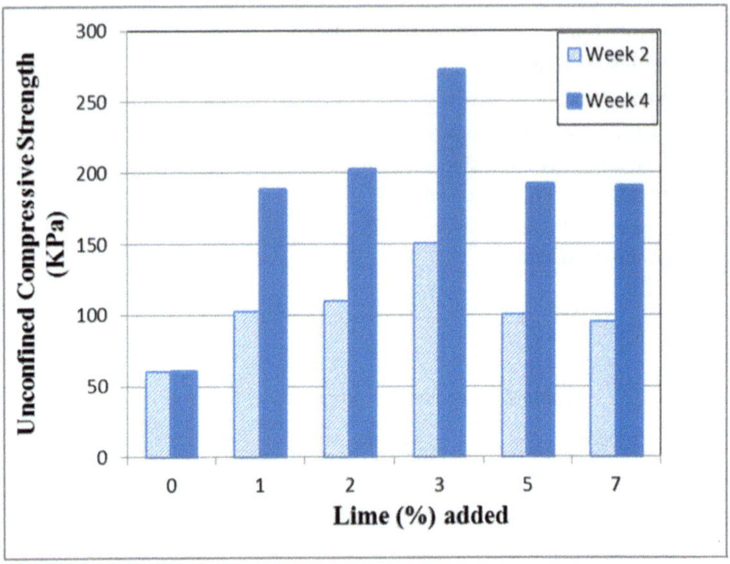

Fig. 1. Selection of optimum lime dosage

5 CBR on Natural Soil

On curing and testing of the natural soil as - soil lime, soil enzyme, and soil enzyme lime mixtures at optimum dosages and at respective OMC, it was observed that the CBR values increased in comparison to the CBR values of untreated soils. Figure 4 describes this improvement. It is seen that the CBR values increased up to 5 times when treated with lime alone and up to 3 times when treated with enzyme alone under dry conditions. An increase of up to 4 times in the case of lime stabilization (Modarres and Nosoudy 2015) and 15 times in the case of enzyme stabilization (Rubens and Sheldon 2002) have been observed in other studies. When treated with enzymatic lime, an increase of more than 6 times was seen in dry CBR test. What should, moreover, be pointed out is that after two-week curing, while only less than two times the improvement was obtained with lime and enzyme stabilization, nearly 4 times improvement was achieved in dry CBR under enzymatic lime stabilization. Similar observations were also found in soaked samples. Thus the results of the current study verify that enzymatic lime has significant benefits as compared to the lime stabilization and enzyme stabilization. The time saved, construction materials saved and economic benefits (Eujine et al. 2016) can be presumably high.

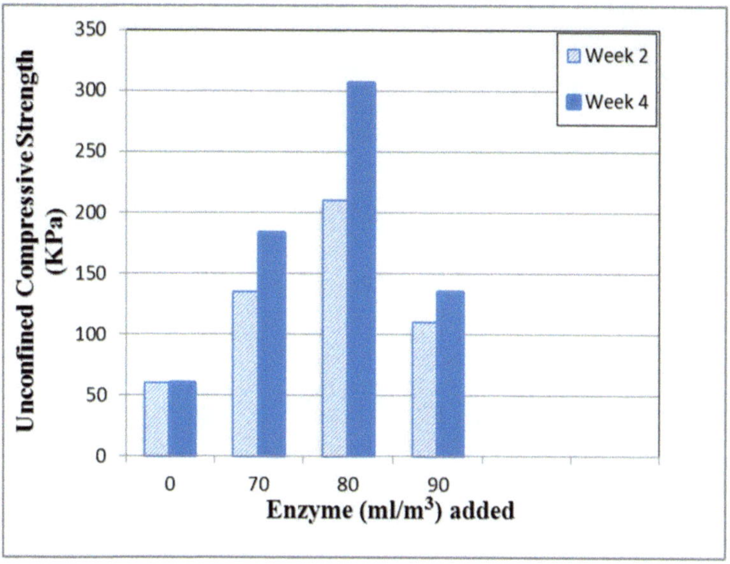

Fig. 2. Selection of optimum enzyme dosage

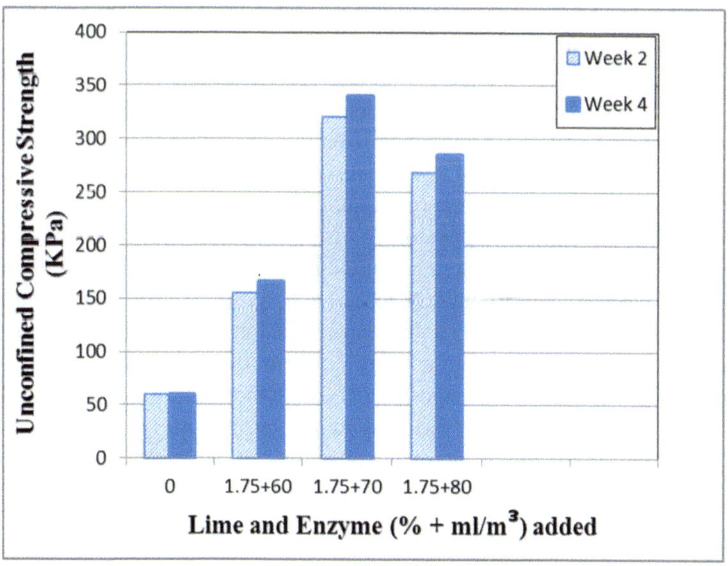

Fig. 3. Selection of optimum enzymatic lime dosage

The Kaolinitic natural soil has responded quite well to all three additives used in the study. The lime in the presence of enzyme has enhanced the soil properties better than in when stabilized with lime alone. It is assumed that the biopolymer-induced aggregation of clay particles has formed a clay-polymer interconnected network through

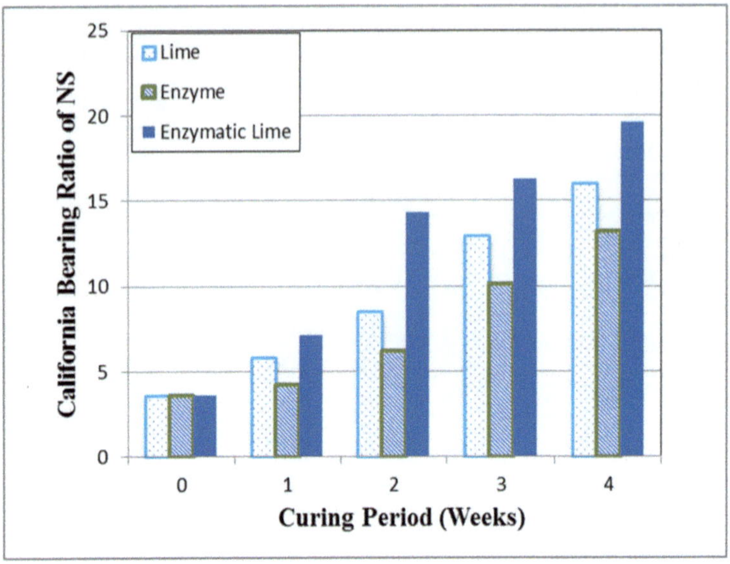

Fig. 4. Improvement in CBR values of natural soil on curing with stabilizers

cation bridging and hydrogen bonds. In other words enzyme proteins adsorbed by the clay particles, thus modify the clay particle (Pinck 1960). Enzyme primarily attaches to the clay substrate, alters clay molecule and later reattached itself from the modified clay by returning to its original form after the reaction is concluded (Pfeiffer 1954). Here when the newly modified soil is combined with lime the cation bonds may be formed at an accelerated rate, bringing about an increase in strength in shorter duration. This may also be due to the low activation energy required for the modified clay-lime reactions.

An area of concern was the denaturing of the enzyme in the presence of high pH lime and high lime-soil reaction temperatures. Extremely high or low pH/temperature values generally bring about a complete loss of activity for most enzymes. But it is also known that when an enzyme-substrate complex is formed, the enzymes convert the local conditions in the reaction site entirely different to those outside the reaction site (Cooper 2000). This way the changes in pH and temperature do not hinder the clay modification. The same may be assumed in the current study. If the enzyme had become denatured, the marked improvement might not have occurred.

6 Results on Control Specimen

Studies on control specimen were conducted in order to determine the effectiveness of the system when used on soil containing different clay fractions. Figures 5 through 6 show the improvement in CBR value of the lab control specimens when treated with lime, enzyme, and enzymatic lime. The results of CBR tests indicate that like in the case of lime and enzyme stabilization, an increase in clay content enhanced the effectiveness of enzymatic lime as a stabilizer. The presence of more clay particles

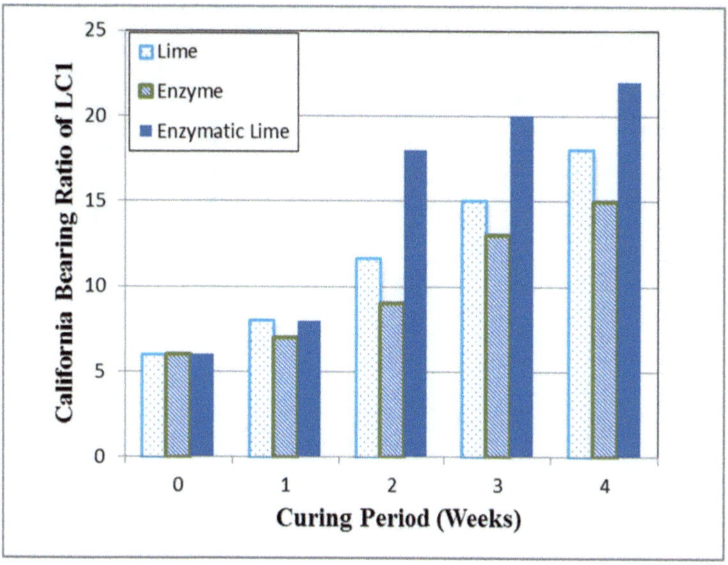

Fig. 5. Improvement in CBR values of LC1 on curing with stabilizers

means more modified clay, more pozzolanic activities with lime and hence higher strength. However, LC2 showed lower CBR gain factor (Sahu 2001; Che Mamat 2013) than LC1.

It can be observed from Fig. 5 that for C-40, enzymatic lime gives a substantial improvement in CBR of treated soil in both dry and soaked conditions. Lime treated soil gave comparatively lower CBR strengths even under dry tests. The strength improvement due to enzyme treatment was much less than enzymatic lime stabilized soils. Although exposure to water has significantly reduced the CBR of all treated soils, enzyme treated soils were affected the most.

In this study highest strength was obtained for C-40, the control specimen with 40% clay fines. It is evident from Fig. 6 that strength gain is not proportional to the clay content. C-60 exhibited lower degrees of strength improvement on treatment with lime, enzyme, and enzymatic lime as compared to C-40. However, the maximum strength improvement was still obtained for enzymatic lime stabilized specimens. Similar to the behavior of enzymatic lime stabilized natural soils, the control specimens also show increased rate of strength improvement and thus a much higher strength within two weeks.

The general trend of CBR test indicates that there is the greatest increase in strength was attributed by the stabilizers when clay content varies from 35% - 50%. When the clay content is too low, adequate strength improvement does not develop due to the majority untreated silt and sand fraction in the soil. Moreover more moderate clay content delays the drying of the soil at the site (NLA 2004). And whenever clay content is too high, the improvement imparted is resisted by the self-flocculated clay (Cao et al. 2012) that prevents cation exchange between admixture and soil. Besides higher

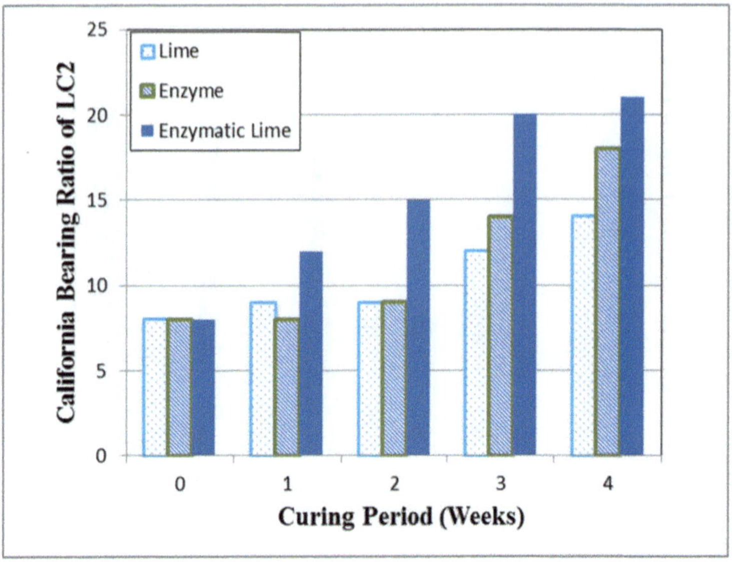

Fig. 6. Improvement in CBR values of LC2 on curing with stabilizers

percentage of clay will dry the soil during curing (Ragassi 1995), displaying lower strengths and brittle behavior. Similarly, during lime stabilization, the optimum proportion of lime addition varies with clay content (Trivedi et al. 2013) and clay mineral. But the use of lime in the presence of enzyme demonstrates that only a minimum amount of lime is required to attain necessary strength modification. The catalyst enzyme thus reduces exposure of high lime percentages to the environment.

Enzymatic lime was found to be effective for soil with a minimum clay content of 20% in this study. And as soils with clay content as a minimum, as 10% were proven to be stabilized with enzymes (Bergmann 2000), this proposed system of stabilization may be applicable to soils with 10% clay fines as well. Physical evidence from the study suggests that enzymatic lime stabilized samples did not become dry on curing, as in the case of lime stabilization. This apparently indicates a very low heat of hydration required during enzymatic lime stabilization. Hence such stabilized roads may be resistant to this abrading action of traffic on highway surfaces. The common issues like potholes, breaking down and softening of the road surface (Rubens and Sheldon 2002) can be avoided, as enzymatic lime tends to make stabilized soil more durable. Since enzymes have already provided dust control to stabilized soils (Shirsavkar and Koranne 2010), the same may be achieved in enzymatic lime stabilized soils as well.

Like in the case of natural soil, enzymatic lime enhanced the rate of strength gain of all stabilized soils specimens. This is very well evident from the data. The mechanism hypothesized is that enzyme modifies the clay and it is the modified clay that interacts with lime, bringing an amplified improvement to soil properties. The same is observed in CBR test results as well. During lime stabilization, the formation of calcium silicate hydrate and calcium aluminate hydrate gels as a result of pozzolanic reaction and their subsequent crystallization to bind the structure together (Tran et al. 2014) takes place after few days of lime addition only (Bergado et al. 1996). But when the enzyme was mixed with lime, not only was the initial delay avoided, a noticeable increment in the rate of strength improvement was also seen.

7 Pavement Design

A flexible pavement was designed based on IRC-37 (IRC 2001). The modified thicknesses for a pavement designed for the stabilized soils assuming medium traffic is given in Table 3. And it is observed that there will be considerable savings in subgrade materials when stabilization is done with enzymatic lime as compared to stabilization with lime and enzyme respectively. The clay fraction in the soil also plays a significant role in determining the effectiveness of stabilizer. In this case, however, the maximum savings were found in the case of natural soil containing 20% clay fines.

Table 3. Modified pavement thicknesses

Sample	Untreated soil	Lime treated	Enzyme treated	Enzymatic lime treated
	Approx. thickness of granular material in mm			
Natural soil	560	220	240	180
LC1	345	185	220	170
LC2	300	230	185	170

8 Conclusions

1. Enzymatic Lime Stabilization of soils can be used in improving the bearing capacity of the subgrade, with noticeable savings on both aggregate and disposal charges
2. A reduction up to 40 mm was noticed on treating the soil with enzymatic lime as compared to soils treated with lime alone – in the case of natural soil. A reduction of
3. Economic advantages of using enzymatic lime stabilization technique include:
   a. Using local soil as much as possible
   b. Avoiding transportation cost
   c. Reduction in construction time
   d. Reduction in construction materials

References

Bell, F.G.: Lime stabilization of clay minerals and soils. Eng. Geol. **42**, 223–237 (1996)

Bergado, D.T., Anderson, L.R., Miura, N., Balasubramaniam, A.S.: Soft Ground Improvement in Lowland and Other Environments. ASCE Press, New York (1996)

Bergmann, R.: Soil Stabilizer on Universally Accessible Trials. United States Department of Transportation and Federal Highway Administration (2000)

BIS.: Determination of Unconfined Compressive Strength, IS 2720 (Part 10): 1991 (1991)
BIS.: Laboratory Determination of CBR, IS 2720 (Part 16): 1987 (1987)
Cao, L., et al.: Flocculations of organisms with clay. In: Proceedings of the International Conference on Applied Biotechnology, Tianjin, China on October 18–19, 2012, 1 (2012)
Che Mamat, R.B.: Engineering properties of Batu Pahat soft clay stabilized with lime, cement, and bentonite for subgrade in road construction. Universiti Tun Hussein Onn Malaysia (2013)
Choudalakis, G., Gotsis, A.D.: Permeability of polymer/clay nanocomposites: a review. Eur. Polym. J. **45**(4), 967–984 (2009). doi:10.1016/j.eurpolymj.2009.01.027
Cooper, G.M.. The Central Role of Enzymes as Biological Catalysts. The Cell: A Molecular Approach, (National Library of Medicine) (2000). http://www.ncbi.nlm.nih.gov/books/NBK9921/
Eren, Ş., Filiz, M.: Comparing the conventional soil stabilization methods to the consolid system used as an alternative admixture matter in Isparta Daridere material. Constr. Build. Mater. **23** (7), 2473–2480 (2009). doi:10.1016/j.conbuildmat.2009.01.002
Eujine, G.N., Sankar, N., Chandrakaran, S.: The engineering behaviour of enzymatic lime stabilised soils. Ground Improv. (2016). doi:10.1680/jgrim.16.00014
IRC.: IRC-37-2001: Guidelines for the design of flexible pavements, 37 (2001)
Isaac, K.P., Biju, P.B., Veeraragavan, A.: Soil stabilization using bio-enzymes for rural roads. In: Proceedings of the Conference by the Indian Road Congress, New Delhi, December 2003 (2003)
Kassim, K.A.: Consolidation characteristics of lime stabilised soil. J. Civ. Eng. **12**(1), 31–42 (2000)
Khalid, N., et al.: The California bearing ratio (CBR) value for banting soft soil subgrade stabilized using lime-pofa mixtures. Electron. J. Geotech. Eng. **19A**, 155–163 (2014)
Little, D.N.: Handbook for Stabilization of Pavement Subgrades and Base Courses with Lime. Kendall/Hunt Publishing Company, Dubuque (1995)
Modarres, A., Nosoudy, Y.M.: Clay stabilization using coal waste and lime—technical and environmental impacts. Appl. Clay Sci. **116–117**, 281–288 (2015). doi:10.1016/j.clay.2015.03.026
Najmiah, N., Razap, B.A.: Effect of Lime on California Bearing Ratio (CBR) of Soft Soils. B.E. Thesis, Universiti Teknologi PETRONAS (2007)
NLA.: Lime-Treated Soil Construction Manual Lime Stabilization & Lime Modification. National Lime Association (2004)
Pfeiffer, J.: Enzymes, the Physics and Chemistry of Life, pp. 171–173. Simon and Schuster, New York (1954)
Pinck, L.A.: Adsorption of proteins, enzymes, and antibiotics by montmorillonite. Clays Clay Miner. 9(1), 520–529 (1960). http://www.clays.org/journal/archive/volume9/9-1-520.pdf
Ragassi, V.: Compressed Earth Block—Manual of Production. Manual of Production: Stabilization, 1 (1995)
Rogers, C.D., Glendinning, S.: Improvement of clay soils in situ using lime piles in the UK. Eng. Geol. 47(3), 243–257 (1997). http://www.sciencedirect.com/science/article/pii/S0013795297000227
Rubens, B., Sheldon, R.M.: Objective Performance Measurement of Actual Road Sites Treated with an Organic Stabilizer. Nature Plus (2002)
Sahu, B.K.: Improvement in California Bearing Ratio of Various Soils in Botswana by Fly Ash. International Ash Utilization Symposium, pp. 22–24 (2001)
Shankar, A.U.R., Rai, H.K., Mithanthaya, R.I.: Bio-enzyme stabilised laterite soil as a highway material. J. Indian Roads Congress **553**, 143–151 (2009)
Sherwood, P.: Soil stabilization with cement and lime. State of the art review. Tech. Rep. Transport Research Laboratory; HMSO, London (1993)

Shirsavkar, S., Koranne, S.: Innovation in road construction using natural polymer. Electron. J. Geotech. Eng. 15, 1614–1624 (2010). http://www.ejge.com/2010/Ppr10.112/Ppr10.112a2r.pdf

Shukla, M., Bose, S., Sikdar, P.: Bio-enzyme for stabilization of soil in road construction a cost-effective approach. IRC Seminar: Integrated Development of Rural and Arterial Road Networks, (December), pp. 1–14 (2003). http://www.avijeetagencies.com/downloads/crrireport.pdf

Tingle, J., et al.: Stabilization mechanisms of nontraditional additives. Transp. Res. Rec. **1989**(1), 59–67 (2007)

Tran, T.D., Cui, Y.J., Tang, A.M., Audiguier, M., Cojean, R.: Effects of lime treatment on the microstructure and hydraulic conductivity of Héricourt clay. J. Rock Mech. Geotech. Eng. **6** (5), 399–404 (2014). doi:10.1016/j.jrmge.2014.07.001

Trivedi, J.S., Nair, S., Iyyunni, C.: Optimum utilization of fly ash for stabilization of subgrade soil using genetic algorithm. Procedia Eng. **51**, 250–258 (2013). doi:10.1016/j.proeng.2013.01.034

Vedula, M., Nath, P., Chandrashekar, B.P.: A Critical Review of Innovative Rural Road Construction Techniques and Their Impacts. National Rural Roads Development Agency, New Delhi (2002)

Numerical Modeling of Heat Production for a Snow-Melting System Using Geothermal Energy in North Dakota

I-Hsuan Ho(✉)

Harold Hamm School of Geology and Geological Engineering,
University of North Dakota, Grand Forks, ND 58202, USA
ihsuan.ho@engr.und.edu

Abstract. This paper is focused on heat production modeling for a proposed snow-melting system using shallow geothermal energy in North Dakota. The paper presents computations for the heat that is required for a pavement snow-melting system based on averaged climatic and precipitation data and using Chapman and Katunich's equation. Based on seasonal heat requirements, finite element analysis using COMSOL was performed to simulate the heat extraction process whereby heat is transferred between the soil and the heat collection pipes in a constant subsurface temperature zone. In addition to the thermal properties of the soil, both the fluid in the pipes and the characteristics of the pipes themselves are shown to be significant factors for the heat extraction process. The results of the numerical analysis also indicate that a high level of thermal conductivity of the soil is desirable for a successful heat collection system. Compared to costly studies that involve full-scale prototypes, simulations via numerical modeling can provide cost-effective insights into the important components of a successful snow-melting system.

Keywords: Heat production · Snow-melting system · Geothermal energy · Finite element analysis (FEA) · Numerical modeling

1 Introduction

According to statistical data obtained by the National Weather Service in 2010, over 200,000 vehicles accidents are caused by weather-related issues annually. Most of these accidents occur in winter. Without doubt, the major problems associated with winter travel are snow and ice on the roads. Hence, if an efficient snow removal system could be developed, especially a system that uses renewable energy, travel in cold regions could be improved dramatically. The effectiveness of a successful snow removal system depends on a combination of several factors, such as climatic conditions, geological conditions, the type of energy used, the availability of heat resources, the design methods employed, and the payback period (Ho and Dickson 2016).

Typical snow removal methods that are currently available include: (a) mechanical removal, (b) chemical (primarily salt) treatments, (c) electric heating, and (d) hydronic systems. However, each of these systems has several disadvantages. For example, removing heavy snow is difficult and can even be dangerous for the operator of the

L. Mohammad (ed.), *Advancement in the Design and Performance of Sustainable Asphalt Pavements*, Sustainable Civil Infrastructures, DOI 10.1007/978-3-319-61908-8_17

snow removal vehicle. Using salt and other chemical treatments to prevent ice buildup involves recurring costs of the chemicals as well as maintenance of the trucks that distribute the chemicals. Salts such as NaCl, $MgCl_2$, and various glycols often are used for such chemical treatments (Ekolist 2001). Their frequent use can weaken and damage asphalt and concrete over time, decreasing the longevity of roads and pavement (Melcher 2001; Ho and Dickson 2016). Furthermore, after the snow melts, the chemical treatment runoff can cause local environmental damage, harming plant life and polluting local water systems. Chemical usage not only damages the environment and roads but also vehicles. Moreover, if the temperature falls below −3.9 °C, chemical treatments cannot melt the snow at all (Liu et al. 2007). Expensive electric coils or water-boiler (hydronic system) heating systems can heat concrete and melt snow and can respond fairly quickly to snowfalls. However, such systems are expensive to install and maintain. In addition, these systems have high recurring energy expenses, can be inefficient in cold climates, and consume large amounts of either electricity or fuel that create additional CO_2 emissions.

In addition to the type of energy supply that is employed, related environmental issues are big concerns. Due to global climate change, many industries are increasingly turning to renewable energy, such as geothermal, wind, and solar energy. In comparison to electric or boiler-based systems, geothermal systems use about 2 to 10% of the energy. The Kyoto Protocol targeted a CO_2 reduction of 4.2% (Olivier et al. 2011) over the next five years. Thus, geothermal energy can serve as a good alternative energy resource to support a heating system.

Some geothermal applications have been initiated in Sapporo, Japan (Nagai et al. 2013), Iceland, Poland (Zwarycz 2002), and Alaska in the United States. However, each area has different climatic conditions, available geothermal resources, environmental concerns, and geological conditions. The weather conditions in the State of North Dakota (the focus of this study) are unique compared to other states in the United States. For example, the energy required to melt snow is 25% more in North Dakota than in Alaska under similar circumstances (Zarling 1995; Rees et al. 2002; Ho and Dickson 2015). The average temperature in Fargo and Grand Forks in North Dakota is −7.2 °C according to the National Oceanic and Atmospheric Administration (NOAA 2015) for the four coldest months from November to February based on data for 2000 through 2015. The temperature of the fluid that must be circulated in pipes to melt snow has been estimated to be 7 °C (Gosnold 2012).

Several studies have focused on geothermal snow-melting systems; however, very little research has been conducted that focuses on such a system specifically for North Dakota. Given North Dakota's unique climate and geological environment, particularly when using low-temperature heating fluids, the objective of this study is to evaluate the feasibility of developing a snow-melting system for pavements in North Dakota using shallow geothermal energy. This paper presents computations for the heat that is required for a snow-melting system in North Dakota based on averaged climatic data and precipitation using Chapman and Katunich's equation (1956). The study focused on heat production modeling using heat collection pipes and a constant subsurface temperature zone. Based on the seasonal heat requirements, finite element analysis (FEA) was used to simulate the heat extraction process using the heat collection pipes. The measured data show that the temperature stabilizes at 7 °C 6 m below the ground surface regardless of the season. The heat is transferred between the soil and the pipes, and both the fluid in the pipes and the characteristics of the pipes themselves are significant for the heat extraction process. The results of the numerical analysis also show that high levels of thermal conductivity are desirable for a successful heat collection system.

2 Parametric Study of Heat Requirements

A parametric study was conducted to determine the heating requirements based on climatic conditions and precipitation for the past 15 years in North Dakota. Several studies of snow-melting systems have been well documented, and many of the proposed designs are based on steady-state conditions. Variables such as the snowfall rate, ambient air temperature, and wind speed often are considered based on weather data compiled over several years, and usually these values represent an average or upper value during a snowfall event. Equation (1) presents the Chapman and Katunich equation (1956) that assumes steady-state conditions and is used to compute the amount of heat per square foot that is needed to melt snow.

$$q_o = q_s + q_m + A_r * (q_h + q_e) \tag{1}$$

where q_o = total heat flux per unit area of the surface, Btu/h ft^2 (W/m^2); q_s = total sensible heat flux, Btu/h ft^2 (W/m^2); q_m = melting load, Btu/h ft^2 (W/m^2); A_r = snow-free area ratio; q_h = sum of the convection and radiation loss, Btu/h ft^2 (W/m^2); and q_e = evaporative loss, Btu/h ft^2 (W/m^2).

In order to use this equation to estimate the heat required to melt snow, a few assumptions must be made regarding the precipitation rate, area ratio of snow coverage, etc. Even the equation itself has some limitations; however, it can still provide a reasonable estimation of the heat required for a snow-melting system. Estimations of the heat requirements are based on the transient weather conditions and climatic conditions, discussed in the following sections.

2.1 Transient Weather Conditions

Most snowfall events happen over the course of several hours, and during that time the precipitation rate, ambient air temperature, humidity, wind speed, and solar radiation can change enough to affect the performance of a snow-melting system. An ideal system would use the least amount of energy possible to melt the snow and ice as soon as the precipitation forms on the surface. Some systems use moisture and temperature sensors that automatically activate in response to freezing conditions and precipitation, whereas others systems are on a timer or manually activated.

When a snowfall event begins, the rate of the snowfall usually increases until the peak snowfall rate is reached. A system can be designed so that it melts the snow as it hits the ground by heating the pavement and preventing the snow from forming on the surface. Preventing snow from sticking to the pavement has several advantages; for example, it insures that driving conditions will be safer throughout the snowfall event, and it prevents the snow from creating an insulating layer that interferes with the melting process. The snowfall rate, temperature, and several other factors affect the amount of energy that is needed to keep the surface free of snow and ice. The water content of the snow also affects the efficiency of the system, as not all amounts (inches) of snowfall require the same amount of thermal energy to melt the snow (Ho and Dickson 2016).

2.2 Climatic Conditions in North Dakota

The climatic conditions constitute the most important influential factor when designing a snow-melting system. The air temperature, rate of the snowfall, frequency of the snowfall, moisture content of the snow, wind speed, and amount of sunlight all control the rate at which the snow will melt. The data for the climatic conditions in North Dakota that were used for the proposed snow-melting system design are from the NOAA. These data were averaged over a 15-year period from 1999 to 2014 (NOAA 2015). A snow-melting system requires climatic condition information that is obtained over the months of November to February of the next year. These winter months typically have the most snowfall and lowest temperatures of the year, when the snow is likely to stick to the pavement and affect transportation safety.

Figure 1 presents the averaged 15-year air temperatures for the cold months from 1999 to 2014. The coldest month-long period in North Dakota is between January of one year and February the following year. Major snowfalls usually start in November of one year and continue (intermittently) through February of the next year. Snow-melting design thus should take into account various climatic conditions to optimize the use of the available geothermal energy. The worst climatic conditions should not be the first priority for a snow-melting system design. However, the principles of the design should consider a system that can accommodate the most critical periods within a year. Other significant climatic factors, such as wind speed, snowfall rate, humidity, and moisture content also need to be considered in the design. Guidelines should be proposed wherein the magnitudes are dependent on the significance of the infrastructure.

2.3 Heat Requirements for Pavement Surface

According to Eq. (1) and Fig. 1, the lowest average temperature, −10.6 °C, occurs in January, and the highest average temperature among the four coldest months is −1.7 °C. Assuming a wind speed of 15 m/s, the free area ratio of snow coverage as 0.85, and the rate of snowfall as 0.1 in./h, the estimated required heat to melt snow is 2,540 BTU/h-m^2. If the temperature used is −1.7 °C and the other climatic conditions remain the same, the estimated heat required is 1,713.6 BTU/h-m^2.

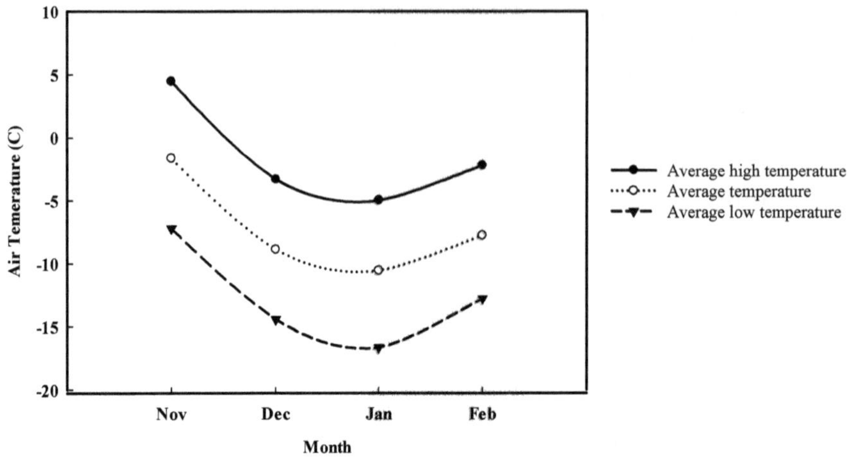

Fig. 1. Average temperatures during cold months in North Dakota.

3 Available Geothermal Resource for Heat Production

The heat production mechanism for geothermal extraction using pipes is complicated because many factors are involved. These influencing factors include the diameter of the pipe, properties of the pipe, flow rate of the fluid in the pipe, the specific heat of the fluid, the thermal conductivity of the soil, etc. Also, the heat transfer mechanisms between the soil and the pipe as well as between the pipe and the fluid are difficult to model. In this study, COMSOL finite element software was employed to simulate the heat production of geothermal energy by modifying an existing heat production model. Several parameters must be input for the FEA, as presented in Table 2. The major governing factors for heat production are controlled by the subsurface temperature, thermal gradient, heat collection pipes (including pipe properties and fluid), and soil properties. Details are given in the following sections. Note that this study considers the heat collection pipe only, not the 'surface' heat pipe that is located immediately below the pavement surface.

3.1 Subsurface Temperature

The available geothermal resource is controlled mainly by a stable subsurface temperature. Figure 2 presents the nationwide subsurface temperature distribution for the United States. The subsurface temperature range in North Dakota is between 5 and 10 °C. Figure 3 presents a series of locally collected data that were measured from a well near Grand Forks in Eastern North Dakota between August 2011 and March 2012. The temperature was found to stabilize at 7 °C with a depth of more than 6 m, regardless of the air temperature. Although the data collected in this area over a 10-year period (2003 to 2012) are limited, they nonetheless provide consistent results, as shown in Fig. 2.

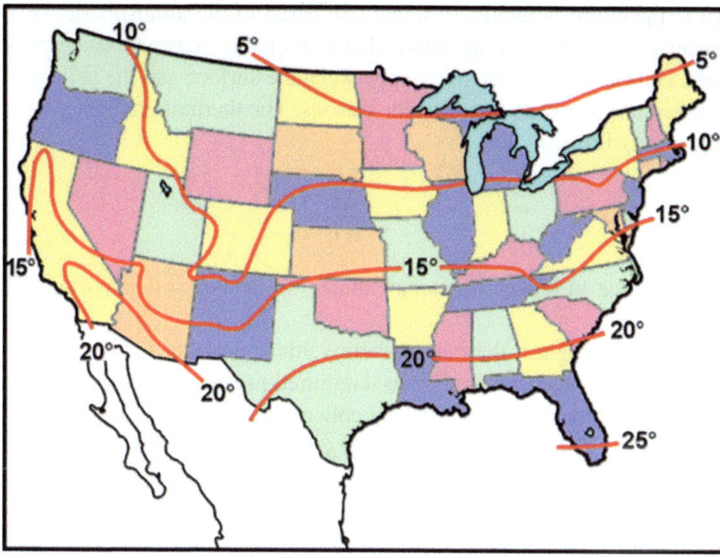

Fig. 2. Subsurface temperature distribution in the United States (after Collins 1925).

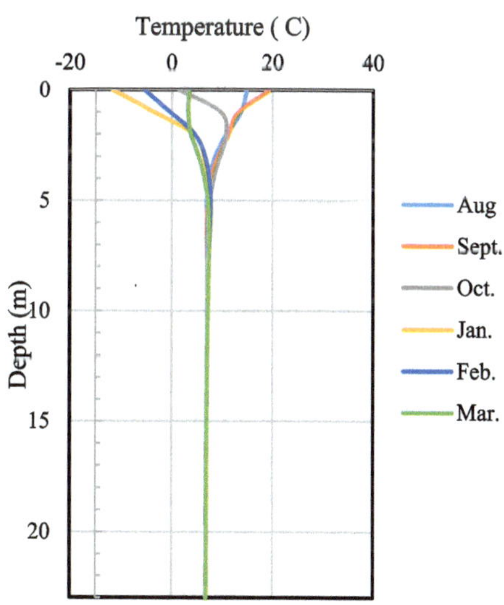

Fig. 3. Subsurface temperature distribution.

3.2 Thermal Gradient

The thermal gradient from the ground surface to the depth of geothermal extraction varies and is subject to seasonal change. In Fig. 3, the subsurface temperature is shown to stabilize at about 7 °C at a depth of 6 m. Although a slight drop in the ground temperature occurs during the coldest months (November to February), the overall subsurface temperature is relatively stable compared to the atmospheric weather conditions because the surface temperature depends on the seasonal changes of the air temperature. Hence, the thermal gradient between the surface and the soil layer that has a stable temperature varies in the different seasons. The thermal gradient, i, is expressed mathematically as Eq. (2):

$$i = \frac{\Delta T}{D} \tag{2}$$

where ΔT = change in subsurface temperature, °C; and D = depth from the ground surface, m.

Accordingly, the highest thermal gradient value appears in January and the lowest appears in March. The thermal gradient is a required parameter for numerical modeling and must be defined for the FEA for a heat collection system. For January, the average temperature used is −10.7 °C and the thermal gradient used in the computation is 2.95. If the average air temperature in November is −1.7 °C, the thermal gradient would be −1.45.

4 Heat Collection Pipes

Figure 4 presents a snow-melting system scenario that uses shallow geothermal energy. In this snow-melting system, the heat is collected through heat collection pipes and circulated to the surface. The heat collection pipes are usually embedded horizontally in the constant temperature layer and the fluid in the pipes is circulated to the surface through a heat pump. The warm fluid delivers heat to the pavement surface to melt the snow and then returns to the ground to collect heat again.

In this simulation, the heat collection pipe was placed 6 m deep and horizontal to the ground surface. As aforementioned, the thermal gradient varies in different seasons. Thus, the thermal gradient used in the numerical model is based on the temperature as the subsurface temperature increases linearly from the ground surface to a depth of 6 m. The thermal gradient, i, is calculated according to Eq. (2).

Based on the heat requirement calculated previously, the range of the required heat is between 1,713.6 BTU/h-m^2 and 2,540 BTU/h-m^2. As long as the heat that is extracted from the shallow geothermal layer can reach the heat required, the system is able to melt the snow on the surface during different months and for different requirements.

The arrangement of the heat collection pipes is based on the pipe properties, soil properties, and boundary conditions. For this study, the heat pump power was assumed to be 4 kW. A total of 48 kWh of heat can be produced in a day. Based on the heat required for a duration of 12 h, the total heat required is 6.03 kWh/m^2 for November and 8.93 kWh/m^2 for January. Thus, the area to be melted for a snow-melting system can be calculated accordingly.

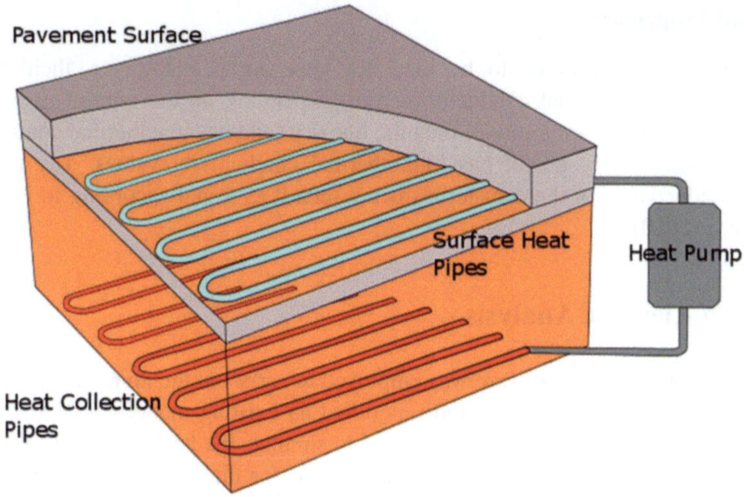

Fig. 4. Profile of a snow-melting system that uses shallow geothermal energy.

4.1 Pipe Properties and Fluid

The heat collection pipe parameters include the pattern of the pipe arrangement, the size of the pipes, and the material of the pipes. In this study, each pipe has a 'snake' shape, as shown in Fig. 5. The PEX pipes used for heat collection are made of cross-linked high-density polyethylene. The thermal conductivity of the pipe material is 0.51 W/mK. The fluid used in the pipes is an ethylene glycol mixture that has a low freezing point of −36.8 °C so that the fluid can prevent the pipe from freezing when the fluid is circulated to the surface. The ratio of specific heat to water is 0.8 under most temperatures. The length of the pipe embedded in the constant temperature layer that is considered for the model is 94 m. The diameter of this pipe is 24.5 mm.

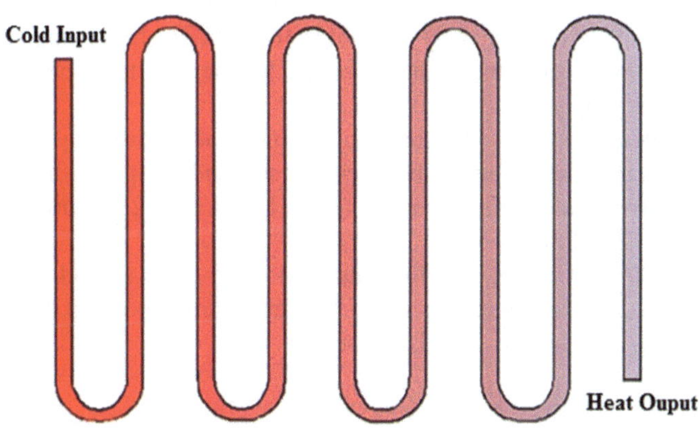

Fig. 5. Snake pattern of heat collection pipe.

4.2 Soil Properties

Geomaterials are considered to be ideal for heat storage. However, their thermal properties, such as thermal conductivity, vary depending on the type of geomaterial used and the degree of saturation. For soils, fine dry sand has low thermal conductivity values and silty sandy gravel has high thermal conductivity values. The range of thermal conductivity for these materials is between 0.15 to 4.4 according to Hamdhan and Clarke (2010).

5 Finite Element Analysis

COMSOL finite element software was employed to carry out the numerical analysis of the heat collection process. This study modified an existing model (COMSOL 2016) called the 'geothermal heat recovery system' to simulate the heat produced from pipes for a snow-melting system. The input temperature for the heat collection pipes was expected to be 0 °C and the outlet temperature for the heat collection pipes was consistent or close to the subsurface temperature of 7 °C. Table 1 provides a summary of the other parameters used in the FEA.

Table 1. Finite element analysis parameters for proposed heat collection system

Diameter of pipe	24.5 mm
Flow rate of pipe	1.5 L/s
Daily heat demand	48 kWh
Power of heat pump	4 kW
Depth of embedment	6 m
Thermal gradient of subsurface	1.5 K/m
Moisture content of soil	100%
Thermal conductivity of soil	1.51 W/mK

The dimensions used for the soil mass are 22 m length, 15 m width, and 9 m depth. The heat supply required for the daily demand is 48 kWh. The heat pump is required to operate 12 h per day during the highest demand for the heat to reach the daily heat demand. The heat supply is 1,3750 BTU/h.

In a heat collection system, the change in the thermal conductivity of the soil will significantly affect the heat transfer between the soil and the pipes. In the FEA, the thermal conductivity values are assumed to be 1.5 W/mK, 2.0 W/mK, 2.5 W/mK, 3.0 W/mK, and 4.5 W/mK.

In the finite element model, the flow rate used is 1.5 L per second. When the heat pump reaches the daily heat requirement, the heat extraction process is stopped, and the flow reduces to 1/10, which is 0.15 L per second.

5.1 Analysis Results

Figure 6 shows the temperature change of the fluid in a pipe at t = 1 day. The flow was maintained at a range of temperatures between 6 °C and 6.6 °C in the stable subsurface temperature zone.

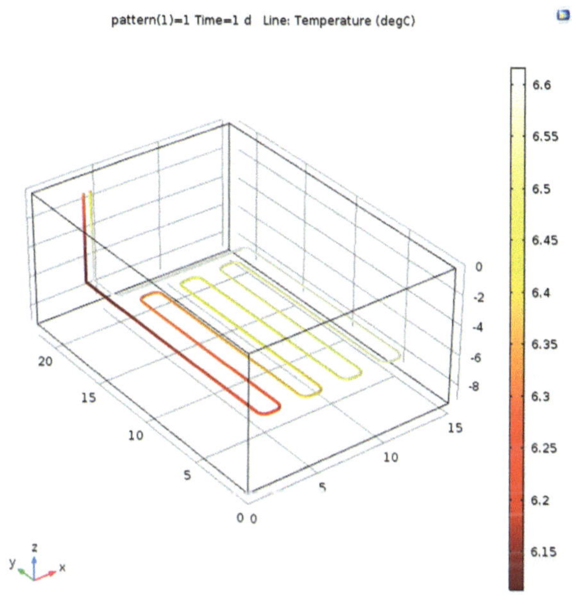

Fig. 6. Temperature change of fluid in pipe for heat collection system at t = 1 day using FEA.

Figure 7 presents a comparison of the outlet temperatures among several types of soils that have different thermal conductivity values. The results indicate that soil with a higher thermal conductivity value is able to maintain a higher outlet temperature at all times during a day than soil with a lower thermal conductivity value. When the heat extraction stops, the low flow rate is able to provide an outlet temperature that is close to the stable subsurface temperature. For soils that have high thermal conductivity values, the temperature of the fluid in the pipe stays high also and is close to the subsurface temperature. In contrast, if a soil has low thermal conductivity values, a relatively lower temperature of the fluid is maintained at the outlet of the pipe during the thermal extraction process.

This paper focuses on heat production using geothermal heat collection pipes in a stable subsurface temperature soil layer. The heat production is based on the low-temperature fluid that circulates through the heat collection pipes. In the heat collection pipes, the inlet temperature of the fluid is assumed to maintain a pavement surface temperature that is equal to or higher than 0 °C. As aforementioned, the heat delivery pipe just below the surface is not considered along with the heat collection pipes in this paper.

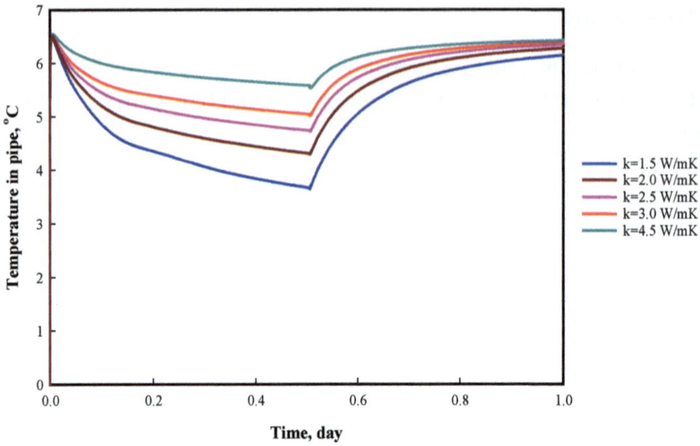

Fig. 7. Temperature change of fluid in pipe for heat collection system at t = 1 day using FEA.

Based on the numerical analysis results, different types of soil and degrees of saturation have a significant effect on thermal conductivity. The thermal conductivity of a geomaterial in turn affects the effectiveness of the heat collection process in a snow-melting system. For practical engineering work, a geomaterial with a high thermal conductivity value is desirable for an efficient heat collection system.

6 Conclusions

In order to extract geothermal energy, the heat collection pipes play a significant role in a snow-melting system. The heat production and the outlet temperature of the fluid in the pipes can be simulated via FEA. Several conclusions can be drawn from this study, as follows:

(1) The soil in which the heat collection pipes are embedded should have high thermal conductivity values.
(2) The heat collection pipes should be embedded in a material that has higher thermal conductivity values to maintain a higher fluid temperature in the heat collection pipes. Practically speaking, the subsurface soil can be replaced by geomaterial that has high levels of thermal conductivity.
(3) Theoretically, the use of low-temperature fluid for a snow-melting system is feasible. However, the area to be melted is subject to the heat requirement.

Acknowledgments. This research was supported by funding from the National Science Foundation North Dakota Experimental Program to Stimulate Competitive Research (ND EPSCoR).

References

Chapman, W.P.: Design of snow melting systems. Heating and Ventilating (April): 95 and (November): 88 (1952)

Chapman, W.P., Katunich, S.: Heat requirements of snow melting systems. ASHAE Trans. **62**, 359–372 (1956)

COMSOL: Pipe Flow Module Application Library Manual (2016)

Federal Highway Administration (FHWA): How Do Weather Events Impact Roads? FHWA Report (2011)

Gosnold, W.: Geothermal Well Data. Grand Forks, North Dakota (2012)

Hamdhan, I.N., Clarke, B.G.: Determination of thermal conductivity of coarse and fine sand soils. In: Proceedings of World Geothermal Congress 2010, Bali, Indonesia (2010)

Ho, I.H., Dickson, M.: Assessment of Pavement Snow-Melting System Using Geothermal Energy in Cold Region. Presented at International Conference for Geo-Energy and Geo-Environmental, GeGe 2015, Hong Kong (2015)

Ho, I.H., Dickson, M.: Assessment of geothermal snow-melting system used in cold region area. Energy Geotechniques. Taylor & Francis, London, 113p (2016)

Lund, J.W.: Pavement snow melting. Geo Heat Center Q. Bullet. **21**(2), 12–19 (2000)

Liu, X., Ree, S.J., Spitler, J.D.: Modeling snow melting on heated pavement surfaces. Part I: model development. Appl. Therm. Eng. **27**, 1115–1124 (2007)

Melcher, K.: Winter road maintenance spreadings in the Czech Republic and in EU countries (2001)

Nagai, N., et al.: Numerical simulation of snow melting using geothermal energy assisted by heat storage during seasons. Heat Transf. Asian Res. **42**(8), 724–744 (2013)

National Oceanic and Atmospheric Administration (NOAA): Weather and Temperature Data. NOAA (2016)

Olivier, J.G.J. et al.: Long-term trend in global CO_2 emissions. 2011 report (PDF), The Hague, Netherlands: PBL Netherlands Environmental Assessment Agency; Institute for Environment and Sustainability (IES) of the European Commission's Joint Research Centre (JRC) (2011)

Rees, S.J., Spitler, J.D., Xia, X.: Transient analysis of snow-melting system performance. ASHRAE Trans. **108**, 406 (2002)

Zarling, J.P.: High capacity intersection thaw system. School of Engineering, University of Alaska Fairbanks, Fairbanks, AK 99775 (1995)

Zwarycz, K: Snow melting and heating systems based on geothermal heat pumps at Goleniow Airport. Poland Transportation Report (2002)

Assessment of Interlayer Bonding Properties with Static and Dynamic Devices

Christiane Raab[1(✉)], Elise Fourquet[2], Omar Abd El Halim[3], and Manfred N. Partl[1]

[1] Road Engineering/Sealing Components, EMPA Swiss Federal Laboratories for Materials Science and Technology, Duebendorf, Switzerland
christiane.raab@empa.ch
[2] Eurovia, Paris, France
[3] Department of Civil and Environmental Engineering, Carleton University, Ottawa, Canada

Abstract. Although static interlayer bond testing has been established for quality control during recent years its significance has been questioned and in the wake of this discussion several more realistic dynamic testing devices have been proposed with the purpose of determining the interlayer bond properties not only at failure but also during service life. The Germany testing device developed at the University of Dresden even found entrance into the European test standards and the device has been commercially available for a couple of years. The paper presents an investigation dealing with the importance of interlayer bond testing for performance evaluation and focuses on a comparison of results from static and dynamic interlayer bond testing. Based on the comparison of static and dynamic bond testing the benefits from dynamic testing will be evaluated and a recommendation for the dynamic testing procedure will be presented.

1 Introduction

During recent years interlayer bonding has gained more and more importance, although it was a long way from testing the bond of a two-layered core manually with a hammer until standard quality control on the road. Lately, more and more testing devices have been developed and modifications of devices were proposed (Raab et al. 2009; West et al. 2005). One of the first developments for a standard shear testing device came from Germany, where University of Karlsruhe (Leutner 1979) proposed a shear apparatus which could be installed into a universal testing machine. Although, this device was only applicable for static testing and the application of a normal force was not possible it was modified and adopted by various other countries and remained one of the most important devices for interlayer bond testing. Regarding the interlayer bond, not only the value at failure as determined in the static test is of importance but the evaluation of bond properties the service life an infrastructure is of importance. In this context several dynamic testing devices have been proposed (Millien et al. 1996; Diakhate et al. 2006; Mohammad et al. 2009). The German testing device developed at

L. Mohammad (ed.), *Advancement in the Design and Performance of Sustainable Asphalt Pavements*, Sustainable Civil Infrastructures, DOI 10.1007/978-3-319-61908-8_18

University of Dresden (Ascher 2007) even found entrance into the European test standards (prEN 12697-48) and has been commercially available for a couple of years. Despite this fact the device has mainly been used in Germany and the experience also in comparison to the static equipment is limited. The present study therefore presents an investigation on interlayer bond testing for performance evaluation and focuses on a comparison of results from static and dynamic interlayer bond testing.

2 Test Devices

2.1 Static Testing Device

For static testing the Layer-Parallel Direct Shear (LPDS) test device (Fig. 1) was used. It is an EMPA (Swiss Federal Laboratories for Materials Science and Technology) modified version of the original equipment developed in Germany by Leutner being more versatile in the geometry and more sophisticated in the clamping mechanism (Leutner 1979; Raab et al. 2009). The modified LPDS test device fits into an ordinary servo-hydraulic Marshall testing machine and allows the testing of cores with a diameter of about 150 mm and rectangular specimens; no normal force can be applied (Raab and Partl 1999). One part of the core (up to the shear plane to be tested) is placed on a circular u-bearing and held with a well-defined pressure by a semicircular pneumatic clamp. The other part, the core head remains unsuspended. Shear load is induced to the core head by a semicircular shear yoke with a deformation rate of 50.8 mm/min, thus producing fracture within the pre-defined shear plane of 2 mm width. The exact testing procedure is also described in Swiss Standard SN 670961 (SN 671961, 2000).

Fig. 1. LPDS static testing device

2.2 Dynamic Testing Device (Cyclic Compressed Shear Bond Testing Device)

The cyclic compressed shear bond testing device was first developed at University of Dresden (Ascher 2007).

The driving force for this development was on the one hand the need to bring the shear phenomena closer to the situation on the road with applying a normal load representing the tire and on the other hand to characterize the bond not only at failure but also during the service life of the pavement.

In the device shown in Figs. 2 and 3 the specimen with a diameter Ø of 100 mm is fixed into half-shells using 2-component epoxy glue. The lower half-shells are mounted on aluminium supports. One of these supports is mounted horizontally and the other vertically movable. The shear force is induced by the vertically movable support, while the horizontally movable support enables the application of normal force and allows adjusting the distance between the lower half-shells. The shearing stress is applied by the hydraulic cylinder of the servo-hydraulic testing machine. The applied load is dynamically displacement-controlled using a sinusoidal signal (tension-compression). The displacement amplitude is temperature dependent and aimed to stay within the linear viscoelastic range. The maximum shear displacements are: 0.03 mm (−10 °C), 0.07 (10 °C), 0.11 (30 °C) and 0.15 (50 °C). There are 2 transducers measuring the shear displacement and 2 transducers measuring the normal displacement of the specimen. A data logging system records loads and displacements during the test. For temperature control the device is placed in a thermostat chamber; the temperature can be chosen between −10 and 50 °C.

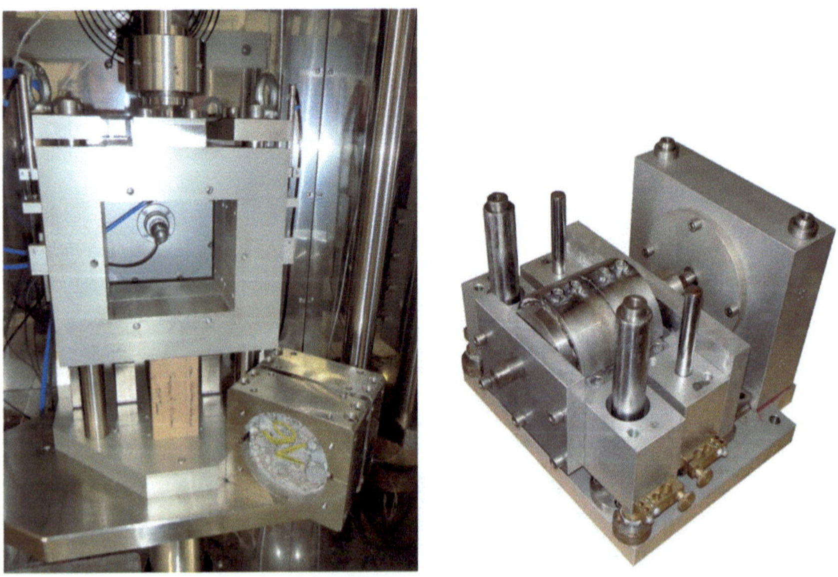

Fig. 2. Dynamic testing device, left: machine and device, right: device open

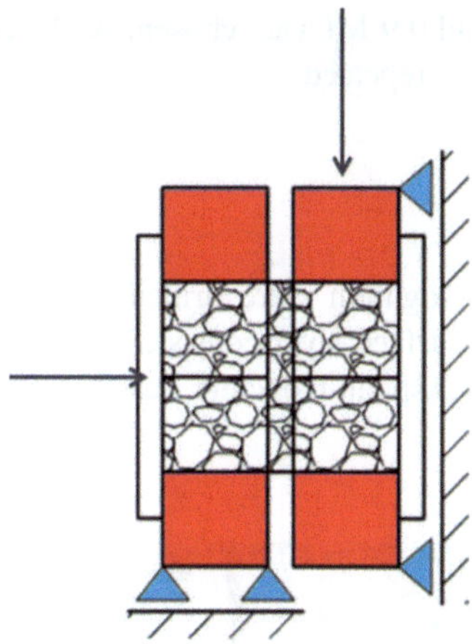

Fig. 3. Dynamic testing device, schematic

Ascher further proposed a testing cycle for different temperatures, frequencies and normal loads as depicted in Fig. 4. For each temperature five normal forces and for each normal force five frequencies are applied. Frequencies of 0.1, 1, 5 and 10 Hz and

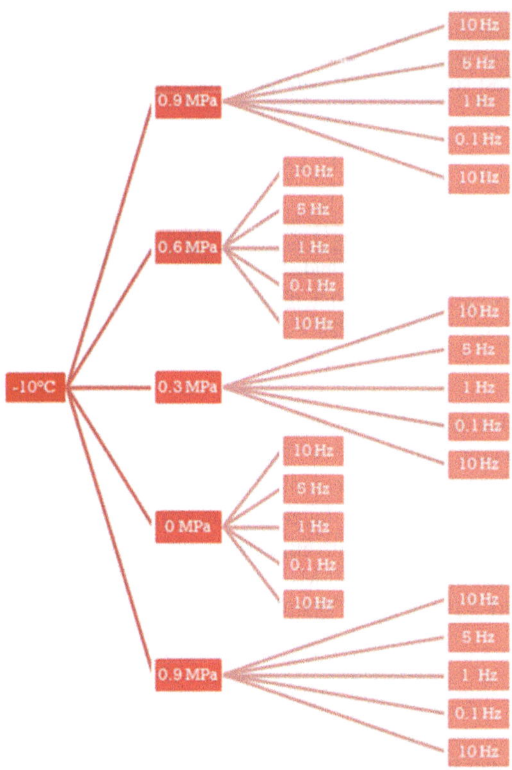

Fig. 4. Testing procedure for dynamic testing

normal loads of 0, 0.3, 0.6 and 0.9 MPa are chosen. At the end of each test, as a way of control the last temperature is repeated.

3 Material

The specimens for the investigation were drilled from a slab taken from a Swiss motorway (see Fig. 5). The surface layer consisted of Stone Mastic Asphalt SMA 11 and the second layer was an Asphalt Concrete Base Course AC 22. The experimental design is given in Table 1.

Fig. 5. Test specimen, left: slab for coring, right: cut 2 layer specimen

Table 1. Experimental design

Layers	Mixture type	Binder grade penetration	Binder content [%]	Air void [vol.%]	Thickness [mm]
Layer 1	SMA 11	55/70 + Trinidad lake asphalt + fibres	5.8	3.2	40
Layer 2	AC 22	Mixelf 10/20	4.8	4.0	50

The core diameter depended on the testing device being 150 mm for the static and 100 mm of the dynamic testing device. The thickness of the surface layer was 40 mm, the base layer was cut to a thickness of 50 mm. For the investigation the interlayer bond between the first and second layer was determined either by static or dynamic testing. Between the layers a cationic emulsion with an application rate of 200 g/mm^2 and a residue bitumen content of 60 M-% had been applied as tack coat according to the standard SN 670 205a-NA/EN 13808:2005.

4 Testing

Testing delivers shear forces from which shear stress τ and shear stiffness S were calculated using the following equations:

$$\tau = F/A \tag{1}$$

where:

F shear force amplitude [N]
A sheared area [mm^2]

$$S = \tau/s \tag{2}$$

where

s shear displacement [mm]

4.1 Static Testing

Regarding static testing the only parameter which was studied was the temperature. Testing was conducted at different temperatures of −10 °C, 10 °C, 20 °C, 30 °C and 50 °C.

4.2 Dynamic Testing

For dynamic testing frequencies of 0.1, 1, 5 and 10 Hz and normal loads of 0, 0.3, 0.6 and 0.9 MPa were chosen. Testing was further conducted at different temperatures of −10 °C, 10 °C, 20 °C, 30 °C and 50 °C.

5 Results

5.1 Static Testing

Figure 6(a) depicts the maximum shear stress calculated from the shear force using Eq. 1 determined from static testing. Every column in Fig. 6 represents the average of 6 specimens.

From the results it becomes clear that the shear stress is highly dependent on the temperature decreasing from 5 MPa at a testing temperature of −10 °C to 0.25 MPa at a testing temperature of 50 °C.

Similar results are depicted for the shear stiffness in Fig. 6(b). Here the values rage from 4.7 MPa/mm at −10 °C to 0.1 MPa/mm at a temperature of 50 °C.

5.2 Dynamic Testing

In order to reduce the amount of data shown in this research shear stiffness was chosen to present the results of dynamic testing. Since dynamic testing is done on the one hand at different frequencies and on the other hand using different normal loads the relationship with temperature cannot be shown in a simple diagram. Figure 7 therefore depicts exemplary the shear stiffness at 1 Hz for different normal stresses and temperatures.

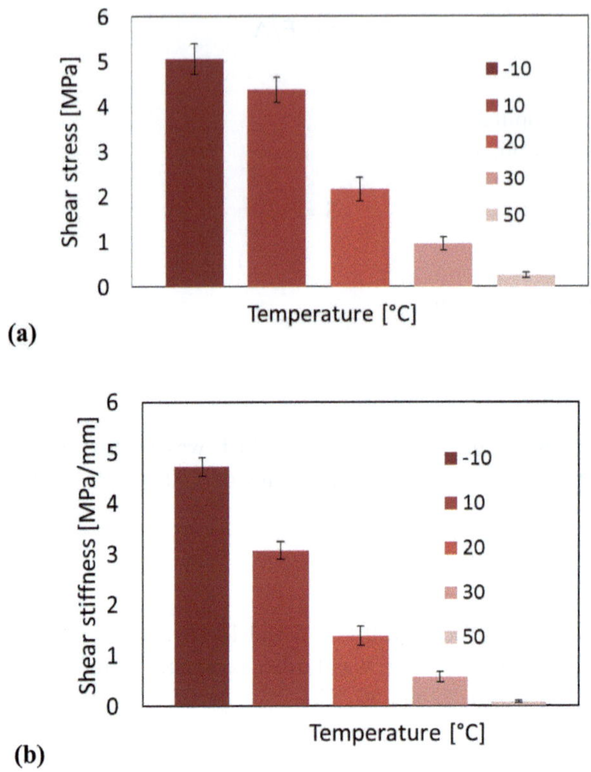

Fig. 6. Static test results, (a) max. shear stress [MPa], (b) max. shear stiffness [MPa/mm]

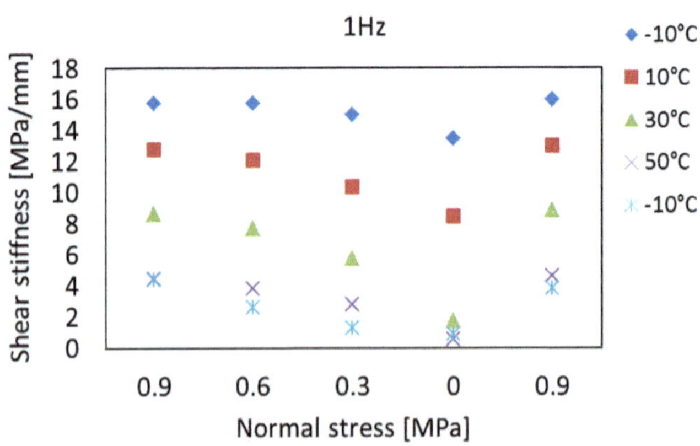

Fig. 7. Dynamic test results, max. shear stiffness [MPa/mm] at a frequency of 1 Hz for different normal stresses

From the results it can be seen that, similar to the static test, shear stiffness decreases with increasing temperature with shear stiffness values ranging from 15.8 MPa/mm to 4.4 MPa/mm for the highest normal stress of 0.9 MPa and from 3 MPa/mm to 1.9 MPa/mm for no normal stress. Further, stiffness increases with increasing normal stress, a phenomenon that has been reported also for static testing when applying normal loads (Raab et al. 2009; Ferrotti et al. 2016).

Figure 8 shows the shear stiffness as a function of temperature and frequency for a normal stress of 0.9 MPa and confirms the influence of temperature on shear stiffness. Furthermore, the dependency of frequency on temperature is clearly visible. At low temperature stiffness remains more or less equal, while at higher temperatures (30 °C to 50 °C) the variation of stiffness with frequency becomes more distinct.

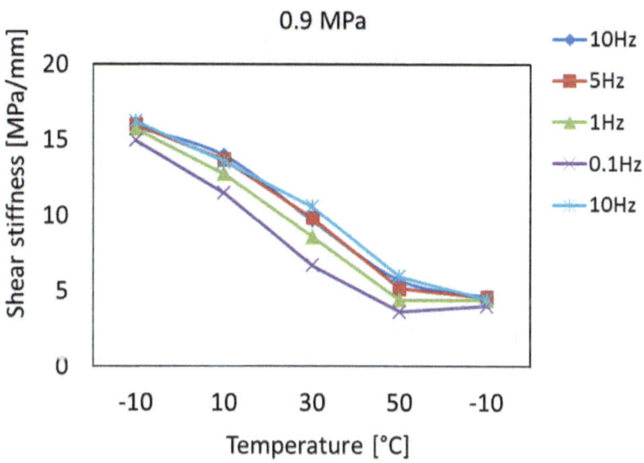

Fig. 8. Dynamic test results, max. shear stiffness [MPa/mm] for different frequencies and for normal stress of 0.9 MPa

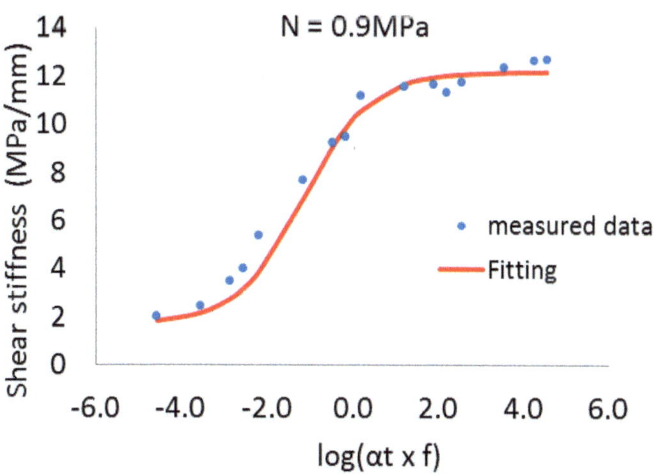

Fig. 9. Master curve of shear stiffness at normal stress of 0.9 MPa

As theory and practice show there is a link between time (frequency) and temperature which enables to construct the so-called Master curve (see Fig. 9; Tables 2 and 3) with the help of the time-temperature-superposition principal. The principle used in this paper is the William-Landel-Ferry (WFL) equation, which leads to the following relationship between the shift factor and temperature:

$$log(a_T) = \frac{p(T - T_0)}{q + T - T_0} \tag{3}$$

where:

T temperature [°C]

T_0 reference temperature [°C] to which all data are shifted

p, q empirical constants of the WLF superposition principle

Finally, the tests results can be fitted using a sigmoidal function:

$$log\ E = d + \frac{a}{1 + e^{b - c\left(\log f + \left(\frac{p(T - T_0)}{q + (T - T_0)}\right)\right)}} \tag{4}$$

where:

E shear stiffness [MPa]

a maximum shear stiffness [MPa]

d minimum shear stiffness [MPa]

b, c shape parameters

Table 2. Table used for calculation of Master curve

DATA			Log freq	Log aT	Log fr * at	fr	E*	Log E	Log E	Error
Temp	Freq	Measured E								
[°C]	[Hz]	[Mpa]					Sigmoidal	Measured	predicted	0.05
−10	10	12.66	1.00	3.55	4.55	3.5E+04	1.2E+01	1.10	1.08	0.00
−10	5	12.63	0.70	3.55	4.25	1.8E+04	1.2E+01	1.10	1.08	0.00
−10	1	12.33	0.00	3.55	3.55	3.5E+03	1.2E+01	1.09	1.08	0.00
−10	0.1	11.74	−1.00	3.55	2.55	3.5E+02	1.2E+01'	1.07	1.08	0.00
10	10	11.33	1.00	1.18	2.18	1.5E+02	1.2E+01	1.05	1.08	0.00
10	5	11.63	0.70	1.18	1.88	7.6E+01	1.2E+01	1.07	1.08	0.00
10	1	11.55	0.00	1.18	1.18	1.5E+01	1.2E+01	1.06	1.07	0.00
10	0.1	11.15	−1.00	1.18	0.18	1.5E+00	1.1E+01	1.05	1.02	0.00
30	10	9.47	1.00	−1.18	−0.18	6.6E−01	9.8E+00'	0.98	0.99	0.00
30	5	9.23	0.70	−1.18	−0.48	3.3E−01	9.1E+00'	0.97	0.96	0.00
30	1	7.66	0.00	−1.18	−1.18	6.6E−02	6.9E+00'	0.88	0.84	0.00
30	0.1	5.37	−1.00	−1.18	−2.18	6.6E−03	3.9E+00	0.73	0.59	0.02
50	10	3.98	1.00	−3.54	−2.54	2.9E−03	3.2E+00	0.60	0.50	0.01
50	5	3.47	0.70	−3.54	−2.85	1.4E−03	2.7E+00	0.54	0.44	0.01
50	1	2.45	0.00	−3.54	−3.54	2.9E−04	2.1E+00	0.39	0.33	0.00
50	0.1	2.03	−1.00	−3.54	−4.54	2.9E−05	1.8E+00	0.31	0.26	0.00

6 Discussion

When comparing the results from static and dynamic testing at first glance the results seem to be similar: increasing interlayer shear stress and shear stiffness with decreasing testing temperature.

At second glance, not only the received values depicted in Table 3 but also the purpose of testing appears to be different.

Table 3. Parameters calculated for the WLF and sigmoidal curve

Sigmoidal				WLF		
d	a	b	c	p	q	T_0 °C
0.23	0.86	−2.33	1.21	13000	110000	20

When static testing represents the result at failure of interlayer bonding, dynamic testing offers a complex evaluation of interlayer bond properties for the service life of the pavement. From the dynamic results master curves can be retrieved which enable to predict the shear stiffness for all combinations of temperature and strains and stresses of interest based on limited testing. Further, the results from dynamic testing are not only valuable for quality control but can be used for various purposes, such as modelling.

Although, dynamic testing seems a valuable tool which offers a variety of possibilities for interlayer bond assessment and evaluation, there still remain a lot of open questions.

Regarding the testing procedure a difference between the first and the last shear stress and shear stiffness test result at −10 °C was found. This shows that the specimens might have been damaged during testing, posing the question for revising the proposed procedure.

Furthermore, the presented evaluation is very limited especially in regard to material. In the evaluation only one kind of asphalt pavement and interface was used which is certainly not enough. Future research will have to investigate different kinds of asphalt pavement layers such as porous or semi dense graded asphalt as well as different kinds of tack coats.

Last not least, dynamic testing proved to be cumbersome and the gluing of specimens offered several problems. Therefore, it is possible that such impairments might influence testing results.

Table 4. Shear stiffness results for static and dynamic testing (normal stress = 0)

Shear stiffness [Mpa][1]		
Temp. [°C]	Static testing	Dynamic testing
−10	4.7	13.4
10	3.1	8.5
20	1.4	
30	0.6	1.8
50	0.1	0.6

7 Conclusions

The conclusions obtained from this study are as follows:

Shear stress and shear stiffness are clearly dependent on temperature leading for static and dynamic test methods to increasing values with decreasing temperature.

Although static and dynamic testing show similar trends regarding the dependency of shear stress and shear stiffness, values are different between both test methods when comparing values for the same testing conditions (temperature and normal load).

The application of a normal stress leads to a more realistic simulation simulating tire forces on the road.

Static and dynamic test do not have the same goals and output: Static testing investigates interlayer bond properties at failure and is considered for quality assessment, while dynamic testing aims at the interlayer bond properties during the service life of a pavement. Dynamic shear stiffness results can further be used for interlayer bond performance prediction (master curve) and modeling purposes.

Although first results are promising, the dynamic testing method presented in this research still needs further evaluation regarding the testing procedure itself as well as regarding the amount and kind of material tested.

Further, interlaboratory studies on testing devices and testing procedures including specimen preparation will be needed for comparison and harmonization of testing devices and methods (Table 4).

References

Ascher, D., Wellner, F.: Untersuchungen zur Wirksamkeit des Haftverbundes und dessen Auswirkungen auf die Lebensdauer von Asphaltbefestigungen [Investigation of the effectiveness of bonding and its influence on the service life of asphalt pavements]. Technical University of Dresden, Germany, Report No. 13589 BR/1, 100 p. (2007)

Diakhate, M., Phelipot, A., Millien, A., Petit, C.: Shear fatigue behaviour of tack coats in pavements. Road Mater. Pavement Des. **7**, 201–222. DOI:ISSN 1468–0629

Ferrotti, G., D'Andrea, A., Maliszewski, M., Partl, M.N., Raab, C., Sangiorgi, C., Canestrari, F.: Interlaboratory Shear Evaluation or reinforced Bituminous Interfaces. In: Canestrari, F., Partl, M.N. (eds.) 8th RILEM International Symposium on Testing and Characterization of Sustainable and Innovative Bituminous Materials. RILEM Bookseries, vol. 11, Springer (2016), ISBN: 978-94-017-7341-6 (Print) 978-94-017-7342-3 (2015)

Leutner, R.: Untersuchungen des Schichtenverbunds beim bituminösen Oberbau [Investigation of interlayer bonding in bituminous pavements]. Bitumen **3**, 84–91 (1979)

Millien, A., Petit, C., Rosier, J.: Comportement au cisaillement des couches d'accrochage dans les chaussée [Shear Bond Behaviour on the Road] rapport interne Laboratoire 3MsGC, Université de Limoges, France. p. 15 (1996)

Mohammad, L.N, Bae, A., Elseifi, M.: Effect of Tack Coat Materials and application rate on the interface shear strength. In: International Conference on Maintenance and Rehabilitation of Pavements and Technological Control (MAIREPAV6), Torino, Sixth Proceedings, Vol. 2, pp. 636–645 (2009)

Partl, M.N., Raab, C.: Shear adhesion between top layers of fresh asphalt pavements in Switzerland. In: Proceedings of 7th CAPSA Conference on Asphalt Pavements for Southern Africa, Victory Falls, Zimbabwe, 29 August–2 September (1999)

Raab, C., Partl, M.N.: Methoden zur Beurteilung des Schichtenverbunds von Asphaltbelägen [Methods for the evaluation of interlayer bonding]. *ASTRA*-Project FA 12/94, Report No. 442, 118 (1999)

Raab, C., Partl, M.N.Abd, El Halim, A.O.: Evaluation of interlayer shear bond devices for asphalt pavements. Baltic J. Road Bridge Eng. **4**(4), 176–195 (2009)

Swiss Standard, Schweizer Norm SN 671961. Bituminöses Mischgut, Bestimmung des Schichtenverbunds (nach Leutner), Verein Schweizerischer Straßenfachleute VSS (2000)

Swiss Standard SN 670 205a-NA/European Standard, EN 13808. Bitumen and bituminous binders – Framework for specifying cationic bituminous emulsions (2005)

West, R.C., Zhang, J., Moore, J.: Evaluation of bond strength between pavement layers. National Center for Asphalt Technology, NCAT Report 05–08 (2005)

Wavelet-Spectrogram Analysis of Surface Wave Technique for Quick NDT Measurement on Surface Layer of Pavement

Sri Atmaja P. Rosyidi(✉)

Department of Civil Engineering, Universitas Muhammadiyah Yogyakarta, Yogyakarta, Indonesia
atmaja_sri@umy.ac.id

Abstract. Reliable assessment of in situ pavements stiffness is an important aspect in effectively managing a pavement system. The aim of this paper is to propose the new procedure, namely the wavelet-spectrogram of surface wave (WSSW) technique for non-destructively measurement of elastic modulus on surface layer of a pavement system. Using two receivers, surface wave propagation on pavement surface was recorded and transformed into in frequency domain by wavelet analysis. For this analysis, a derivative Gaussian wavelet was selected as an appropriate mother wavelet for seismic waveform propagating along pavement surface. Thus, an interactive 2-D plot of time-frequency spectrogram consisting of wave-energy spectrum was simultaneously generated. CWT-filtration method was implemented in order to reduce the effect of noisy signal recorded during measurement. From selected wave spectrogram, the unwrapped phase different spectrum was generated to obtain phase velocity which was performed by least-square linear regression. Finally, the elastic modulus of pavement surface layer was calculated from a modified relationship between phase velocity, Poisson ratio and density of pavement surface layer. The results show that the proposed technique is able to measure in situ elastic stiffness of the surface layer. In addition, the change of the surface layer stiffness is also able to be monitored. The stiffness (elastic modulus) produced by the WSSW technique is classified as a modulus at very low strain level.

1 Introduction

The spectral-analysis-of-surface-wave (SASW) is a well-known of non-destructive techniques (NDT) in pavement evaluation and geotechnical investigation. The SASW employs the surface waves dispersion in order to determine the shear wave velocity corresponding to stiffness of each layer at pavement and soil profile. The method performs three steps of an elaborated data process, i.e., (1) collecting seismic data, (2) constructing an experimental dispersion curve and (3) conducting an inversion process of the dispersion curve for generating a stiffness profile. The method has been improved and utilized in many civil engineering applications, e.g., site characterization (Stokoe et al. 1994), soil density (Kim et al. 2001), pavement characterization (Rosyidi et al. 2007, Yusoff et al. 2015), soil damping measurement in soft soil sites (Rosyidi and Taha 2012), and asphaltic pavement measurement (Hazra and Kumar 2014, Rosyidi 2015).

L. Mohammad (ed.), *Advancement in the Design and Performance of Sustainable Asphalt Pavements*, Sustainable Civil Infrastructures, DOI 10.1007/978-3-319-61908-8_19

In inversion process, the advanced mathematical approach is required to produce the stiffness profile from a dispersion curve. Reliable inversion procedure using stress-wave propagation theories, i.e., the transfer matrix method, the dynamic stiffness matrix method and the finite difference method should be employed. All the methods require an initial profile model consists of a set of homogeneous layers extending to infinity in the horizontal direction. In each layer of profile, the information such as a thickness, a shear wave velocity, a Poisson's ratio (or compression wave velocity), and a mass density are assigned. Based on the initial profile, a theoretical dispersion curve is calculated using one of these wave propagation theories. The theoretical dispersion curve is then compared with the experimental dispersion curve. If the two dispersion curves do not match, the initial profile (number of layers, layer thickness, shear wave velocity, or any combination) is adjusted, and another theoretical dispersion curve is calculated. The trial-and-error procedure is repeated until the two curves match, and then the associated assumed profile is considered the real profile. In addition, when the SASW method is performed on irregular stiffness profile, i.e., pavement structure, the trial-and-error procedure becomes a difficult analysis and takes longer time in data processing. Moreover, several researchers have reported on difficulties related to surface wave measurements at pavement sites. Most of these difficulties are reported to originate from the influence of higher modes of propagation (Al-Hunaidi 1992, Tokimatsu et al. 1992, Ganji et al. 1998; Ryden et al. 2004).

The seismic surface wave method is not able to separate different modes of propagation on a pavement system and thus measures a superposition of all propagating waves at the specific receiver locations. This superposed effect, often termed apparent phase velocity or pseudophase velocity, changes with distance and has forced the evaluation of the data to take into account the position of the receivers and the superposition of different modes for the inversion of experimental dispersion curves. Ryden et al. (2004) proposed a new approach in seismic pavement testing where the different modes of propagation are separated, thereby potentially clarifying some of the noted difficulties with the SASW method applied to pavement testing. Their approach is based on the multichannel analysis of surface wave (MASW) data processing technique. However, due to complexity in surface wave analysis, application of these methods on pavement evaluation is still relatively limited. On the other hand, in pavement evaluation system, the need of accurate, cost-effective and non-destructive evaluation is becoming ever important because the rehabilitation and management of roads is becoming increasingly difficult due to the increasing number of aging roads and limited budgets. As well, for practical purpose, pavement engineer usually needs a quick and relatively effortless analysis for determining the structural condition of pavement surface layers.

In this paper, a new procedure in a surface wave technique which is using continuous wavelet transform combined with the elastic stiffness formulation for obtaining the phase velocity is introduced. This method is namely the wavelet-spectrogram of surface wave (WSSW) technique. The time-frequency decomposition of continuous wavelet transform (CWT) on seismic signals is employed to characterize the phase information of phase different spectrum. It provides a reliable information of wave spectrum in the pavement profile. An algorithm on phase different calculation in this analysis aims to avoid the use of a complex inversion algorithm to obtain the elastic modulus of the pavement surface layer. In the SASW method, data analysis and processing in frequency domain has been carried out by fast Fourier transforms (FFT). However, due to Fourier transform works by expressing any arbitrary periodic function of time with period as sum a set of sinusoidal, this analysis becomes inconsistent and is unable to preserve the time dependence and describe the evolutionary spectral characteristics of non-stationary processes (Rosyidi et al. 2009). Wavelet analysis is becoming an effective tool for analyzing localized variations of power within a time series. By decomposing a time series into time-frequency spectrum (TFW), one is able to determine both the dominant modes of variability and how those modes vary in time. A typical result from a case study is presented herein for the structural assessment of an existing asphalt concrete (AC) pavement in Purwakarta, West Java, Indonesia.

1.1 Continuous Wavelet Transform

The continuous wavelet transform (CWT) has been used in many studies in geophysics (Foufoula-Georgiou and Kumar 1995) and civil engineering application (Rosyidi et al. 2009). The continuous wavelet transform (CWT) technique is an alternative tool for localizing the interested frequency of seismic signal processing particularly in non-stationary problems. Wavelets dilate in such a way that the time support changes for different frequency. When the time support increases or decreases, the frequency support of the wavelet is shifted toward high or low frequencies, respectively. Therefore, as the frequency resolution increases, the time resolution decreases and vice versa. The characteristic of time-frequency resolution creates the CWT technique useful for non-stationary seismic analysis.

A wavelet is defined as a function of $\psi(t) \in L^2(\Re)$ with a zero mean, which is localized in both time and frequency. By dilating and translating the wavelet $\psi(t)$, it can be used to produce a family of wavelets as:

$$\psi_{\sigma,\tau}(\tau) = \frac{1}{\sqrt{\sigma}}\psi\left(\frac{t-\tau}{\sigma}\right) \tag{1}$$

where σ is the dilation parameter or scale and τ is the translation parameter ($\sigma, \tau \in \Re$ and $\sigma \neq 0$). The wavelet has also various wavelet shape used for signal analysis which is called the mother of wavelet, i.e., Gaussian, Morlet, Paul and Mexican Hat. An appropriate selection of wavelet shape signal analysis depends on the seismic waveforms.

The CWT is defined as the inner product of the family wavelets $\Psi_{\sigma,\tau}(t)$ with the signal of $f(t)$ which is given as:

$$F_W(\sigma,\tau) = \langle f(t), \psi_{\sigma,\tau}(t)\rangle = \int_{-\infty}^{\infty} f(t)\frac{1}{\sqrt{\sigma}}\bar{\psi}\left(\frac{t-\tau}{\sigma}\right)dt \tag{2}$$

where $\bar{\psi}$ is the complex conjugate of ψ, $F_W(\sigma,\tau)$ is the time-scale map.

In this study, the mother wavelet of the Gaussian Derivative (GoD) was used. The real component of the GoD wavelet in the time and frequency domains is defined as follows:

$$\psi_0(t) = \frac{(-1)^{m+1}}{\sqrt{\Gamma\left(m+\frac{1}{2}\right)}} \frac{d^m}{d\eta^m}\left(e^{-t^2/2}\right) \tag{3}$$

$$\hat{\psi}_0(s\omega) = -\frac{i^m}{\sqrt{\Gamma\left(m+\frac{1}{2}\right)}}(s\omega)^m\left(e^{-(s\omega)^2/2}\right) \tag{4}$$

where m is the wave number and Γ is the Gamma function. The complex wavelet is generated by the addition of a Heaviside function in the frequency domain. This wavelet decays with the square root of the gamma function. The GoD has wavelet's derivative order that can be varied in order to get the best resolution of the waveform.

2 Research Method

2.1 Field Measurement

In field measurement of the WSSW method, a set of ball bearing with weight from 5 to 15 g was used as an impact source to generate seismic waves on the surface layer of pavement. Generated seismic waves were then detected using two high frequency accelerometers. Consequently, the signals were recorded using an ADT analog-digital acquisition connected to a notebook computer for post processing (Fig. 1). In this study, two configurations of mid-point receiver spacing and the receiver-source spacings were employed in order to sample different depths of pavement surface. The configuration of mid-point receiver spacings was described in Fig. 2. The receiver spacing (d_2) was obtained as a distance which the length is less than and/or equal to the thickness (H) of pavement surface layer. Whereas, the distance between a source and first receiver (d_1) was set as equal to receivers spacing (d_2). Due to the interested pavement area in measurement was asphaltic surface layer, the short receiver spacings of 5, 10 and 15 cm with a high frequency source were only used. In the test, repetition procedure (forward and backward) until at least 4 to 6 times in each receiver spacing measurement should be employed in order to minimize the effect of internal phase shift between receivers and to enhance a good average of received signals.

Beside the WSSW test, the SASW measurement was also carried out at same sites of existing road pavement in Purwakarta, West Java, Indonesia. The pit test was also conducted to acquire the information of a pavement profile. From the pit test, the pavement structure consists of an asphalt concrete (AC) layer (18 cm), crushed stone of base course (10 cm), sub-base course (30 cm) over a subgrade layer.

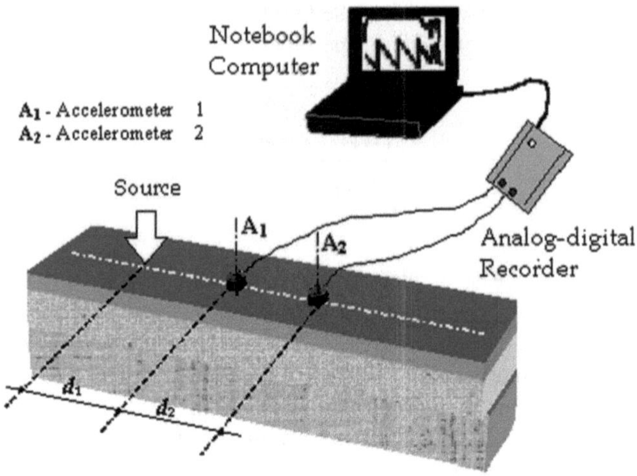

Fig. 1. Experimental set up of WSSW measurement on a pavement structure

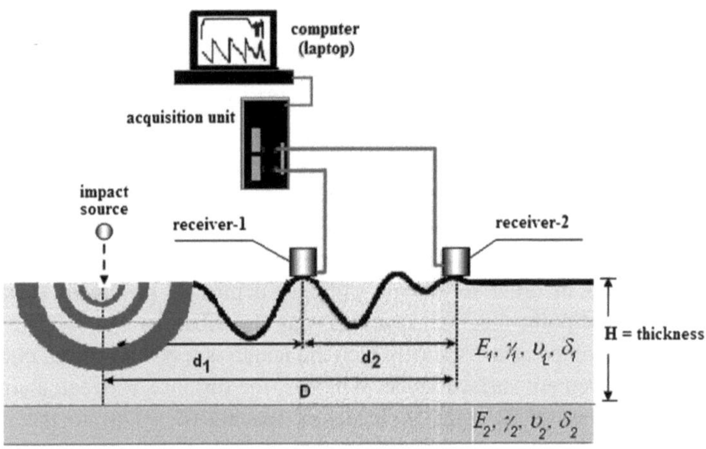

Fig. 2. Mid-point receivers configuration of WSSW measurement on pavement structure

2.2 Data Analysis and Process

A proposed procedure of the WSSW method consists of seismic data analysis using continuous wavelet transform and elastic modulus calculation of elastic linear model on pavement surface layer which is described as follows:

1. The seismic wave data are collected from field measurement using a configuration of the mid-point receiver spacings.
2. The time-frequency spectrum analysis based on continuous wavelet transform of Gaussian Derivative (DoG) mother wavelet is carried out for two signal waveforms recorded from field measurement. By decomposing a time series of seismic

waveform into time-frequency (TF) spectrogram, the dominant modes of variability and how those modes vary in time can be determined very well.

3. Phase information of the transfer function (phase spectrum) are then determined from both TF spectrograms. The data informs the time difference of wave propagating from first to second receiver. A mathematical expression for calculating phase spectrum from TF spectrogram is defined by (Rosyidi and Taha 2012):

$$H(f) = \frac{Y(f)}{X(f)} \approx \frac{W^{Y}_{f(u,s)}}{W^{X}_{f(u,s)}} = \frac{\int_{-\infty}^{\infty} Y(t)\frac{1}{\sqrt{\sigma}}\psi * \left(\frac{t-\tau}{\sigma}\right)dt}{\int_{-\infty}^{\infty} X(t)\frac{1}{\sqrt{\sigma}}\psi * \left(\frac{t-\tau}{\sigma}\right)dt} \quad (5)$$

where,

$X(f)$ = spectrum input of signal, $X(t)$, from first receiver,
$Y(f)$ = spectrum output of signal, $Y(t)$, from second receiver,

$$W^{Y}_{f(u,s)} = \int_{-\infty}^{\infty} Y(t)\frac{1}{\sqrt{s}}g\left(\frac{t-u}{s}\right)e^{-i\xi(t-u)} \quad (6)$$

$$W^{X}_{f(u,s)} = \int_{-\infty}^{\infty} X(t)\frac{1}{\sqrt{s}}g\left(\frac{t-u}{s}\right)e^{-i\xi(t-u)} \quad (7)$$

From Eq. 7, the phase spectrogram in time-frequency domain can be obtained by:

$$H(u,s) = \frac{W^{XY}_{f}(u,s)}{W^{XX}_{f}(u,s)} = \frac{\left|W^{XY}_{f}(u,s)\right|e^{i(\theta_Y(a,b)-\theta_X(a,b))}}{W^{X}_{f}(u,s)^{*} \times W^{X}_{f}(u,s)} \quad (8)$$

Thus, phase different is obtained from the ratio of the imaginary to real part of the phase spectrogram which is expressed as:

$$\phi = \tan^{-1}\left(\frac{\Im H(u,s)}{\Re H(u,s)}\right) \quad (9)$$

4. The coherence function is used to visually inspect the quality of signals being recorded in the field and have a real value between zero and one in the range of frequencies being measured. The value of one indicates a high signal-to-noise ratio (i.e., perfect correlation between the two signals) while values of zero represents no correlation between the two signals. The coherence function is a ratio of the output power caused by the measured input to the total measured output which is defined as:

$$\gamma^2(f) = \frac{G_{yx}(f) \cdot G^*_{yx}(f)}{G_{xx}(f) \cdot G_{yy}(f)} \tag{10}$$

5. A linear relationship between the phase different and frequency from the transfer function spectrum is then derived. The phase velocity is examined as a function of distance from the slope value (m). This relationship can be written as:

$$\phi = \left|\frac{360\mathrm{D}}{\mathrm{V_{ph}}}\right| f = mf \tag{11}$$

By Eq. 11, one can easily determine V_{ph} by performing a least-square linear regression over the frequency range in the transfer function spectrum and the slope of the best-fit line (m) can be obtained. The phase velocity of surface wave propagation is independent of the wavelength for up to a wavelength approximately equal to thickness of the uppermost layer. The range of wavelength to be used can be estimated from the phase velocities (V_{ph}) of the material anticipated at the site:

$$\lambda = \frac{V_{ph}}{f} \tag{12}$$

where f is the frequency.

6. If one simply generates high frequency waves and assumes that properties of the uppermost layer are uniform, the dynamic elastic moduli of the pavement materials can easily be determined as follows:

$$\mathrm{E} = 2\frac{\gamma}{g} V_S^2 (1+\mu) = \frac{\gamma}{g} |\mathrm{KV_{ph}}|^2 \tag{13}$$

$$K = (1.13 - 0.16\mu)\sqrt{\frac{2(1-\mu)}{(1-2\mu)}} \tag{14}$$

where E is the dynamic elastic modulus, respectively, V_S is the shear wave velocity, V_{ph} is the phase wave velocity, g is the gravitational acceleration, γ is the total unit weight of the material and μ is the Poisson's ratio.

3 Results and Discussion

3.1 Stiffness and Elastic Modulus

The WSSW measurements were carried out at 12 sites on existing asphalt pavements. A typical result of received signals from field measurement was shown in Fig. 3. The body waves (Primary and Secondary wave) and surface wave (Rayleigh wave) are clearly shown in both signal recordings. Therefore, the Gaussian Derivative (GoD) continuous wavelet transform (CWT) was employed on signals for generating a time-frequency (TF) plot. Both CWT plots may overcome on identification problem of the spectral characteristics of non-stationary signals measured in two receivers. The typical CWT spectrogram for received signals with an improved time-frequency resolution is shown in Fig. 4. From Fig. 4, two main energy events at different frequency bands were also clearly detected which may result in both low and higher mode of seismic waves. It is able to observed that lower frequency energy was found in the range of 2.8–16 kHz in CWT spectrogram. This spectrum range is identified as surface wave signals. The level and frequency range of these signals were determined by independent-measurement of high level energy in surface wave propagation (up to 60% of total wave energy). Thus, the other energy event is identified as wave mode from direct and reflected body waves which was found in more than 20 kHz. The dominant wave energy found in CWT spectrogram at interested frequency range of surface waves can be clearly captured. The TF of GoD-CWT provides good resolutions at high frequencies of signals. It is also effective in the detection of frequency bandwidth of wave groups using various derivation order of this mother wavelet (GoD).

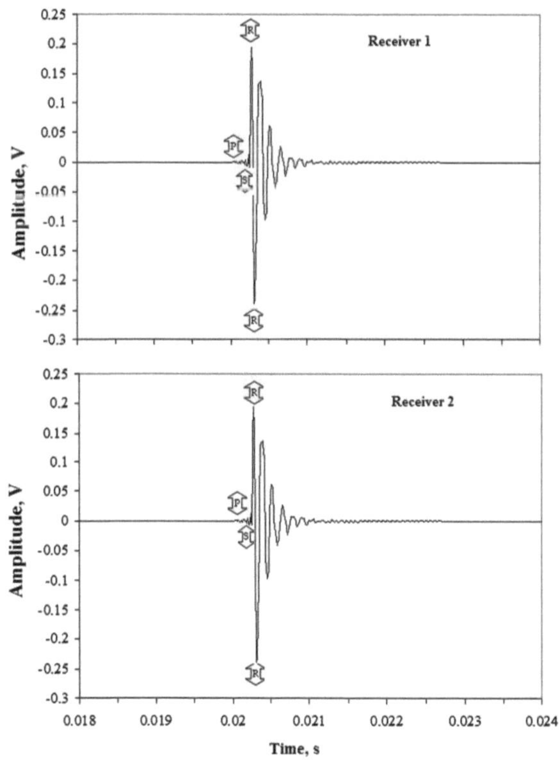

Fig. 3. Signal from WSSW measurement on a pavement structure

Based on both TF spectrograms (Fig. 4), the phase different between two signals for every wave frequency were then calculated. As a result, a transfer function spectrum on wrapped phase different curve was obtained as given in Fig. 5. The phase data

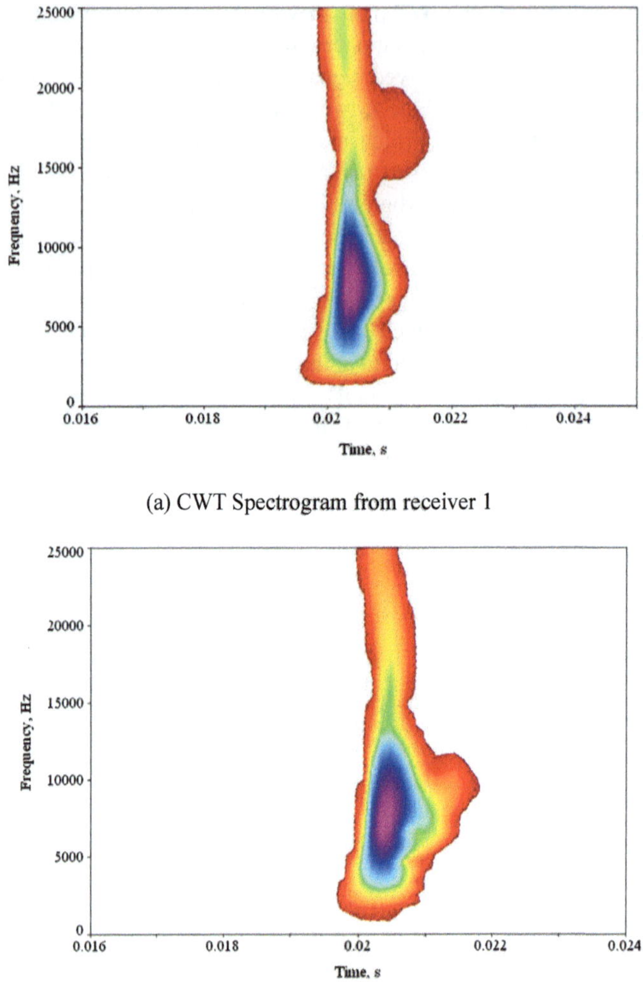

(a) CWT Spectrogram from receiver 1

(b) CWT Spectrogram from receiver 2

Fig. 4. Time-frequency plot of received signals from measurement

oscillates between $-\pi$ and π radian (−180 and 180 degrees). This is the standard illustration of spectrum presenting phase data because the detail variation in data can be observed in a small space.

Figure 5 also shows that the phase different curve reveals a smooth trend of variation in phase with frequency up to a frequency of 25 kHz. It indicates the high-frequency surface waves were detected representing the high stiffness of the asphalt concrete surface layer on a pavement structure. The quality of phase data is also controlled by the coherence function. As shown in Fig. 6, the phase data up to frequency of 20 kHz have the value of coherence magnitude above 0.98.

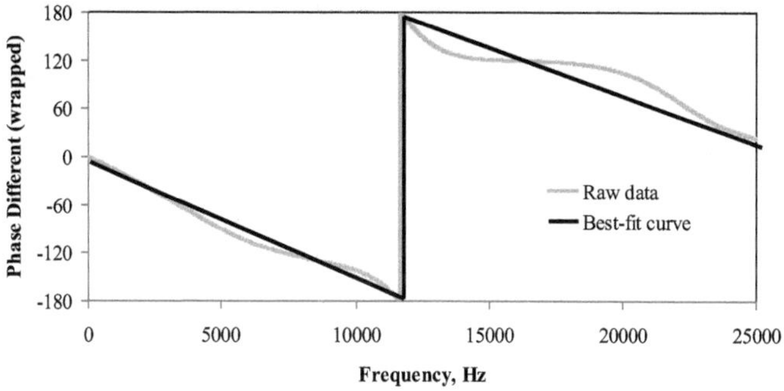

Fig. 5. Comparison between raw data and best-fit curve of wrapped transfer function spectrum from measurement

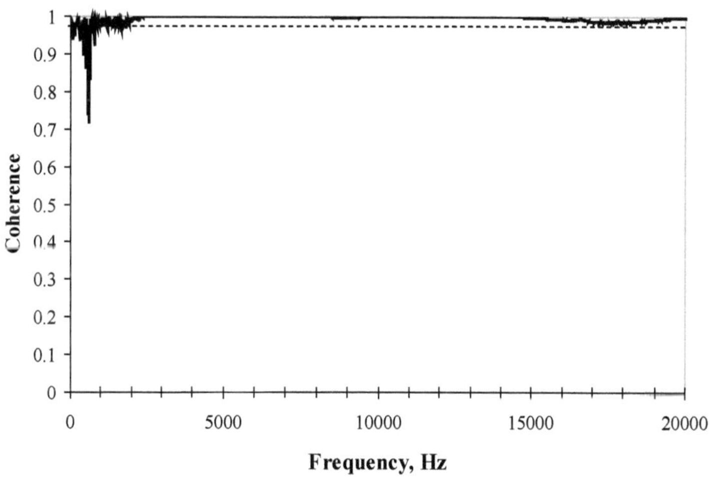

Fig. 6. Coherent function spectrum of received signals from measurement

In order to obtain the modulus value of the surface layer, the smoothed fitting process of weighing function was used. The fitted curve between the raw and the smoothed phase spectrum is shown in Fig. 5. Thus, the phase data is unwrapped by adding the number cycles to each phase. The unwrapped of the raw phase spectrum is also shown in Fig. 7. The unwrapped phase spectrum is smoothed by the linear regression as the best fit curve to the raw data. The slope of the line is more or less constant with frequency. A line is fitted to the curve in the range of the frequency corresponding to wavelengths shorter than the thickness of the surface layer. Slope value of the line can be used to determine the elastic modulus of the surface layer of the pavement profile using Eqs. 11, 13 and 14.

From Fig. 7, the slope (m) of best-fit curve is found to be 0.0140. Consequently, the phase velocity can be calculated using Eq. 13 which is found to be 1028.57 m/s. Based on the phase velocity, field configuration data and material parameters, such as of

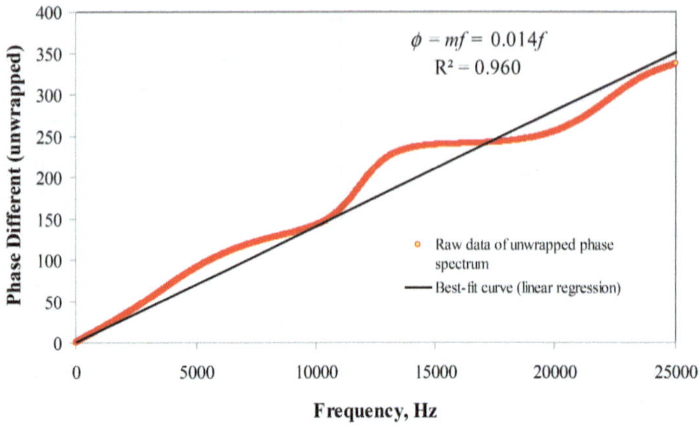

Fig. 7. Comparison between raw data and best-fit curve of unwrapped transfer function spectrum from measurement and slope analysis to obtain m value

receiver spacing (d_2) of 5 cm, Poisson's ratio of asphaltic layer of 0.25 and unit weight for pavement material (AC) of 2,200 kg/m^3, the elastic modulus of asphaltic (AC) surface layer is obtained as 845,662,040.80 kg/m^2 or 8456.62 MPa.

The result shows that the elastic modulus of AC layer can be easily determined using the WSSW technique. However, the value of elastic modulus presented is relatively high. It is due to the seismic technique measures the dynamic stiffness at very low strain level (less than 103%). In this level, the material modulus behavior can be assumed as a constant and have only a maximum value. To illustrate the usefulness and sensitivity of this approach in testing the surface layer, changes in the stiffness of the existing surface and the overlay layer of the pavement were measured in situ. Figure 8 shows that the different stiffness of an existing and overlay surface layer from the measured profile can be investigated well. Based on these results, it can be summarized that the stiffness changes in the surface layer were easily, non-destructively, and fast measured by the WSSW technique.

3.2 Validation with the SASW Method

In order to validate the results from the WSSW test, the spectral-analysis-of-surface-wave (SASW) analysis was conducted at same locations of a road pavement. In this method, a set of transient impact sources was used to generate surface wave energy that propagates horizontally near the surface layer of the pavement. The phase differences of signal data were obtained from the cross-power spectrum. Thus, the phase information was then unwrapped to produce the dispersion curve of the phase velocity versus wavelength. An inversion process was then iteratively employed to confirm the experimental dispersion curve from the theoretical model established. A 3-D stiffness matrix model (Rosyidi 2007) was employed in the SASW inversion analysis. Final stiffness profile was obtained after 16 times of iteration with the root-mean-square error (RMS) of 35.47 m/s or average deviation of about 5.92%.

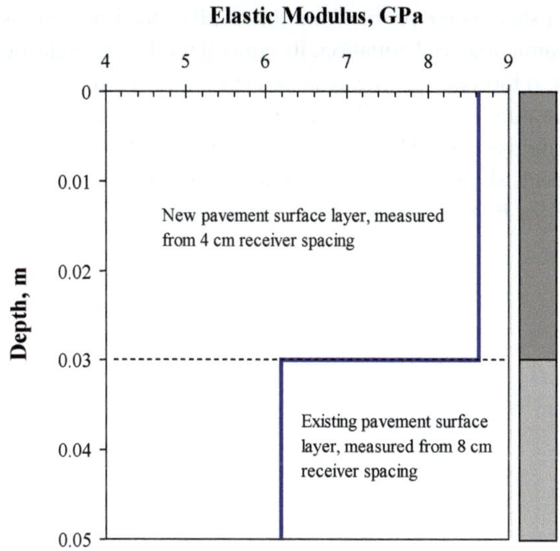

Fig. 8. Comparison between elastic modulus of an overlay and existing surface layer at pavement roads, Indonesia

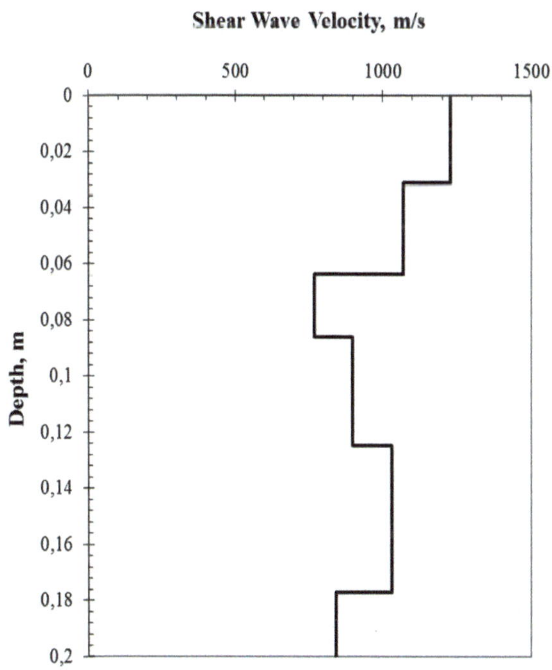

Fig. 9. The shear wave velocity profile from inversion of the experimental dispersion curve

The equivalent shear wave profile from the result of the inversion is shown in Fig. 9 and using the dynamic material equation, its equivalent dynamic elastic modulus profile is given in Fig. 10. A modulus profile as shown in Fig. 10 is only given to 10 cm depth of pavement structure due to the asphaltic layer was indicated at this level. A good agreement of elastic modulus resulted from the WSSW and the SASW method is also presented in Fig. 10. The result shows that the difference between both methods is calculated at 0.01% and 1.14% for first and second layer of pavement surface layer, respectively.

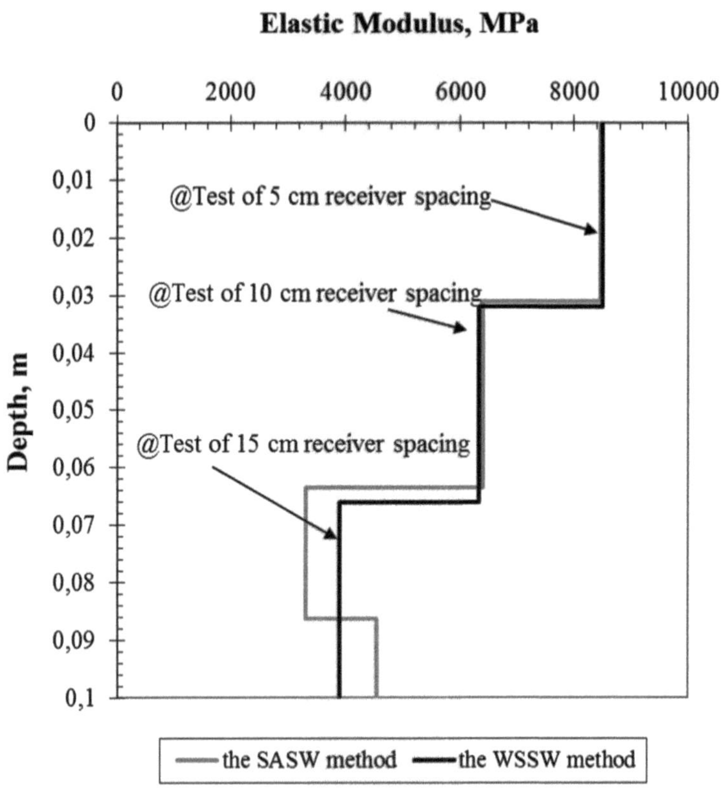

Fig. 10. The dynamic elastic modulus of the pavement profile from the SASW method and its comparison with the WSSW method

The falling weight deflectometer (FWD) method was also employed to validate the elastic modulus of pavement surface layer from the WSSW test. The elastic modulus of pavement surface layer obtained from both measurements is shown in Fig. 11. The elastic modulus obtained from the WSSW is higher comparison with value determined by the FWD. As mentioned earlier, the modulus measured at very low strain levels associated with surface wave method is in maximum value and it is independent from strain amplitude. In addition, the high frequency used in the seismic result in higher values of stiffness for pavement material. In the case of FWD, the modulus was measured at frequency of around 30 Hz (Fig. 12).

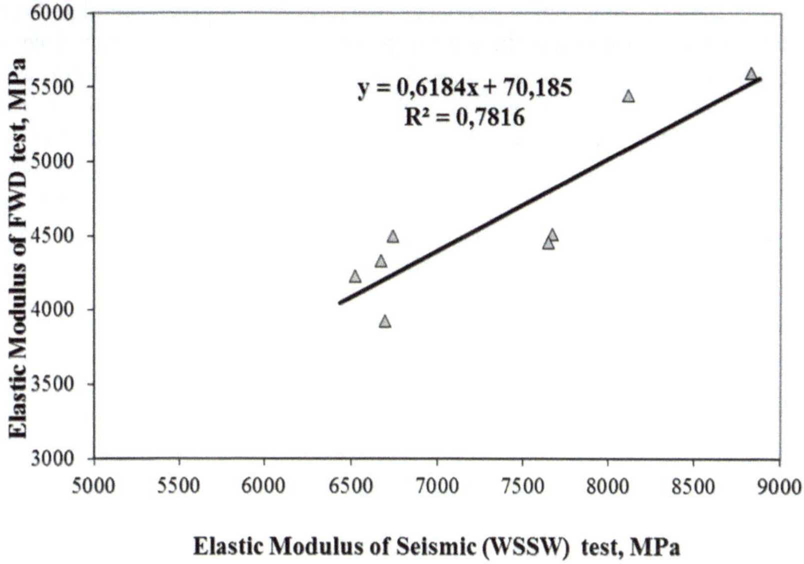

Fig. 11. Comparison elastic modulus of surface layers from the SASW test compared to the FWD test at Purwakarta State Road, Indonesia

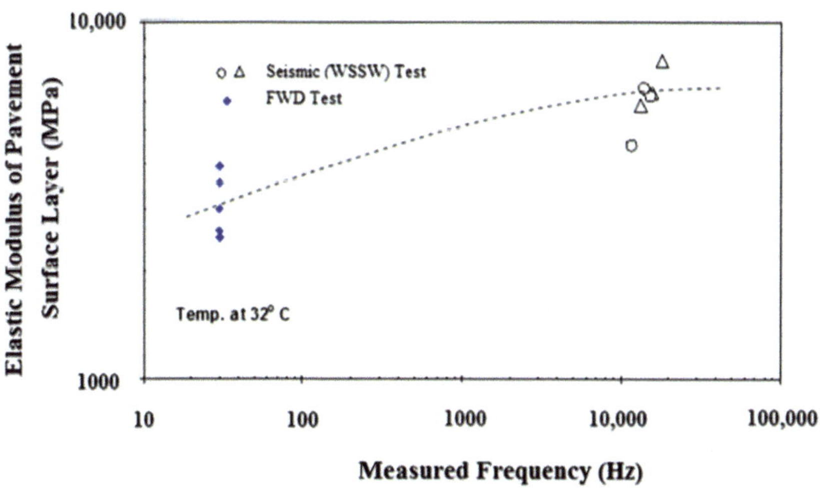

Fig. 12. Influence of frequency on the small strain elastic modulus of asphalt concrete at measured temperature of 32 °C at Purwakarta Road, Indonesia

4 Conclusions

The dynamic elastic modulus profile obtained from the proposed technique of the wavelet-spectrogram analysis of surface waves (WSSW) were presented in this paper. This technique improves the conventional SASW measurement testing and was used for investigation in the complex pavement structure. In this paper, the identification, denoising and reconstruction of the wave response spectrum from seismic surface wave propagation using time-frequency analysis of continuous wavelet transforms is presented. The spectrogram could be used to clearly identify the various events of interest mode of the seismic surface waves. By the simple calculation on phase spectrum from the surface wave data, the elastic modulus of the surface layer can be obtained without the complex calculation of the inversion process. The calculation is easy and can be simply implemented. This technique is also a very sensitive non-destructive testing (NDT) to monitor the change of the modulus of the existing surface and overlay layers.

Acknowledgments. This work is part of research project funded by Ministry of Research, Technology and Higher Education, Indonesia and the Universitas Muhammadiyah Yogyakarta (UMY). Their supports are gratefully acknowledged. We would also like to thank Dr. Siegfried (Puslitbang Jatan, Bandung) and research assistants for their assistance during field works.

References

Al-Hunaidi, M.O.: Difficulties with phase spectrum unwrapping in spectral analysis of surface waves nondestructive testing of pavements. Can. Geotech. J. (1992). doi:10.1139/t92-055

Foufoula-Georgiou, E., Kumar, P.: Wavelets in Geophysics. Academic Press, Cambridge (1995)

Ganji, V., et al.: Automated inversion procedure for spectral analysis of surface waves. J. Geotech. Geoenviron. Eng. ASCE (1998). doi:10.1061/(ASCE)1090-0241(1998)124:8(757)

Hazra, S., Kumar, J.: SASW testing of asphaltic pavement by dropping steel balls. Int. J. Geotech. Eng. (2014). doi:10.1179/1938636213Z.00000000051. Taylor & Francis

Kim, D.S., et al.: Evaluation of density in layer compaction using SASW method. Soil Dyn. Earthq. Eng. (2001). doi:10.1016/S0267-7261(00)00076-2. Elsevier

Rosyidi, S.A.P., et al.: Signal reconstruction of surface waves on SASW measurement using Gaussian derivative wavelet transform. Acta Geophys. (2009). doi:10.2478/s11600-009-0015-8. Springer

Rosyidi, S.A.P.: Comparison between 2-D and 3-D stiffness matrix model simulation of SASW inversion for pavement structure. Civil Eng. Dimens. J. Civil Eng. Sci. Appl. **9**(1), 42–48 (2007)

Rosyidi, S.A.P.: Simultaneous in-situ stiffness and anomalies measurement on pavement subgrade using tomography surface waves technique. Proced. Eng. (2015). doi:10.1016/j.proeng.2015.11.057. Elsevier

Rosyidi, S.A.P., Taha, M.R.: Wavelet spectrogram analysis of surface wave technique for dynamic soil properties measurement on soft marine clay site. Seism. Waves Res. Anal. (2012). doi:10.5772/27530. Intech

Rosyidi S.A.P., et al.: Development of VS-CBR-DCP empirical model for determining dynamic stiffness of pavement base layer using SASW. In: Proceedings of International Conference on Advanced Characterisation of Pavement and Soil Engineering Materials, pp. 895–902 (2007)

Ryden, N., et al.: Multimodal approach to seismic pavement testing. J. Geotech. Geoenviron. Eng. ASCE (2004). doi:10.1061/(ASCE)1090-0241(2004)130:6(636)

Stokoe II, K.H., et al.: Characterization of geotechnical sites by SASW method. In: Wood, R.D. (ed.) Geotechnical Characterization of Sites, pp. 15–26. Oxford and IBH Publishing Co., New Delhi (1994)

Tokimatsu, K., et al.: Effects of multiple modes on Rayleigh wave dispersion characteristics. J. Geotech. Eng. ASCE. (1992). doi:10.1061/(ASCE)0733-9410(1992)118:10(1529)
Yusoff, N.I.M., et al.: Measurements of the Elastic Modulus of Pavement Subgrade Layers Using the SASW and FWD Test Methods. The Baltic Journal of Road and Bridge Engineering. Technika, Vilnius (2015). doi:10.3846/bjrbe.2015.22

Thermal Energy Harvesting from Asphalt Roadway Pavement

Utpal Datta(✉), Samer Dessouky, and A.T. Papagiannakis

Department of Civil and Environmental Engineering,
University of Texas at San Antonio, San Antonio, TX, USA
utpal.buet@gmail.com,
{samer.dessouky,at.papagiannakis}@utsa.edu

Abstract. This study is aimed at developing energy harvesting system for electrical power generation from asphalt pavement roadways. Energy harvesting technologies from roadways is a new research arena which involves technologies that capture the wasted heat energy in pavements, accumulate and store it for later use without depletion of natural resources. The temperature of pavement surface rises up to 50 °C to 55 °C in the summer (depends on geographic location) due to the effect of the solar radiation. Soil temperature is consistently stabilized at 27 °C creating a thermal gradient profile across pavement layer structure. This Thermal gradient is a potential source of energy that can be harvested. The proposed system collects heat energy from the pavement surface and transfers to thermoelectric generators module along the pavement side. Field demonstration, laboratory testing and finite element simulation suggested that electric energy is approximately 10 mW per thermoelectric module (6.4 cm × 6.4 cm dimension). At scalable system, the produced energy can be used for LED street illuminating or empowering roadway monitoring devices inserted within the pavement in off-grid areas. Furthermore. Changes in the response of these sensors can be used as a means of monitoring the health of the pavement layers, where these sensors are installed. This paper demonstrates finite element simulation on the capability of the proposed system and preliminary findings on the output power generation.

1 Introduction

1.1 Background

Sustainable energy harvesting is one of the challenges in generating green power. There are several energy sources available for this purpose, such as photovoltaic, wind and geothermal. An emerging technology for sustainable energy harvesting is thermoelectricity, whereby temperature gradients in a structure are used to directly power thermoelectric generators (TEGs). Asphalt pavements are ideally suited for installing such technology because they readily heat up during the day especially in low latitudes. Example of peak asphalt concrete temperatures for the LTPP Section 48-0001 at Austin, Texas [1] is shown in Fig. 1. It can be observed that surface temperatures remain in the 40 °C to 55 °C range for most of the year, except for the period between the end of November and the beginning of March. On the other hand, roadside soil

L. Mohammad (ed.), *Advancement in the Design and Performance of Sustainable Asphalt Pavements*, Sustainable Civil Infrastructures, DOI 10.1007/978-3-319-61908-8_20

temperatures at a depth of 180 mm to 200 mm from the surface varies from 18 °C to 34 °C year round (Fig. 2) [2]. In such cases, there is clearly a 20 °C to 22 °C thermal gradient available throughout most of the year. This thermal gradient can be used to drive thermoelectric power generation modules (typically 5 mm in thickness) to produce electricity. This paper describes the development of such system.

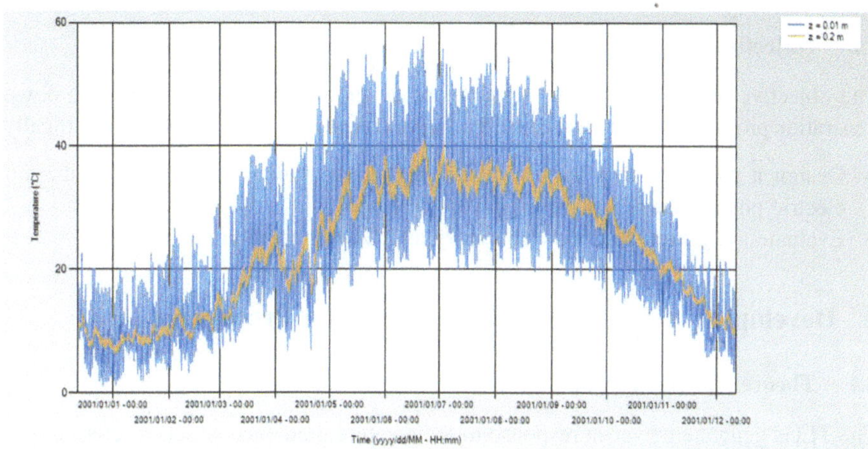

Fig. 1. Pavement temperature profile at Austin, Texas at depth of 10 mm and 200 mm using TEMP software

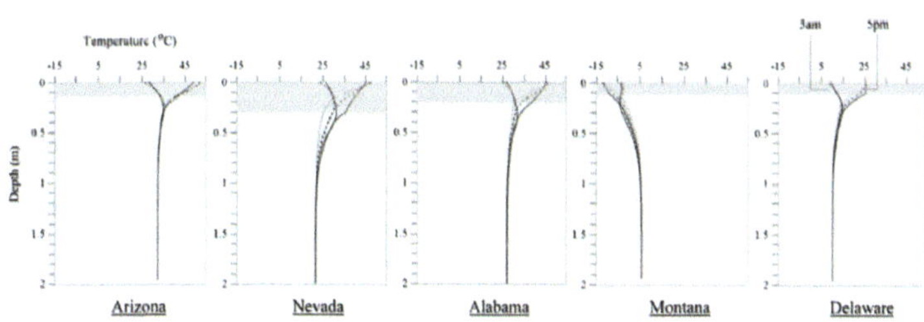

Fig. 2. Temperature profile in pavement systems across 2-m depth at different states [2]

Energy harvesting systems from temperature gradients within a pavement is a novel concept. Limited literature describes efforts to develop such systems through computer simulations and laboratory testing [3–5]. One of the problems of these systems is the location of the thermoelectric element within the asphalt concrete layer. This exposes the TEGs to traffic hazards and compromises their longevity. This not only makes TEGs less accessible, but also weakens the pavement structure itself. Other limitations are the heat dissipation that inadvertently get conveyed across the TEG leading to a reduction in system efficiency over time.

In fact, a thermoelectric power generation system should be installed deep into the pavement structure and away from the driving lanes to be accessible from roadside. The prototype described in this paper was designed to have the TEG below the pavement surface, use conductive elements to transfer the heat at lower depth, has a heat sink to maintain lower soil temperature and be installed along roadside to be accessible without interfering traffic.

1.2 Objectives

The objective of this paper is to describe the development of a thermoelectric power generation prototype from temperature gradients in asphalt pavements. More specifically:

- Design a prototype that harvests the temperature gradients available for thermoelectric power generation. From pavement layers.
- Evaluate the prototype under laboratory and field temperature conditions.

2 Developing a Thermoelectric Generation Prototype

2.1 Theoretical Background

The TEGs generate power in response to temperature gradients, which is referred to as the Seebeck effect [6]. There are four engineering parameters which define the performance of a TEG namely; ΔT, Temperature gradient (difference) between the hot and cold side of the TEG; V, Voltage generated by TEG due to temperature gradient; I_L, electric current drawn by the TEG and P_L, output power produced by the TEG. The generated voltage is given by:

$$V = N\left(0.0002 \times 1.004^{\Delta T}\right) \times \Delta T \tag{1}$$

where N is the total number of thermoelectric elements (574 for a 64 mm × 64 mm module or 254 for a 40 mm × 40 mm module; twice the number of thermocouples), and $\Delta T = T_{Hot} - T_{Cold}$, where T_{Hot} is the heat flux temperature of the hot side and T_{Cold} is the cold side surface temperature [7]. To maximize the output voltage and power, an external resistant is used with the TEG. When an external resistance, R_L, is applied to the TEG, the output voltage that appears across that load, V_L, is less than V and is a function of the external and the internal resistance R_{int} of the TEG. The output voltage, current and power are given by:

$$V_L = \frac{VR_L}{(R_L + R_{int})} \tag{2}$$

$$I_L = \frac{V_L}{R_L} \tag{3}$$

$$P_L = V_L I_L \tag{4}$$

2.2 Conceptual Design

The thermoelectric energy harvesting prototype consists of four basic components (Fig. 3):

- A Z-shape thermally isolated copper plate.
- A thin square TEG module sandwiched between the copper plate and bottom heat sink.
- A heat sink connected to the bottom of the TEG.
- A digital multimeter/voltmeter to monitor temperature and output voltage and power.

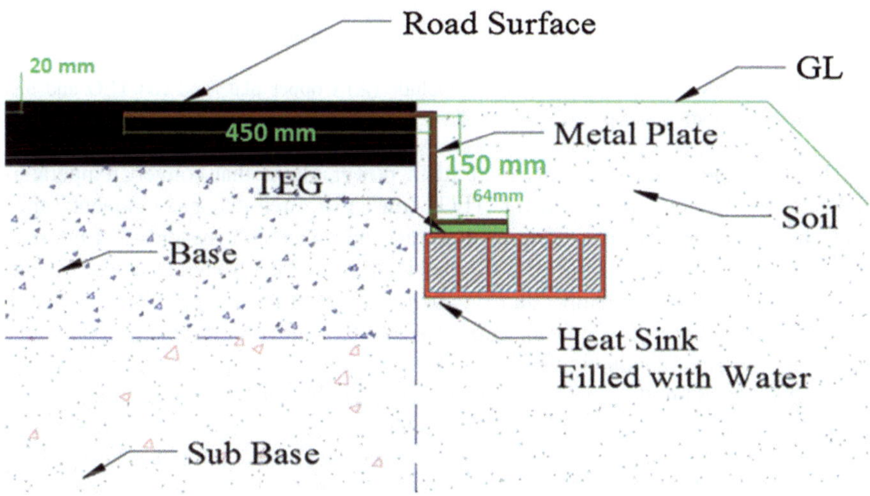

Fig. 3. Schematic of the thermoelectric energy harvesting prototype in pavement (GL: Ground level)

The copper plate allows transferring the heat from top of pavement surface down to the top of TEG module, while the heat sink provides a relatively constant temperature over time, while dissipating the heat that gets conveyed across the TEG. The more temperature gradient (ΔT) it ensures; the more output electrical power is produced. To maximize the prototype efficiency several models of thermal harvester were analyzed through finite element (FE) analysis. Finally form the output of the FE an optimum model was selected for laboratory testing. Finally, a field testing of the prototype was examined at in situ pavement conditions.

The copper plate is an elongated Z-shape metal (Fig. 4a). It is segmented in three parts; A *Heat collector* which is inserted at 20 mm depth from the top pavement surface near the edge. A *Heat conveyor* which is inserted into the near soil up to 180 mm depth from the surface. This segment is properly insulated using one layer of 2.5 mm thick Poly Vinyl Chloride (PVC) sheets. A *Heat Transfer Plate* which transfers the thermal heat into the top of the TEG module. The outer surface of the heat conveyor is covered by a PVC sheet to reduce heat loss (Fig. 4b). The bottom surface of the heat conveyor plate is attached to the TEG's upper surface with highly conductive thermal paste.

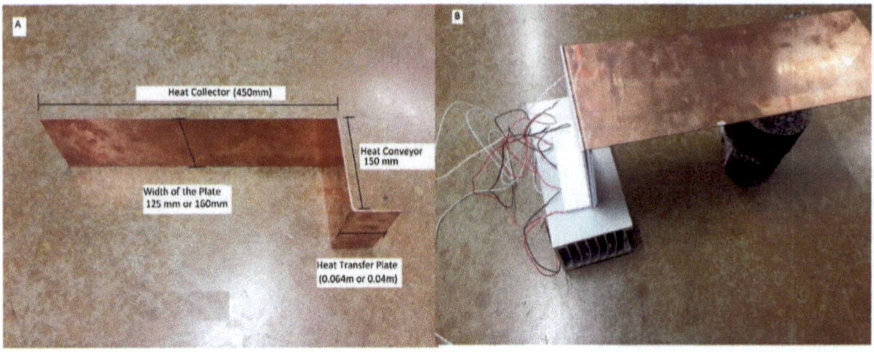

Fig. 4. The harvester prototype components include (A) Copper plate and (B) TEG and heat sinks attached to the copper plate

The TEGs are solid state devices used in power plants that convert temperature gradients directly into electric power. TEGs consist of three major components: thermoelectric materials, thermoelectric modules and thermoelectric systems that interface with the heat source (Fig. 5) [8, 9]. Thermoelectric materials generate power directly from heat by converting temperature gradient into electric voltage. These materials must have both high power factor (i.e., mWatts/°C) and low thermal conductivity. Having low thermal conductivity ensures that no significant amounts of heat from the warm side of the TEG flows into the cooler side, thus reducing the temperature gradient driving thermoelectricity. Traditionally, the main three semiconductors known to have both low thermal conductivity and high power factor were bismuth telluride (Bi2Te3), lead telluride (PbTe), and silicon germanium (SiGe) [6]. A thermoelectric module is a circuit containing thermoelectric materials that generate electricity from heat directly. A thermoelectric module consists of two dissimilar thermoelectric materials joined at their ends. There are two types, namely, n-type (negatively charged) and p-type (positively charged) semiconductors [6]. In this study two different types TEGs were used; Two of 64 mm × 64 mm modules and Four 40 mm × 40 mm modules.

The TEG generates power as a result of the temperature gradient between its top and bottom surfaces. Inadvertently, some the heat conveyed from the top to the bottom

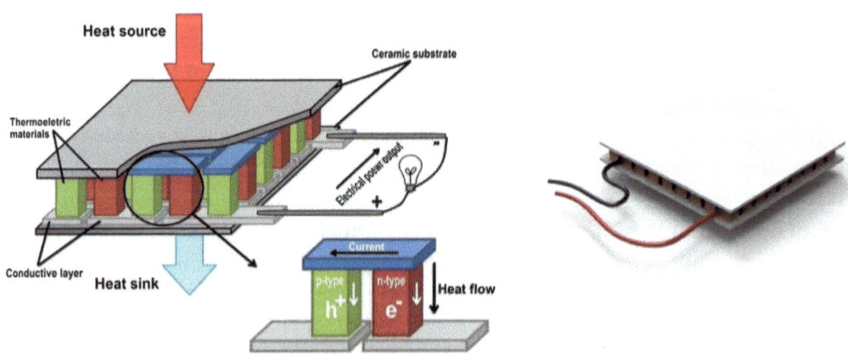

Fig. 5. Schematic diagram of the internal components of TEG [8, 9]

is minimizing the temperature gradient and output voltage. If this heat is not dissipated effectively from the bottom side, it will minimize the efficiency of the prototype. Therefore, a custom designed heat sink was developed. The aluminum heat sink used in the prototype is 300 mm (H) × 150 mm (W) × 115 mm (D) (Fig. 4b). The inner chambers of the heat sink can be filled with a heat dissipating fluid, such as water. After sealing it properly, the heat sink is glued under the bottom surface of the TEGs with highly conductive thermal paste. The heat sink maintains the TEGs bottom surface temperature as close as the surrounding soil. Hence a steady temperature gradient is ensured to maximize the power generation.

With the help of a K type thermometer data logger the temperature data of the temperature data of the road surface (T_1), heat transfer plate (T_2), surface of heat sink (T_3) and soil (T_4) was recorded. The available thermal gradient is measured as $\Delta T_0 = T_1 - T_4$ and the effective/recovered thermal gradient is measured as $\Delta T = T_2 - T_3$. TEGs are connected in series. Their positive and negative poles are extended with connecting wires up to the ground surface. The output voltage V and current I is measured with a given resistance value. The output power is calculated by multiplying voltage by current (Eq. 4).

3 Finite Element Analysis

The aim of the FE analysis, conducted using ABAQUS, was to design materials and dimensions of the thermoelectric prototype. In order to get the maximum possible temperature gradient and output power from the prototype, following key parameters were examined, i.e., effective materials for the harvester components, design depth of the TEG, dimensions of the heat transfer plate as well as number of TEG units to be used. The FE analysis involved the following assumptions:

1. The prototype is analyzed as one-dimensional heat transfer problem with steady state flow.
2. For each iteration, the pavement surface and soil temperature used as boundary conditions are assumed constant.
3. The prototype is thermally insulated i.e., no heat loss occurred during heat conduction.
4. The effective yield period for heat harvesting is considered to be 8 h in a day (i.e., 10:00 AM to 6:00 PM).

The temperature gradient between pavement surface and soil at designed depth is the available source of thermal energy. Prior to the FE analysis, the pavement temperature profile was analyzed by TEMPS software [11] to numerically quantify the thermal gradient available in pavement to effectively operate an energy harvesting prototype. For this analysis, weather data in Austin, Texas (i.e., mean hourly temperature, mean hourly wind speed, and hourly solar radiation) were used from LTPP program (Fig. 1). The temperature profile analysis indicates that except November to February the pavement temperature at 20 mm from surface varies in the range of 40 °C to 58 °C and at a depth of 200 mm it varies from 18 °C to 35 °C. In this study, the 18 °C to 36 °C was consider as adjacent roadside soil temperature at 200 mm depth (Fig. 1).

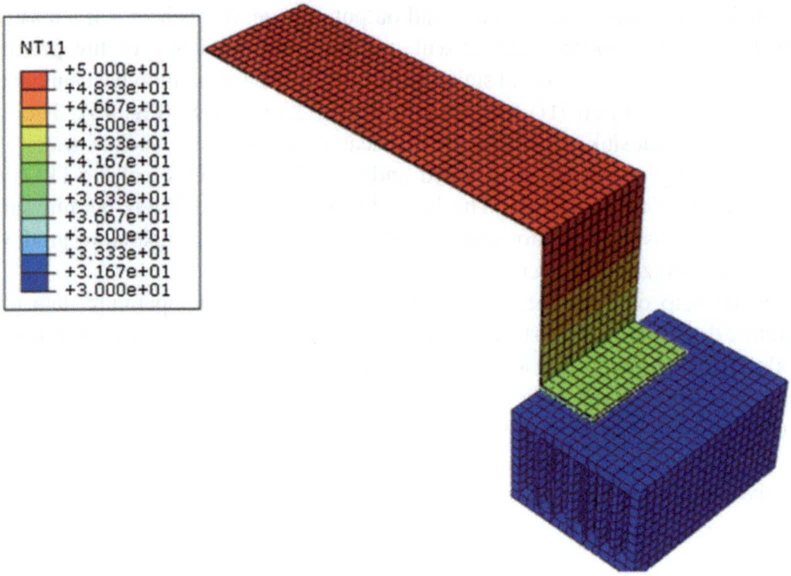

a) FE model of prototype 1

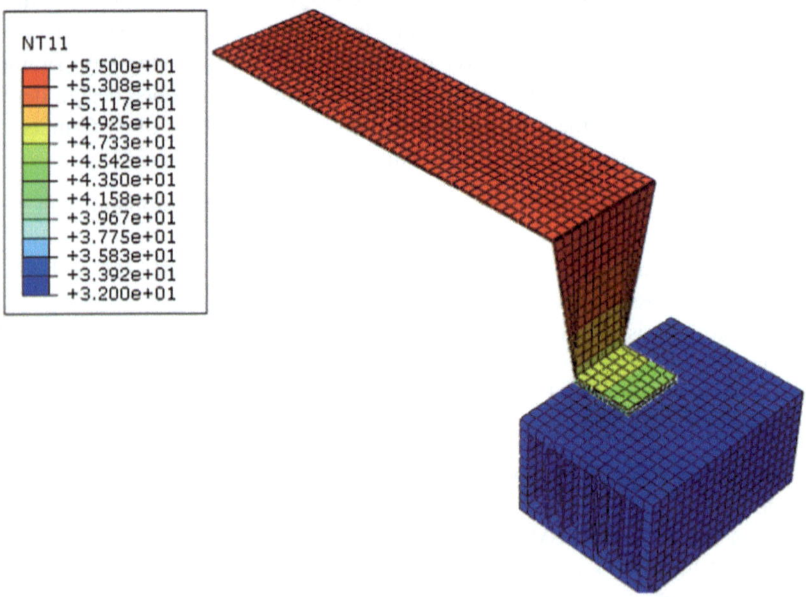

b) FE model of prototype 2

Fig. 6. FE analysis results for (a) energy harvesting prototype 1 and (b) energy harvesting prototype 2

The FE model in Fig. 6a, b consists of a Z-shape metal plate with thermal boundary conditions of 50 °C or 55 °C at top left side (e.g., pavement). A one dimensional steady state flow analysis was conducted to determine the temperature at the bottom right side (soil) of the model. The model was used to study the above mentioned key parameters to justify the optimum design elements for effective energy harvesting prototypes. The ultimate goal is to transfer that gradient between top and bottom surface of the TEG (i.e., through the heat transfer plate and heat sink connected to the TEG). With respect to the thermal properties used in the FE, the TEGs have thermal conductivity of 32 W/m.k [10], while copper plate and aluminum heat sink have thermal conductivity of 385 and 205 W/m.k [9], respectively."

In order to determine design depth (at which heat transfer plate, TEGs and heat sink will be placed), for a specific set of boundary conditions, several models with different depths were analyzed. A depth of 50 mm, 100 mm, 180 mm, 225 mm and 300 mm where considered to balance between installation cost and effectiveness of the energy harvester. The FE analysis indicates that the optimum design depth is 180 mm. (Figure 7). Going deeper into the soil reduces ΔT. The reason is with the increase of depth the heat conduction length increases resulting in reduction of thermal gradient.

For road surface temperature 50 °C and soil temperature 26 °C, three different commercially available metal conducting plates were considered namely; steel, aluminum and copper. The FE analysis shows that the copper plates yield the best output (Table 1).

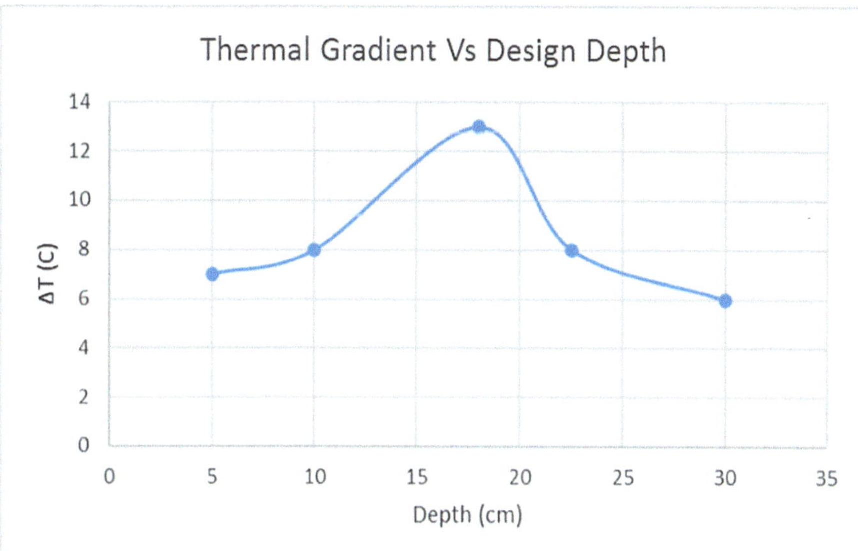

Fig. 7. FE analysis for determining the design depth of the prototypes

Table 1. FE analysis results on selecting optimum materials for Z-shape metal plate

Materials	Road surface temp. (°C)	Soil temp. (°C)	Temp. on heat transfer plate (°C)	Temp. on heat sink (°C)	Avg. recoverable temp. gradient (°C)
Copper	50	26	41–43	28–29	13
Aluminum	50	26	32–35	28–29	5
Steel	50	26	31–33	28–29	3

The FE analysis recommends the following three models after analyzing their efficiency. For the first selected model was with 2 TEGs (64 mm × 64 mm) whose design dimensions are: *Heat collector* with dimensions of 450 mm × 125 m, *Heat conveyor* with dimensions of 150 mm × 125 mm and *Heat Transfer Plate* with dimensions of 64 mm × 125 m. For the second chosen model with 4 TEGs (40 mm × 40 mm) the design dimensions are: *Heat collector* with dimensions of 450 mm × 160 mm, *Heat conveyor* with dimensions of 150 mm × 160 mm and *Heat Transfer Plate* with dimensions of 40 mm × 160 mm. For the third selected model was with 1 TEG (64 mm × 64 mm) whose design dimensions are: *Heat collector* with dimensions of 450 mm × 125 m, *Heat conveyor* with dimensions of 150 mm × 1/2 (125 + 64) mm and *Heat Transfer Plate* with dimensions of 64 mm × 64 mm (Table 2).

Table 2. FE result on determining the optimum dimensions of the Z-shape metal plate

Metal plate dimensions (mm × mm)	Road surface temp. (°C)	Soil temp. under 180 mm depth (°C)	Temp. on heat transfer plate (°C)	Temp. on heat sink (°C)	Avg. recoverable temp. gradient (ΔT) (°C)
(a = 450 × 64, b = 150 × 64, c = 40 × 64)*	50	30	38–41	30–31	9
(a = 450 × 160, b = 150 × 160, c = 40 × 160)	50	30	40–43	30–31	11
(a = 450 × 64, b = 150 × 64, c = 64 × 64)	50	30	39–41	30–31	10
(a = 450 × 125, b = 150 × 125, c = 40 × 125)	50	30	43–45	30–31	13
(a = 450 × 150, b = 150 × 1/2 (125 × 64), c = 64 × 64)	50	30	45–46	30–31	15

*a = dimension of heat collector, b = dimension of heat conveyor, c = dimension of heat transfer plate

Average soil temperature and pavement surface temperature of yield period in different months (April to June) were considered as boundary condition during the analysis. It is found that for a temperature of 22 °C (difference between pavement surface and soil at designed depth) the output gradient between top and bottom surface of the TEGs are 11 °C (Table 3).

Table 3. FE analysis results on output thermal gradient with various road and soil temperature

Boundary conditions			Input to thermo-electric generators		
Road surface temperatures (°C) (Fig. 6a)	Soil temp. under 180 mm depth (°C) (Fig. 6a)	Available temp. gradient (ΔT) (°C)	Temp. on heat transfer plate (°C)	Temp. on heat sink (°C)	Avg. recoverable temp. gradient (ΔT) (°C)
45	27	18	36–39	27–29	10
50	30	20	40–43	30–31	11
55	32	23	43–45	32–34	11

4 Laboratory and Field Testing

4.1 Prototype Fabrication and Laboratory Testing

The laboratory testing involved the evaluation of the prototype with 2 larger size TEGs (Fig. 4). The dimension of the heat sink was 300 mm × 150 mm × 115 mm. The Z-shape copper plate dimensions for each prototypes were chosen based on the FE results as mentioned previously. The outer surface of the heat conveyor is covered by a PVC and the bottom surface is fully attached to the TEG with highly conductive thermal paste. The road surface temperature condition was controlled using a heated water tub. The soil condition temperature was controlled using a heat sink filled with water at room temperature (Fig. 8). The heat conducting plate was fully inserted into the heated water tub. Temperature data was collected by a data logger and the output voltage, current and power were measured by a power meter.

Fig. 8. Laboratory test setup of the prototype and power measuring device

Tests were run at four different water tub temperature conditions (representing pavement) of 45 °C, 50 °C, 55 °C, 60 °C and heat sink temperature (representing soil) ranges from 26 °C and 28 °C. Using the data logger, the temperature was collected over a period of 40 min at four spots in the prototype; water tub, heat transfer plate at top of TEG, heat sink at bottom of TEG and room temperature. Results showed that the imposed temperature gradient was in the range of 20 °C and the output temperature gradient from the lab test was measured to be between 11 °C and 13 °C, (Fig. 9a). For both prototypes, the output voltage was in the range of 500 mV to 700 mV and the output current was in the range of 22 mA to 25 mA. Hence, the resulting output power was in the range of 11 mW to 22 mW (Fig. 9b). In summary the potential average rate of power generation for this thermoelectric energy harvesting is 1.67 mW/°C (i.e., an average output of 20 mW for ΔT of 12 °C).

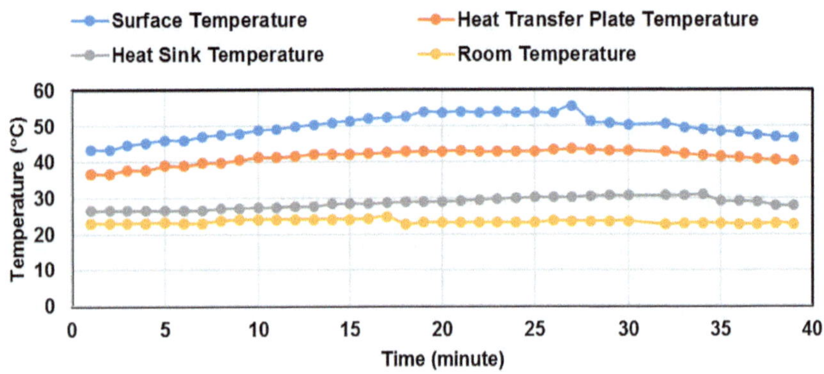

a) Prototype temperature profile in laboratory

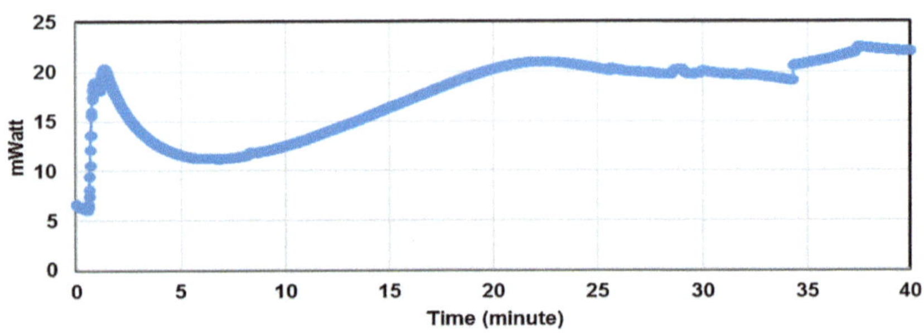

b) Prototype power output

Fig. 9. Laboratory test results showing (a) temperature profile at different components and (b) Power output for the two TEG prototype

4.2 Field Testing

The thermoelectric energy harvesting prototype developed in the lab were field tested on the pavement of the University of Texas at San Antonio campus. A 500 mm × 175 mm × 37.5 mm strip of asphalt pavement was removed at the edge of the asphalt pavement roadway to insert the energy harvester. A 900 mm × 450 mm 450 mm pit was dug in the soil adjacent to the road where the prototype was installed to accommodate the data acquisition system (Fig. 10). A four channel temperature data logger (AZ Instruments 4 Channel K Type Thermometer SD Card Data Logger) was used to collect the temperature data. A power meter (Gossen METRAHIT) was used to measure output voltage, current and power. The TEGs are connected in series and their positive and negative poles are extended with connecting wires up to the ground surface. The output voltage V and current I is measured with external resistance of 8.3 Ω.

Fig. 10. Field installation and data collection

The data was collected during the day between 1:00 AM and 6:00 PM for both prototypes for the period between April 2016 and July 2016. Figure 11a, b shows that the average thermal output gradient observed from the prototype was in the range of 5 °C to 7 °C and Fig. 11c represents the power output up to 10 mW in June 6, 2016. The average output voltage was in the range of 480 mV to 650 mV and output current was in the range of 10 mA to 18 mA and output power generated was in the range of 5 mW to 16 mW (Table 4).

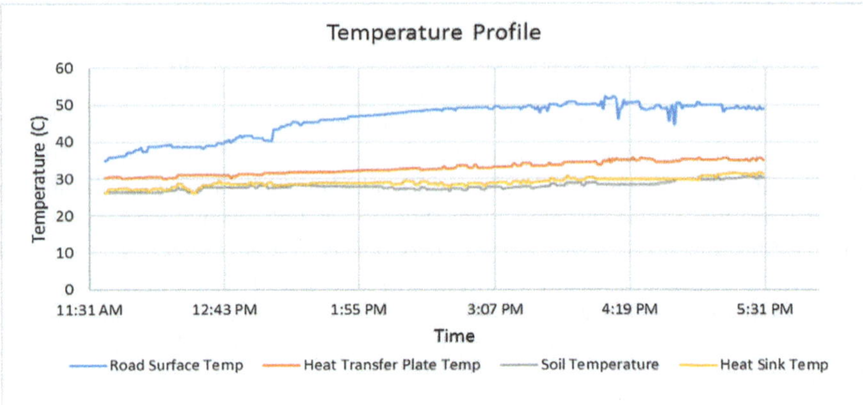

a) Prototype temperature profile in the field

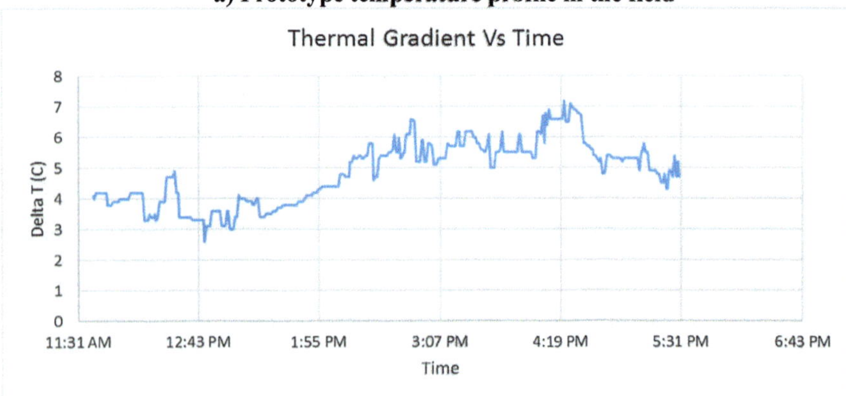

b) Prototype thermal gradient

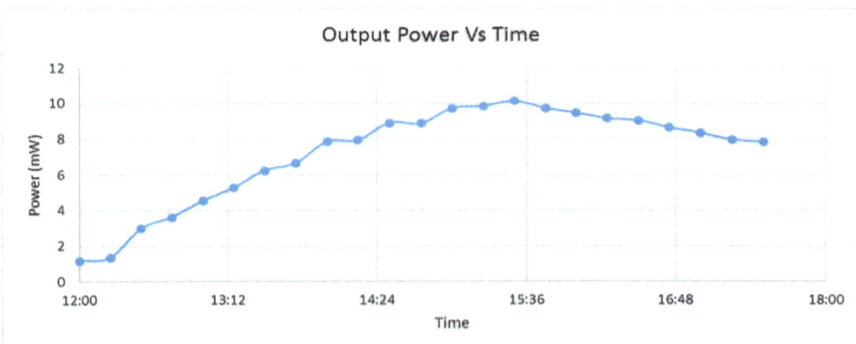

c) Prototype power output

Fig. 11. Field results during day time showing (a) temperature profile, and (b) output power from 2 TEGs on 06 June 2016

Table 4. Field test power output for the two (64 mm × 64 mm) TEGs prototype

Monthly average	Boundary conditions			Input to thermo-electric generators			Electricity output measurements		
	Road surface temp. (°C)	Soil temp. at 180 mm depth (°C)	Available temp. gradient (ΔT) (°C)	Heat transfer plate temp. (°C)	Heat sink top temp. (°C)	Recoverable temp. gradient (ΔT) (°C)	Voltage (mV)	Current (mA)	Power (mW)
Apr	47	26.3	18	34.3	28.2	6.1	520	16	8.3
May	52.3	29.4	20	37.7	30.1	7.6	650	22	14.3
Jun	54.7	32	23	39	31.9	7.1	630	18.6	11.7
Jul	57	34	23	41	34.2	6.8	590	17.3	10.2

5 Cost Analysis

For the prototypes with the two and four TEGs, the cost of the materials excluding installation was $190 and $94, respectively. Based on the experimental data collected and the temperature FE analysis, the estimated annual power output of one of these prototypes is approximately 170 and 80 kWh, respectively (Average household energy consumption per day is 30 kWh). In terms of monetary value, assuming $0.20/kWh, the total revenue is $34 and $16 per year, respectively.

6 Conclusions

Thermal harvesting prototypes using the temperature gradients within asphalt pavements appear to have the potential for sustainably generating energy roadside. The energy harvesting process is green and environmental friendly. It is not obtrusive and it does not interfere with the traffic. Furthermore, if the conductive plate top is installed 20 mm below the asphalt concrete pavement surface, the harvester would not interfere with periodic pavement overlay activities.

Numerical analysis using FE was used to quantify the components design of the prototypes. Laboratory and field testing was conducted to validate the functionality of the prototypes. Under the conditions tested, a strip of average 5-in. length along traffic directions can generate an average of 10 mW of electric power continuously over a period of 8 h a day. The recoverable power density per square feet pavement section is 16 mW (approximately) and the daily energy accumulation of the prototype is 0.5 kWh.

The produced power may look very small for conventional electronic devices. But it can be used for powering pavement health monitoring and roadway communication devices which needs power as low as 3.0 mW [12]. Especially in off-grid areas where it will cost millions of dollar to power these devices from on-grid electricity. Also it can be an alternate source of self-powered electronic road signs or markings and street illuminating with LEDs with high performance and efficient power consuming as low as 1.0 mW [13] in remote areas. With the help of proper electronic devices, the output power can be stored by charging capacitor or batteries for continuous supply of electricity.

The currently available TEGs produce greater power output in higher temperature flux (greater than 300 °C) in which they were developed for. For the proposed harvesting prototype, the heat flux in pavement temperature is considerably low for these devices. So for customized TEGs which yield better power output in lower input temperature (50 °C), the prototype efficiency could considerably improve.

References

1. Pavement temperature. https://infopave.fhwa.dot.gov. Accessed 17 May 2016
2. Hall, M., Dehdezi, P., Dawson, A., Grenfell, J., Isola, R.: Influence of the thermophysical properties of pavement materials on the evolution of temperature depth profiles in different climatic regions. J. Mater. Civ. Eng. 32–47 (2012). doi:10.1061/(ASCE)MT.1943-5533.0000357
3. Wu, G., Yu, X.B.: Computer-aided design of thermal energy harvesting system across pavement structure. Int. J. Pavement Res. Technol. **6**(2), 73–79 (2013). doi:10.6135/ijprt.org.tw/2013.6(2).73
4. Liang, G., Li, P.: Research on thermoelectrics transducers for harvesting energy from asphalt pavement based on seebeck effects (2015). http://www.crcnetbase.com/doi/abs/10.1201/b18415-78. Accessed 16 July 2016
5. Park, P., Choi, G.S., Rohani, E., Song, I.: Optimization of thermoelectric system for pavement energy harvesting (2014). http://www.crcnetbase.com/doi/abs/10.1201/b17219-220. Accessed 2 June 2016
6. Thermoelectric generator. https://en.wikipedia.org/wiki/Thermoelectric_generator. Accessed 23 May 2016
7. Modules. http://www.txlgroup.com/tegs.html. Accessed 25 May 2016
8. Longtin, J.P.: Thermoelectrically powered sensing for nuclear power plants. http://long2.eng.sunysb.edu/NEUP.html. Accessed 6 June 2016
9. Copper and Aluminum thermal conductivity. http://hyperphysics.phy-astr.gsu.edu/hbase/Tables/thrcn.html. Accessed 23 May 2016
10. Ceramic substrate. http://global.kyocera.com/prdct/fc/list/tokusei/denndou/index.html. Accessed 23 May 2016
11. TEMPS software. http://www.arc.unr.edu/Software.html#TEMPS Accessed 12 Jan 2016
12. Strain sensors. www.omega.com/techref/pdf/straingage_measurement.pdf. Accessed 30 Jan 2016
13. LED lights. https://www.thorlabs.com/thorproduct.cfm?partnumber=LED375L. Accessed 10 June 2016

Monitoring the Impact of Micro Cracks Healing Cycles on the Deformation of Asphalt Concrete Under Repeated Loading

Saad Issa Sarsam(✉) and Hanan Kadim Husain

Department of Civil Engineering, College of Engineering,
University of Baghdad, Baghdad, Iraq
saadisasarsam@coeng.uobaghdad.edu.iq

Abstract. The micro cracks healing property in asphalt concrete is considered as a good start for understanding and controlling the durability issue of the pavement. However, the impact of repeated micro crack healing cycles have not been investigated thoroughly. In this work, cylindrical asphalt concrete specimens of 100 mm diameter and 63 mm height have been prepared in the laboratory using Marshall method. Specimens were divided into two groups. The first group was subjected to repeated indirect tensile stresses, while the second group was subjected to repeated double punch shear stresses (both at 25 °C) to initiate micro cracks within the specimens using controlled stress mode of loading for 0.1 s followed by rest period of 0.9 s for specified load cycles. Afterword, Specimens were subjected to external heating in an oven at 60 °C and allowed to heal for two hours, conditioned at 25 °C for two hours, then subjected to another course of repeated tensile or shear stresses. The healing process was continued for two successive courses of loading and heating. The deformation of the specimens was monitored through continuous video capture. The impact of asphalt content, type of test, and healing cycles on the permanent deformation have been analyzed. It was concluded that higher deformation of 16% could be detected under tensile stresses as compared to that under shear stresses. The healing cycles causes (7–30) % and (16–37) % reduction in permanent deformation under repeated (ITS) and repeated (PSS) respectively.

Keywords: Micro crack · Healing · Shear · Tensile · Asphalt concrete · Repeated loading

1 Introduction

Asphalt concrete pavement is subjected to excessive wheel load repetitions and environmental impact throughout its design life, this will initiate micro cracks in the pavement layer, which adversely affect its elastic behavior and serviceability (Sarsam 2016; Sarsam and Barakhas 2015). Asphalt concrete mixture is known to have self-healing capability when proper heating or rest periods are allowed. The self-healing capability is related to the viscosity of the asphalt cement, which increases with increasing healing time, temperature and when the crack size is very small (Garcia et al. 2011; Sarsam 2015). Healing may occurs due to temperature increases within or

L. Mohammad (ed.), *Advancement in the Design and Performance of Sustainable Asphalt Pavements*, Sustainable Civil Infrastructures, DOI 10.1007/978-3-319-61908-8_21

surrounding the pavement while in the micro-cracking range, it increases the useful life of the pavement (Ajam et al. 2016). Repeated loading test and resilient modulus determination have been implemented by Roque et al. (2012) to evaluate the healing behavior of asphalt mixtures. A decrease in resilient modulus with an increasing number of load cycles is indicative of accumulation of micro-damage during the damage phase. Recovery of resilient modulus during the healing phase is indicative of damage recovery or healing. As demonstrated by Barrasaa et al. (2014), asphalt concrete mixes deteriorates under repeated traffic loading and the stiffness decreases. However, due to the healing effect asphalt mixes demonstrate strength recovery. Another study by Sarsam (2015) stated that Crack healing rate had been determined in terms of the recovered stiffness, deformation, and increment in fatigue life. Beam specimens were subjected to load repetitions at 20° C until 50% of the stiffness was retained, then beams were stored in an oven for two hours at 85° C, cooled to 20° C, then subjected to another cycles of load repetitions. It was concluded that Steric hardening appeared to play a significant role in a mixture's response during the loading and healing portions of laboratory tests. The healing phenomena could be utilized to overcome the negative impact of crack initiation. As reported by Sarsam and Rahem (2009), Su et al. (2016), it was believed that crack healing may retain the resilient properties of asphalt concrete. Many researchers had claimed that high tensile contact stresses generated on the road surface close to the tire edges may cause the appearance of small cracks. Before ageing, some of these small cracks may disappear by the kneading action of the tires and the healing effect of asphalt, while surface stiffness is low as a result of high temperature (Little et al. 2000). At the level of asphalt pavements, cracks which are observed at the pavement surface can be either bottom-up cracks which develop due to tensile strains at the bottom of the asphalt layers or top-down cracks at the pavement surface due to tensile and shear stresses, climatic effects and ageing (Kim et al. 2003). At macro level, healing is thought to occur in two ways. One way is that some of the micro cracks can be healed during the rest periods between two axle passages. Another possibility is that micro crack healing happens during summer when the temperature is high. This implies that micro cracks developing during the winter can be healed during a hot summer (Jian 2012).

The aim of this investigation was to assess the impact of healing cycles, asphalt content and loading stress type on the deformation of asphalt concrete. The influence of repeated indirect tensile stress ITS and the repeated double punch shear stress PSS on the permanent deformation will be investigated.

2 Materials and Methods

2.1 Asphalt Cement

It was obtained from Dora refinery, the physical properties are listed in Table 1.

Table 1. Physical properties of asphalt cement

Test	Result	Unit	SCRB specification
Penetration (25 °C, 100 g, 5 s) ASTM D 5	43	1/10 mm	40–50
Ductility (25 °C, 5 cm/min) ASTM D 113	156	Cm	≥ 100
Softening point (ring & ball) ASTM D 36	49	°C	50–60
Flash point (cleave land open cup) ASTM D 92	341	°C	≥ 232
After Thin-Film Oven Test ASTM D-1754			
Retained penetration of original, % ASTM D 946	31	1/10 mm	<55
Ductility at 25 °C, 5 cm/min, (cm) ASTM D-113	147	Cm	>25
Loss in weight (163 °C, 50 g, 5 h)% ASTM D-1754	0.175	%	–

2.2 Coarse and Fine Aggregates

Both type of aggregates have been obtained from Al-Nibaee quarry, Table 2 illustrates the physical properties of coarse and fine aggregates. Table 3 demonstrates the chemical and mineralogical composition of coarse aggregates.

Table 2. Physical properties of Al-Nibaee coarse and fine aggregates

Property	Course aggregate	Fine aggregate
Bulk specific gravity (ASTM C 127 and C 128)	2.610	2.631
Apparent specific gravity (ASTM C 127 and C 128)	2.641	2.6802
Percent water absorption (ASTM C 127 and C 128)	0.423	0.542
Percent wear (Los-Angeles Abrasion) (ASTM C 131)	20.10	–

Table 3. Chemical composition of Al-Nibaee coarse aggregates

Chemical compound	% Content
Silica, Sio_2	82.52
Lime, CaO	5.37
Magnesia, MgO	0.78
Sulfuric anhydride, SO_3	2.7
Alumina, Al_2O_3	0.48
Ferric oxide, Fe_2O_3	0.69
Loss on ignition	6.55
Total	99.09
Mineral composition	
Quartz	80.3
Calcite	10.92

2.3 Mineral Filler

Ordinary Portland cement was implemented as mineral filler in the asphalt concrete mixture, the Physical properties and chemical composition of Portland cement are shown in Table 4.

Table 4. Physical properties and chemical composition of Portland cement

Chemical composition	% Content
Silica, Sio_2	21.52
Lime, CaO	62.50
Magnesia, MgO	3.76
Sulfuric Anhydride, So_3	1.6
Alumina, Al_2O_3	5.63
Ferric Oxide, Fe_2O_3	3.32
Loss on Ignition	1.3
Total	99.63
Physical properties	
% Passing Sieve No.200 (0.075 mm)	98
Apparent specific gravity	3.1
Specific surface area (m^2/kg)	3.55

2.4 Selection of Asphalt Concrete Combined Gradation

The selected gradation in this work follows the (SCRB 2003) Specification for dense graded wearing course, with 12.5 (mm) nominal maximum size of aggregates. Table 5 show the selected aggregate gradation.

Table 5. Gradation of aggregate for wearing course.

Sieve size (mm)	19	12.5	9.5	4.75	2.36	0.3	0.75
Selected gradation	100	95	83	59	43	13	7
SCRB specification	100	90–100	76–90	44–74	28–58	5–12	4–10

2.5 Preparation of Hot Mix Asphalt Concrete

The aggregate was first washed, dried to a constant weight at 110°C, then sieved to different sizes, and stored in plastic containers. Coarse and fine aggregates were combined with mineral filler to meet the specified gradation. The combined aggregate mixture was then heated to a temperature of (150°C) before mixing with asphalt cement. The asphalt cement was heated to a temperature of (150°C) to produce a kinematic viscosity of (170 ± 20) centistokes. Then, asphalt cement was added to the heated aggregate to achieve the desired amount, and mixed thoroughly using mechanical mixer for two minutes until all aggregate particles were coated with thin film of asphalt cement. Marshall Size specimens were prepared in accordance with (ASTM D1559 2009) using 75 blows of Marshall hammer on each face of the specimen. The optimum asphalt content was determined as per the procedure above to be 4.9% by weight of aggregates. The prepared Marshall Size Specimens were divided into two sets, the first set was subjected to the repeated indirect tensile stress test at 25°C, while the second set was subjected to repeated double punching shear stress test at 25°C. Additional asphalt concrete specimens have been prepared using asphalt cement of 0.5% above and below the optimum asphalt content. Specimens have been tested in duplicate, and the average value was considered for analysis. Figure 1 shows part of the prepared specimens.

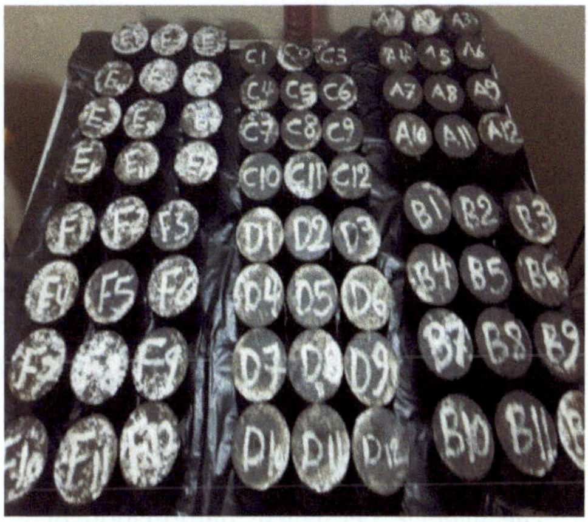

Fig. 1. Part of the prepared specimens

2.6 Monitoring the Deformation of Asphalt Concrete Under Repeated Indirect Tensile Stress

The test was conducted according to AASHTO (2013). The Pneumatic repeated load system (PRLS) shown in Fig. 2 was implemented, asphalt concrete specimens were subjected to repeated indirect tensile stress for 20 min at 25°C to allow the initiation of micro cracks. Such timing and test conditions was suggested by Sarsam and AL-Shujairy (2015). Compressive repeated loading was applied on the specimen which

Fig. 2. PRLS system

was centered on the vertical diametrical plane through two parallel loading strips (12.7 mm) wide. Such load assembly applies repeated indirect tensile stress (ITS) on the specimen in the form of rectangular wave with constant loading frequency of (60) cycles per minutes. A heavier sine pulse of (0.1) sec load duration and (0.9) sec rest period is applied over the test duration. Figure 3 demonstrates the repeated (ITS) test setup. Before the test, Specimens were stored in the chamber of the testing machine (PRLS) at room temperature of (25 ± 1°C), dial gage of the deformation reading was set to zero before test starts, and the pressure actuator was adjusted to the specific stress level equal to 56 kPa. A digital video camera was fixed on the top surface of the (PRLS) to capture dial gage reading. The test was continued for 20 min, upon completion of test, the recording was terminated. The Specimens were withdrawn from the testing chamber and stored in an oven for 120 min at 60° C to allow for crack healing initiation. Specimens were returned to the testing chamber, conditioned for 120 min at 25° C, and then subjected to another cycle of repeated indirect tensile stress at 25° C for 20 min. The deformation of the specimens under repeated indirect tensile stress and the number of load repetitions were also captured using video camera. The procedure of heating, conditioning and testing of the specimen was repeated for a second cycle of healing. The impact of healing on the accumulated permanent deformation was assessed. The average permanent deformation of two samples of each asphalt cement percentage were calculated and considered for analysis as recommended by Sarsam (2016). A Total of 18 specimens were prepared and tested.

Fig. 3. Repeated indirect tensile test

2.7 Monitoring the Deformation of Asphalt Concrete Specimens Under Double Punch Shear Stress

The second set of asphalt concrete specimens were subjected to repeated double punch shear stresses (PSS) for 20 min at 25°C to allow the initiation of micro cracks. Compressive repeated loading was applied on the specimen which was centered between the two plungers of 25.4 mm diameter as per the procedure described by Sarsam and AL-Zubaidi (2014). Such load assembly applies compressive load which was resisted by the aggregate particles interlock of the specimen through shear resistance. The stress on the specimen is in the form of rectangular wave with constant loading frequency of (60) cycles per minutes. A heavier sine pulse of (0.1) sec load duration and (0.9) sec rest period is applied over test duration. Before the test, Specimens were stored in the chamber of the testing machine at room temperature (25 ± 1°C), dial gage of the deformation reading was set to zero before test start and the pressure actuator was adjusted to the specific stress level equal to 56 kPa. A digital video camera was fixed on the top surface of the (PRLS) to capture dial gage reading. Figure 4 demonstrates the PSS test setup. The test was continued for 20 min, upon completion of test, the recording was terminated. The Specimens were withdrawn from the testing chamber and stored in an oven for 120 min at 60°C to allow for crack healing initiation. Specimens were returned to the testing chamber, conditioned for 120 min at 25°C, and then subjected to another cycle of repeated punching shear stress at 25° C for 20 min. The deformation of the specimens under repeated punching shear stress and the number of load repetitions were captured using video camera. The procedure of heating, conditioning and testing of the specimen was repeated for a second cycle of healing. The impact of healing on the accumulated permanent deformation was assessed. The average permanent deformation of two samples of each asphalt cement percentage was calculated and considered for analysis as recommended by Sarsam (2015). A Total of 18 specimens were prepared and tested. Figure 5 demonstrates the asphalt concrete specimens after repeated ITS test, while Fig. 6 shows the specimens after repeated PSS test.

Fig. 4. Repeated punching shear test

Fig. 5. Specimens after repeated ITS

Fig. 6. Specimens after repeated PSS

3 Results and Discussion

3.1 Impact of Healing Cycles on Deformation Under Repeated Tensile Stresses

Figures 7, 8 and 9 demonstrates the variation in the deformation under repeated (ITS) for various asphalt percentages. It indicates that at 1000 load repetitions, the permanent deformation decreases by 3% after the first healing cycle, while the reduction reaches 17% after the second healing cycle for 4.4% asphalt content. When asphalt content increases to 4.9%, that implementation of healing cycles had significantly reduces the permanent deformation by 17.2% and 30% for one and two healing cycles respectively. When asphalt content was increased to 5.4%, the reduction in permanent deformation was 7% and 23% after one and two healing cycles respectively.

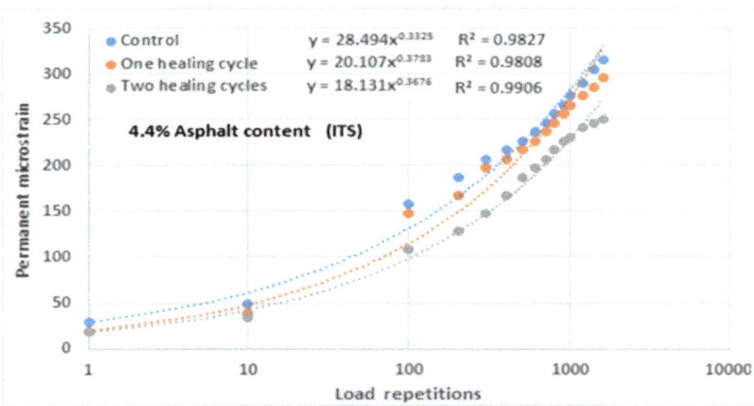

Fig. 7. Deformation at 4.4% asphalt under ITS

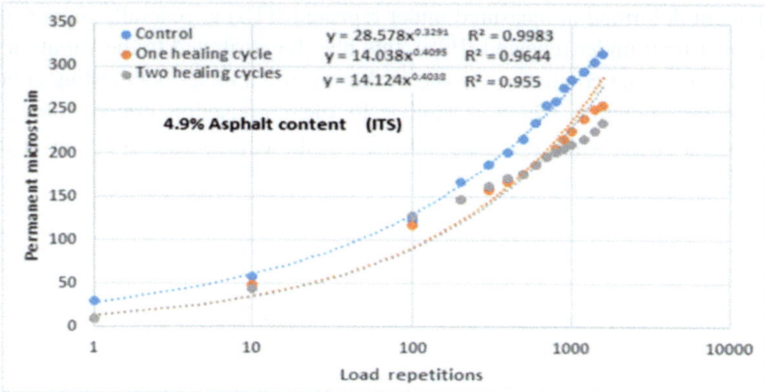

Fig. 8. Deformation at 4.9% asphalt under ITS

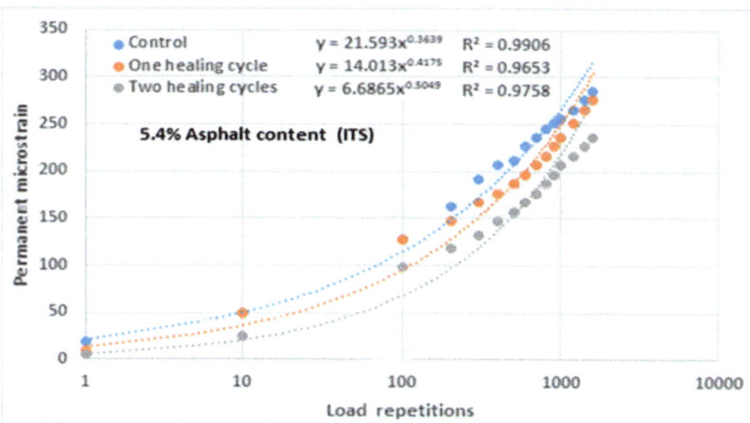

Fig. 9. Deformation at 5.4% asphalt under ITS

3.2 Impact of Healing Cycles on Deformation Under Repeated Punching Shear Stresses

Figures 10, 11 and 12 shows the variation in the permanent deformation under repeated (PSS) for various asphalt percentages. It can be noted that at 1000 load repetitions, the permanent deformation decreases by 16% and 37% after the first and second healing cycles respectively for 4.4% asphalt content. When asphalt content increases to 4.9%, that implementation of healing cycles had significantly reduces the permanent deformation by 25% and 30% for one and two healing cycles respectively. When asphalt content was increased to 5.4%, the reduction in permanent deformation was 29% and 37.5% after one and two healing cycles respectively. It can be noted that the permanent deformation measured under repeated (ITS) is generally higher by 16% than that measured under repeated (PSS), this may be attributed to the variation in the testing techniques, the confinement issue of the specimen under both testing condition, and the testing direction which is similar to the compaction direction in case of punching shear stress while it is perpendicular to the compaction direction in case of repeated (ITS).

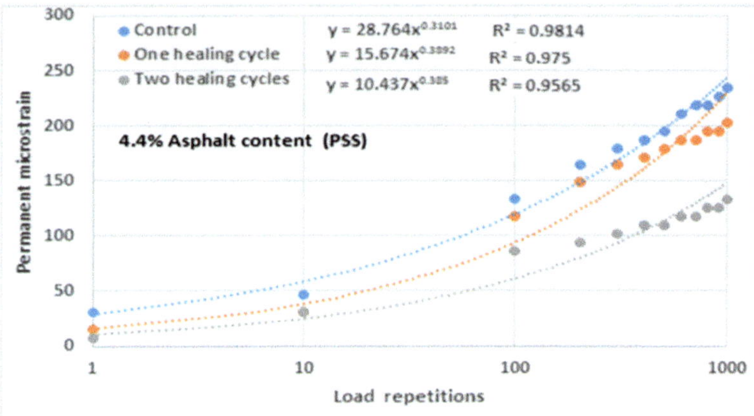

Fig. 10. Deformation at 4.4% asphalt under PSS

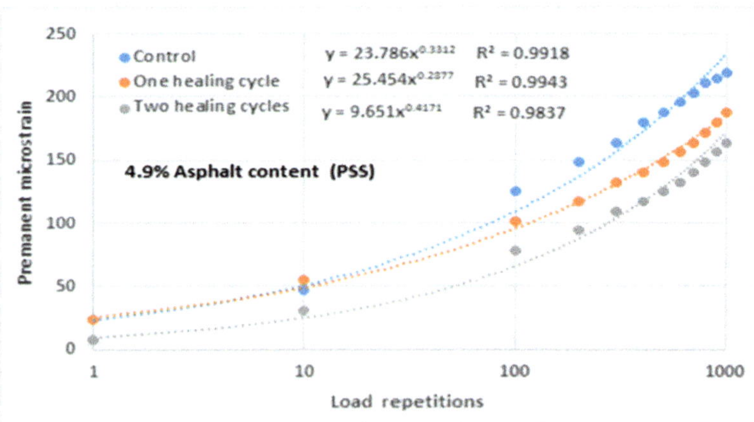

Fig. 11. Deformation at 4.9% asphalt under PSS

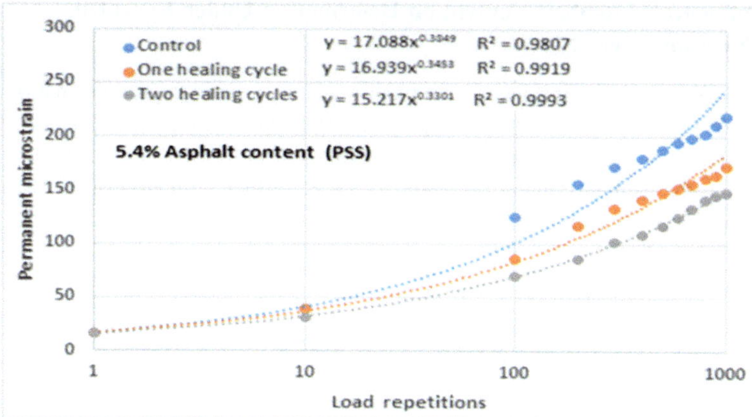

Fig. 12. Deformation at 5.4% asphalt under PSS

3.3 Influence of Asphalt Content on Deformation Under Repeated Tensile Stresses

Figure 13 illustrates the influence of asphalt content on permanent deformation under repeated (ITS) after the second healing cycle. At 1000 load repetitions, as the asphalt content increases from 4.4% to 4.9%, the permanent deformation decreases by 8%, while when the asphalt content increases to 5.4%, further reduction of 14.5% in permanent deformation could be detected.

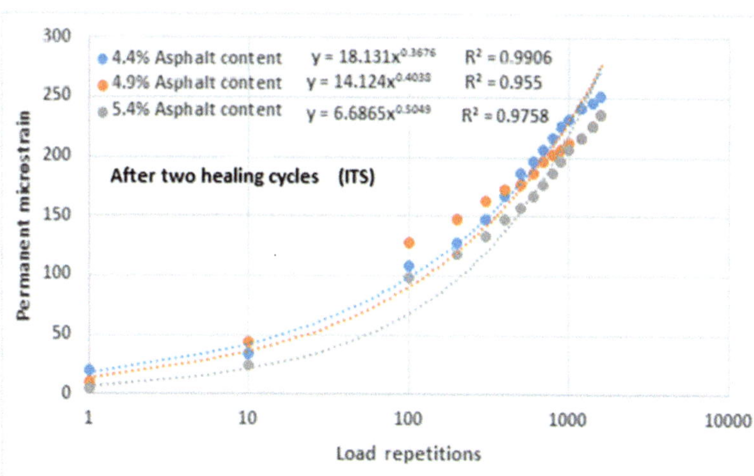

Fig. 13. Impact of asphalt content under ITS

3.4 Influence of Asphalt Content on Deformation Under Repeated Punching Shear Stresses

Figure 14 illustrates the influence of asphalt content on permanent deformation under repeated (PSS) after the second healing cycle. At 1000 load repetitions, as the asphalt content increases from 4.4% to 4.9%, the permanent deformation decreases by 12%, while when the asphalt content increases to 5.4%, further reduction of 23.5% in permanent deformation could be detected.

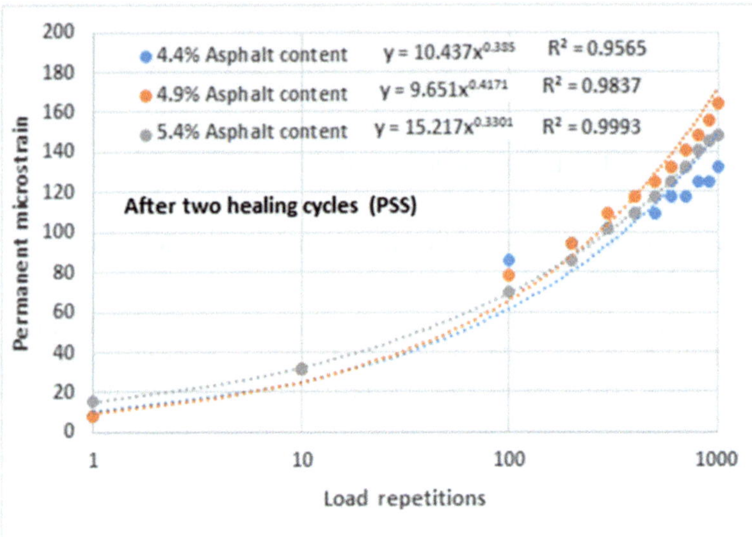

Fig. 14. Impact of asphalt content under PSS

4 Conclusions

Based on the limited testing program, the following conclusions could be drawn

1. At 1000 of (ITS) load repetitions, the permanent deformation decreases by (3 and 17) %, (17.2 and 30) %, and (7 and 23) % after the first and second healing cycle for 4.4, 4.9, and 5.4% asphalt content respectively.
2. At 1000 of (PSS) load repetitions, the permanent deformation decreases by (16 and 37) %, (25 and 30) % and (29 and 37.5) % after the first and second healing cycles for 4.4, 4.9, and 5.4% asphalt content respectively.
3. The permanent deformation measured under repeated (ITS) is generally higher by 16% than that measured under repeated (PSS).
4. The permanent deformation under repeated (ITS) decreases by (8 and 14) % when asphalt content increases from (4.4 to 4.9) % and (4.9 to 5.4) % respectively.
5. The permanent deformation under repeated (PSS) decreases by (12 and 23.5) % when asphalt content increases from (4.4 to 4.9) % and (4.9 to 5.4) % respectively.

References

AASHTO: Standard Specification for Transportation Materials and Methods of Sampling and Testing. American Association of State Highway and Transportation Officials, 14th Edition, Part II, Washington, DC (2013)

Ajam, H., Lastra-González, P., García, P.: Self-healing of dense asphalt concrete by two different approaches: electromagnetic induction and infrared radiation. In: Proceedings, 8th RILEM International Conference on Mechanisms of Cracking and De-bonding in Pavements, vol. 13, pp. 241–246. Springer, Netherlands (2016)

ASTM: Road and Paving Materials. Annual Book of ASTM Standards, Volume 04.03. American Society for Testing and Materials, USA (2009)

Barrasaa, R., Lópeza, V., Montoliua, C., Ibáñezb, V., Pedrajasc, J., Santarénd, J.: Addressing durability of asphalt concrete by self-healing mechanism. In: Proceedings, XVIII Congreso Panamericano de Ingeniería de Tránsito, Transporte y Logística (PANAM 2014), Social and Behavioral Sciences 162, pp. 188–197 (2014)

Garcia, A.H., Schlangen, E., van de Ven, M.F.C.: Asphalt mastic self-healing. In: Proceeding of the 3rd International Conference Self-Healing Materials, pp. 299–300. Bath, UK (2011)

Jian, Q.: Self-Healing of Asphalt Mixtures-Towards a Better Understanding of the Mechanism, PhD. Dissertation, TU Delft (2012)

Kim, Y., Little, D.N., Lytton, R.L.: Fatigue and healing characterization of asphalt mixtures. J. Mater Civil Eng. ASCE **15**(1), 75–83 (2003)

Little, D., Lytton, R., Si, Z., et al.: Crack Phenomenology: Formation and Healing—Task K Findings, Interim Report. Texas Transportation Institute, College Station, TX (2000)

Roque, R., Simms, R., Chen, Y., Koh, C., Lopp, G.: Development of a Test Method That Will Allow Evaluation and Quantification of the Effects of Healing on Asphalt Mixture. University of Florida, Project UF00084223, Final report (2012)

Sarsam, S.I.: Sustainability of asphalt pavement in terms of crack healing phenomena: a review. In: Trends in Transport Engineering and Applications, Volume 3, Issue 2, STM Journals (2016)

Sarsam, S.I.: Crack healing potential of asphalt concrete pavement. Int. J. Sci. Res. Knowl. **3**(1), 001–012 (2015)

Sarsam, S., Rahem, F.: Cracking and Crack Healing in Asphalt Concrete Pavement, Graduation Project Report. Department of Civil Engineering, University of Baghdad, Iraq (2009)

Sarsam, S., Barakhas, S.: Influence of load repetitions and heating on micro crack healing of asphalt stabilized subgrade soil. Int. J. Mater. Chem. Phys. **1**(3), 399–405 (2015)

Su, J., Yang, P., Wang, Y., Han, S., Han, N., Li, W.: Investigation of the self-healing behaviors of microcapsules/bitumen composites by a repetitive direct tension test. Materials **9**(600), 2–18 (2016). doi:10.3390/ma9070600

State Commission of Roads and Bridges SCRB: Standard Specification for Roads & Bridges. Ministry of Housing & Construction, Iraq (2003)

Sarsam, S.I., AL-Shujairy, A.M.: Assessing tensile and shear properties of recycled sustainable asphalt pavement. J. Eng. **21**(6) (2015)

Sarsam, S.I., AL-Zubaidi, I.: Resistance to deformation under repeated loading of aged and recycled sustainable pavement. Am. J. Civil Struct. Eng. **1**(2), 34–39 (2014). Sciknow Publications Ltd. USA

A State-of-the-Art Review of Different Conditions Influencing the Behavioral Aspects of Flexible Pavement

Piyush G. Chandak[1(✉)], Anand B. Tapase[2], Sabir S. Sayyed[1], and Abdulrashid C. Attar[3]

[1] Annasaheb Dange College of Engineering and Technology, Ashta, India
{pgc_civil, sss_civil}@adcet.in
[2] Karmaveer Bhaurao Patil College of Engineering, Satara, India
tapaseanand@gmail.com
[3] Rajarambapu Institute of Technology, Sakhrale, India
abdulrashid.attar@ritindia.ed

Abstract. The paper provides a state-of-the-art review of different conditions influencing the behavioral aspects of flexible pavement. The conditions which influence the pavement behavior are loading intensities and tyre pressure intensity, environmental conditions, surface and subsurface temperature, seepage, thickness combinations of different component layers, material properties, etc. Current pavement design and analytical procedures are discussed and questioned. The life of the road is theoretically designed for repetition of standard axle load, wherein, negligence in consideration of overloaded vehicles and high tyre inflation pressures even up to 1 MPa is noticed. The latest developments in the design and analytical procedures of pavements are highlighted. An overview of finite element modeling efforts involving different aspects is also presented. A combined effect of various actual conditions in the field affecting the pavement performance needs to be studied in details, therefore, there is a demand for an application of analytical tool which can accommodate the details of the complex system. In this connection, it should be noted that the versatile finite element solution technique holds a bright promise. Therefore, it is proposed to discuss at length the application of the finite element method towards the design of the flexible pavements.

Keywords: Flexible pavement · Finite element analysis

1 Introduction

Direct or indirect application of empirical relationship is the backbone of current pavement design procedures which are based on long-term experience and field tests such as AASHO Road Test [27, 34]. A good pavement design is one that provides the expected performance with appropriate economic consideration. The material characterisation of different pavement layers and traffic loading are the primary inputs into the mechanistic model. These approaches are reasonably accurate for design purposes in early years but the scenario has been changing due to rapid change in axle load and its

L. Mohammad (ed.), *Advancement in the Design and Performance of Sustainable Asphalt Pavements*, Sustainable Civil Infrastructures, DOI 10.1007/978-3-319-61908-8_22

configuration which has to be considered for its worst effect while designing flexible pavements. The enhanced study is necessary to find fatigue and rutting distress developed in different layers of flexible pavements due to the combined effect of different environmental parameters (like seepage, temperature etc.) and overloading with inflated tire pressure. Flexible pavements are subjected to various conditions which affect the pavement response, but it is practically not possible and even feasible to make a combination of all above parameters in a field, which creates a need to develop a model which considers a multifaceted system comprising of characteristics of the various environmental condition. FEM is a multidisciplinary tool which can analyze or evaluate the combined effect of above parameters which will be helpful to predict the behavioral aspects of flexible pavement. Day by day increasing the flow of commercial vehicles along with excessively loaded trucks beyond double its capacity and changes in environmental factors has been responsible for reducing the life of pavement [26]. In the present paper various improvements in the design procedures and its application to the present day condition are discussed in details.

2 Literature Review

Empirical methods, limiting shear failure methods, limiting deflection methods, regression methods based on pavement performance of road tests (like AASHTO) and mechanistic-empirical methods are the five basic categories on which the flexible pavement design methods are dependent [14]. The disadvantage of an empirical approach like CBR method of design is that the outcome is based on particular set of environmental, material, and loading conditions. The design is not validated to the other conditions. The major approach for designing a pavement along with stability and safety is the riding comfort wherein the methods based on limiting shear failure approach fails to address. Pavement distresses in the form of potholes, fatigue in the form of cracking and rutting are caused due to excessive stresses and strains developed in the pavement which is not seen in the methods which are based on the limiting deflection approach. The design equations based on AASHTO method which is a result of road test can be applied only to the conditions for which it was formulated [2, 21]. For its application to other conditions extensive modifications based on theory or experience is needed. The Shell method and the Asphalt institute method are the most promising mechanistic-empirical methods which use strain at critical locations to control permanent deformation. The ability to predict the type of distress is the advantage of M-E design approach over the other approaches.

Traffic and loading, materials, environment, and failure criteria are four broad categories in which design factors are distributed. The axle loads, the number of load repetitions, wheel spacing, tyre-pavement contact area and vehicle speed should be included in traffic and loading factors. In mechanistic design procedure, the mention traffic inputs are used in the structural model except for the number of repetition which is used in distress model.

Temperature and seepage both affect elastic moduli of the various layers which ultimately affects the performance of pavement [14, 27]. As in a case of elastic moduli which is temperature dependent the design of pavement should be done considering the dynamic behavior of the various material, increasing traffic, and uneven climatic conditions incorporated during analysis. Along with temperature and precipitation, in the particular areas wherein environmental factors like frost penetration, freezing index, the nearness of ground water table are the factors which need to be considered in the analysis. The climatic models including the heat transfer model, etc. For determining the temperature distribution with respect to space and time, the moisture equilibrium model for determining final moisture distribution in the subgrade, and infiltration-drainage model for predicting the degree of saturation of granular bases can be used in mechanistic design procedure [14]. Zuo et al. [39] observed seasonal temperature and water content variations and evaluated their effects on predicted pavement life. The results of the parametric study showed that the temperature averaging period, the temperature gradient in the asphalt, and the timing and duration of a wet base and subgrade conditions affect the estimation of pavement life. To determine the responses of pavement such as stresses, strains, and displacement in critical components inputs should be given in the form of basic material properties like modulus and poison's ratio in structural models. Those responses are then used in the various failure criteria's in the form of distress model including fatigue cracking, rutting, and low temperature cracking, from which the reliability level and their final design are obtained.

Pavement responses such as stress, strains, and deflection caused due to load, material properties, climatic condition is determined in mechanistic design, which is then empirically correlated to the pavement performance. A number of such mechanistic pavement response models have been developed over the years, ranging from Boussinesq's one-layer model to multi-layer elastic theories to finite element models. Boussinesq [14] formulated a simple formulation for determining stresses, strains and deflections of a homogeneous, isotropic, linear elastic half-space, with Young's modulus and Poisson's ratio subjected to a static point load. Boussinesq's equations were originally developed for a static point load. Later, Boussinesq's equations were further extended by other researchers for a uniformly distributed load by integration [37]. Boussinesq's closed form solution is a very simple and useful approach to mechanistic pavement analysis. Yoder and Witczak [37] suggested that Boussinesq theory can be used to estimate subgrade stresses, strains, and deflections when the base and the subgrade have similar stiffness. Burmister [14] developed a solution for two layers and subsequently for a three layers system. Several layered analytical elastic models have been developed which are generally based on Burmister. Finite element analysis methods have been developed to model flexible pavement responses. Duncan et al. [7, 14] first applied the finite element method for the analysis of flexible pavements.

Helwany et al. [13], illustrated the usefulness of finite element method by discretising a three layer pavement system with the right boundary at a distance of about 8 times the loaded radius subjected to different types of loading. Two-dimensional followed by three-dimensional analysis is performed after validating the analysis with reported case histories to measure the critical performance parameters. Modified Duncan model is used to represent the nonlinear behavior of granular base layer. 2-D axis-symmetrical finite element analysis is performed using the computer program DACSAR. The analysis is further elaborated to 3-D analysis using Nike-3D computer program.

Park et al. [33] focused on the behavior of low volume roads under nonlinear finite element model (FEM). Single tyre load of 40 KN [4] was used for analysis of side boundary. For this study nonlinear finite element model with stress, dependency is developed. The single tyre load used for analyses side boundary is 10 times tyre radii and at 2 m depth and resilient modulus is noted at 3 different locations i.e. at the surface, near and edge of loading point. Further, this study can be extended by adding, more layers (increasing thickness) of AC and including base and sub base. This model can be very helpful for selection of proper materials with proper characteristics which further improves the analysis process. By comparing the developed nonlinear finite element model with a linear model, it was concluded that nonlinear FEM was more suitable to calculate reduced tension in the bottom half of unbound base layers. Upon further comparison with field tests, it was observed that nonlinear stress-dependent finite element method was reliably closer to the observed measurements. Proper selection of materials is also necessary to improve the prediction of the behaviors using developed stress dependent nonlinear FEM.

Saad et al. [5] reported the use of three-dimensional finite element analysis (FEA) for elastoplastic models with ADINA program. The work focuses on the analysis of conventional flexible pavement foundation systems, dynamic response, corresponding to fatigue, strain at the bottom of the asphalt concrete layer and rutting strain at the top of subgrade material. A convenient FEA model was developed considering the domain size and traffic load simulation, which was then examined using sensitivity analysis to interpret the influence of foundation parameters on pavement response. A primary analysis was performed with a view to compare the effect of the elastoplastic material as foundation material with the linear elastic material as the foundation materials. From the analysis, it is noticed that the elastoplasticity caused an increase in rutting strain, maximum tensile strain, and maximum vertical surface deflection. The model sensitivity analysis considering elastoplastic materials, highlights that subgrade material has little or no impact on fatigue strain but reduces rutting strain remarkably. Regardless of subgrade quality, the analysis performed by Saad et al. [6] illustrates that a strong base quality significantly decreases fatigue, strain for thick and thin bases as compared to the weak base. While the reduction in base thickness showcased an increase in fatigue, strain while rutting strain constantly increased.

Siddharthan et al. [24] assessed field pavement responses for a variety of loadings in different environmental conditions by various off-road vehicles three-dimensional moving load analysis, which investigated contact stress distribution by bulky off-road vehicle tyres and compared with computed and actual pavement behavior. Thin and thick pavements were tested for three types of empty and loaded vehicles. Field testing for relatively thin and relatively thick flexible pavement was tested under three types of empty and loaded off-road vehicles. Each pavement section was instrumented with pressure cells, deflection gages, and strain gauges at the bottom of the asphalt–concrete layer. The generated finite-layer approach could handle complex surface loadings such as multiple loads and non-uniform tyre-pavement contact stress distributions.

The beneficial effect of high modulus geosynthetic materials into the foundation of pavement is documented in the study conducted by Saad et al. [6] for the improvement in the critical parameters. It was concluded from the study that the geogrid reinforcement when placed at bottom of the AC layer, rutting strain reduced up to 48% and was independent of base thickness and subgrade quality. In the thin base pavement of 152.4 mm it was observed that rutting strain reduced in the range of 2–34%. When placed at one-third from base bottom rutting strain value decreases significantly, but can be increased by using the sound base material.

Diefenderfer et al. [8] developed a model for predicting the pavement temperature, which is beneficial to those who need to determine the pavement temperature profile in order to calculate in situ pavement engineering characteristics. By using ambient temperature trends from historical records, these models can predict anticipated future pavement temperatures and thus aid researchers in determining the amount of time that pavements are subjected to critical temperatures.

Immanuel and Timm [16] used layered elastic analysis to compare predicted vertical stress in the base and subgrade layers to field measured vertical pressures obtained from the National Center for Asphalt Technology (NCAT) Test Track. The authors found that the predicted pressure was only a reasonable approximation up to vertical pressures of 82 kPa in the base and 48 kPa in the subgrade. Wu et al. [36] calculated the vertical stress at the bottom of base layer using the multilayer elastic program ELSYM5 and compared it to data from the Louisiana ALF. The study found the calculated vertical stresses to be two to eight times higher than the measured field values.

Masad et al. [32] focused on comparative analysis of using isotropic and anisotropic models in flexible pavement response along with an evaluation of permanent deformation and fatigue cracking using NCHRP 1-37A [20] design guide. Anisotropic properties were assumed for unbound base and subbase layers to measure surface deflection which reduced unrealistic tensile stresses which were assumed in the isotropic model in the granular base. Finite element predictions were comparatively analyzed with AASHO road test measured deflections.

The anisotropic nonlinear model depicted longer fatigue lives which reduces the need for large shift factors while establishing a relationship between empirical laboratory calculations to field conditions. The study highlights total permanent deformation in asphalt layers to be more considering anisotropic properties in NCHRP 1-37A model. For the anisotropic, model high tensile stresses were calculated in the lower regions of asphalt layers and total rutting for the model was more than an isotropic model. The Mechanistic-Empirical Pavement Design Guide (MEPDG) recommends using the multilayer elastic program JULEA, which is a modified version of WESLEA, and linear models to compute flexible pavement responses [19, 20]. Pavement analysis procedures that are derived from the theory of elasticity are based on simplifications of the real condition.

Mulungye et al. [22] generated a two-dimensional four layer model using ANSYS/ED to evaluate structural performance of finite element method considering factors like tyre pressure, wheel configuration and axle load variations of a transportation truck. The results obtained from the model analysis were verified with in situ full-scale test data. Andrew et al. [3] propose a simulation model that utilizes transient, two-dimensional finite difference method to calculate the temperature distribution in asphalt pavements in reaction to hourly thermal environmental conditions and to calculate associated thermal stresses. The study uses the stiffness index to evaluate the thermal stress distribution in asphalt pavements and initial thermal stress maps during the seasonal and diurnal freeze and thaw cycles. The generated numerical model permitted an hour-by-hour calculation of the pavement temperature and thermal stress distribution in two-dimensional cross section. Pavement tilt angle had a substantial impact on the pavement temperature distribution while precipitation and evaporation had a cooling effect on the pavement surface temperatures. Lower thermal stresses were observed when higher thermal conductivity layers were placed on the pavement top surface. For a more realistic prediction of pavement temperatures, the numerical model can be further extended to consider the effects of snow and mushy zones.

Su et al. [18] studied evaluation of shear stress in the pavement, so as to know the actual influence of tyre-pavement contact pressure on pavement shear stress in asphalt mix years are applied and studied and then the analysis is done and on semi-rigid asphalt pavement. For analysis in 3D FE model, single axle load with dual tyres was applied Shear stress value comes maximum at tyre edge. The maximum shear stress occurs as 60 mm below the tyre edge. Also, it is noticed from the work that vertical loading and tyre pressure are also the factors which are responsible for producing shear stresses.

Ziari et al. [12] analyses the use of steel slag in asphalt mixture to replace fractional fine aggregates. Steel slag is a material with an environmental issue which cannot be decomposed nor incinerated. These materials are brittle in nature but rich in carbon and silicon, which makes it strong enough to be replaced for fine aggregates to save the environment. Significant outcomes from the study show that the maximum size up to 4.75 mm and an optimum replacement was 10%. The various parameters such as strength index, high-temperature stability, and water stability are up to mark and improve creep stiffness when broken slag is used.

Tarefdar et al. [23] studied the six existing pavement section's design guides and parameters like mean, maximum likelihood, median, the coefficient of variation, and density distribution function of subgrade strength R value were determined. The outputs obtained were then compared in reliability and thickness. Probabilistic procedure yield lesser reliability values than AASHTO. The procedure based on the coefficient of variation of R value deals properly with the subgrade variability. Increasing the minimum R-value for sub-excavation does not necessarily provide an exact solution to meet design reliability. To mitigate different distresses, alternative designs were suggested for the existing pavement thicknesses by modifying material and subgrade properties.

Ranadive et al. [25] illustrates that axisymmetric analysis of flexible pavement is carried out by a computer program (ANSYS), and different performance parameters of pavement were studied for varying conditions of thickness. The increase in thickness of the base course and sub-base course layer does not help to reduce stress and deflection as compared to asphalt concrete layer in which it is observed that there is a substantial reduction in stress as there is an increase in thickness of asphalt concrete. Also, it is reported that the maximum deflection of flexible pavement under wheel load varies with an increase in thickness of the different layers of the component. It reduces when BC layer thickness increases, while increases when base-course and sub-base course thickness increases. As stiffer materials are employed in the upper layers, the noticeable reduction in subgrade stress and deflection is observed. For any given subgrade soil type, this allows a reduction of the thickness of the stiffer layer over a similar thickness of unbound granular material to satisfy the requirements of an allowable subgrade distress or limiting deflection criteria. The Interface between asphalt concrete layer and base course layer affects the distribution of stress in the lower layers of the road structure. Analytical methods can predict the performance of flexible pavement, which helps to bypass costly field experiments; hence it is beneficial to society. Sahoo et al. [35] studied the effect of nonlinearity in a granular layer on critical pavement responses of low volume roads. From the sensitive analysis, a subgrade depth of 2000 mm and length of 1800 mm in the longitudinal direction is determined. To model the non-linearity of granular material, the author used Drucker- Prager plasticity model. From reported work, it is found that the vertical strain on top of subgrade and central deflection increases by 9% and 7% respectively when compared to the values obtained using linear elastic analysis. The author recommends consideration of non-linearity in granular layers for more accurate modeling of pavement.

Rahman et al. [29] focuses on pavement response subjected to various traffic factors like different axle configuration, tyre imprint areas, and inflation pressure. Three-dimensional finite element analysis regarding fatigue and permanent deformation using ABAQUS software was performed. Three layers of pavement were considered as asphalt surface, granular base and subgrade which were assumed to respond linearly and elastically to the static load applied. Interaction properties were assumed as having frictionless contact between two adjacent layers. ABAQUS software was used to measure horizontal tensile and vertical compressive strains. From the considered study it was summarized that tyre imprint area needed to be a rectangle with two semicircles on either side and was considered to be a region of more extensive research work.

Wang et al. [11] analyzed the effects of truck tyre types based on tyre geometries and specifications obtained from tyre manufacturers. A two-dimensional finite element analysis was conducted to investigate truck tyre types affected the near surface pavement response. A 2-D finite element based models were generated using tyre geometries and structure along with tyre pavement interaction models. Tyre-pavement contact stress, top down, cracking and instability rutting based interaction models were also generated.

Although noise factor was not considered in the study of tyre pavement interaction models which was limited to the study of tyre contact stress distributions only. Another major limitation of the study was that the only radial truck tyre and static loading condition were only considered. The material property was considered as linear elastic (no wear or permanent shape change) for single load application. ADINA version 8.3 was used to generate a 2-D model using finite element analysis.

For same axle load dual and wide based tyres generated very close contact stress to the recommended inflation levels while super, the single tyre had highest contact stress. Maximum shear stress and principal stress were much higher for super single tyre than dual and wide based tyres which indicated that the super single tyre might cause more damage to the near surface pavement than other two combinations.

The IRC: 37-2012 [15] guidelines for the design of flexible pavement recommends using the IITPAVE, which is a modified version of FPAVE developed under the research scheme R-56 for layered system analysis.

Gogoi et al. [30] studied to investigate if there is any co-relation in rutting and fatigue distress. Tensile strength and vertical strain on subgrade are used to find out fatigue and rutting stresses values respectively. For analysis purpose, data was collected from federal highway agency (FHA), the USA for the same month of any given year where no maintenance activity had been conducted. It was observed that initially fatigue and rutting increased, but afterward, fatigue cracking level reached up to 5%, rutting increases appreciably. The study does not give any cause or effect of these stresses but proposes two regression equation which can be used to forecast their progression. The work can be further extended to know whether their progression is interrelated during their formation which can be useful for the design of pavements and distress progression models.

Kranthi Kumar et al. [17] studied the use of reclaimed asphalt pavement (RAP) in hot bituminous and cold bituminous mixes as surface or a base layer. RAP obtained from the National Highway near Rajkot of Gujarat state was a laboratory experimented with a combination of RAP and virgin aggregates. The obtained RAP portion was mixed with by both hot as well as cold bitumen mixes procedure and tested for different mix proportions with virgin aggregates.

Pavement performance for both mixes was carried out using Mechanistic-Empirical Pavement Design Guide (MEPDG). RAP proportion for various mixes was varied from 0 to 40% and evaluated for properties like resilient modulus of Bituminous Concrete (BC) mixes, phase angle, indirect tensile strength (ITS), total rutting, bottom top cracking, international roughness index (IRI) for base course material.

Inferences were drawn by laboratory testing of different mixes and it was concluded that up to 20% RAP be used in BC and DBM layers with VG30 bitumen routinely. As per MEPDG calculations, BC mixes with 20% RAP may give equal or better results than a mix of fresh aggregates and VG30 bitumen in terms of rutting, cracking and IRI. Percentage of stone dust variations from 10 to 20% for 3 and 4% bitumen emulsion respectively does not affect resilient moduli and ITS. Resilient moduli and ITS were found to be higher for RAP mixes with 4% bitumen emulsion than 3% bitumen emulsion. RAP can be extensively used for both cold and hot mixes.

Sinha et al. [1] Illustrates the usefulness of finite element method to study the performance of a flexible pavement with different types of local materials in its sub base. Three types of naturally occurring materials, namely; course sand, conventional subbase material, stone dust and four types of industrial waste materials; Blast furnace slag, granulated blast furnace slag, Linz-Donawitz slag and fly ash were used. In this work multilinear elastoplastic hardening model in ANSYS was used and the effect of type of subbase on the life of the pavement is evaluated. In the study, the right boundary was placed at 110 cm from the outer edge of the loaded area, which is more than 7 times the radius 150 mm of the applied load. A uniform pressure of 575 kPa was applied to a circular contact area having a radius of 150 mm causing a single axle load of 40.80 KN.

Tapase and Ranadive [28, 34], reported the usefulness of two-dimensional finite element analysis to study the effect of variation in thickness of different component layers on the critical parameters. They noted that the tensile strain at the bottom of the bituminous layer (BL) and compressive strain on top of the subgrade decrease with an increase in the thickness of BL, which ultimately results in an increase of fatigue and rutting lives.

3 Observations and Future Research Direction

A literature survey showed that considerable progress has been made in the development of response models for mitigating various parameters influencing pavement performance, ranging from Boussinesq's one-layer model to multi-layer elastic theories to finite element models. However, some areas of designing pavement are still not addressed properly.

Some of the debatable issues are discussed below:

Out of the various factors affecting the performance of the pavement, the combined effect of temperature, seepage and loading condition is missing. As the analysis is based on a number of assumptions, only a single parameter at a time is addressed as reported in a number of literature. Conventional construction materials like aggregates are becoming progressively scarce on account of environmental concerns as well as legal restrictions on quarrying while the construction activity has expanded phenomenally. This has shifted focus from the large scale use of conventional aggregates to use of local, recycled and engineered marginal aggregates in construction. So here the scope is to use alternative materials for conventional materials like aggregate and bitumen, which can be partially or completely replaced by alternative materials or waste materials like fly ash, granulated blast furnace slag, plastic, rubber, e-waste, etc. which are creating disposal problems. So here the scope is to analyze the suitability of waste materials in different component layers of flexible pavement.

From the literature, it is noticed that emphasize is given to the factors responsible for the deterioration of pavement in the form of fatigue and rutting of pavement which is theoretically related to CBR values of the subgrade. So here need is to estimate the pavement life in the form of a number of standard axles or cumulative standard axles which can be correlated with the fatigue life and rutting life as per the criteria recommended by various codes in practice like IRC: 37-2012 [15] for the actual subgrade condition. Effect of temperature on pavement, during day time and night time as well as due to seasonal changes, can be studied in details. Seepage through subgrade results in a decrease in the life of the pavement, similarly, the effect of water table on swelling and shrinkage property of subgrade can be studied. The effect of excess wheel loading, modern tyre pressures cannot be neglected. However, from the past studies and through the recommendations of the Indian Road Congress (IRC) and AASHTO code of practice, wheel load, tyre pressure and their combinations can be shortlisted which can be correlated with actual field conditions.

Current pavement design and analytical procedures rely mostly on the direct and indirect application of empirical approach [26, 27] wherein most of the analytical models are based on linear elastic theory. Even though the approach is simple and is to analyze it does not reflect the realistic approach to the nonlinear behavior of underneath granular layers and viscoelastic behavior of bituminous layer. To incorporate the complex behavior of the present pavement system along with modern loading configuration more realistic analysis is required hence there is need of 3D modeling.

2D Finite element models including axis-symmetric, plain strain, plain stress are used effectively for the evaluation of critical pavement responses on various condition which are influencing the pavement behavior. An appropriate combination of 2D and

3D finite element analysis will not only give the realistic result but also help in achieving the optimum computational time.

Material constitutive models such as linear-elastic, nonlinear-elastic [10], viscoelastic and elastoviscoplastic models as per the requirements and environmental conditions should be employed to describe the behavior of the materials in different layers of the pavement. In the current design procedures value of elastic moduli is considered constant throughout its design life which is impossible. Elastic moduli is a dynamic entity which depends on environmental factors.

Also, the work can be extended to enhance the scope to calculate various entities like the effect of the forces arising out of frost and thawing cycles, the entities such as distribution of seepage pressures, gradients and exit gradients, etc. Even though researchers presented a number of improvements in the design and analytical procedures, still there is a clear need for developing a more rigorous analytical tool which will enhance the above-mentioned scope and can be extended to study the combined effect of various parameters simultaneously. This calls for more intensive research in this area in the coming years for the design of sustainable flexible pavements.

Hence the paper describes an outline for future research which is under progress using versatile finite element method.

4 Projected Methodology for Finite Element Analysis

The selected observations from the literature study will be considered for analysis including the viscoelastic behavior of bituminous layer, nonlinearity in granular base layers. Initially, it is proposed to carry the rigorous two-dimensional finite element analysis of the parameters selected for analysis using a multipurpose software ANSYS. Further, the three-dimensional analysis will be carried on the identified area of more interest from the two-dimensional analysis for more accurate prediction of results. In finite element analysis, a continuum is discretized into finite elements, over which the material properties and governing relationships are applied and expressed in terms of unknown values at corners. Set of equations is formed by considering the above continuum with appropriate loading and constraints. A solution of these equations gives the approximate behavior of the continuum considered for analysis [9, 38]. Figure 1 shows a general methodology adopted for flexible pavement design. The pavement configuration including the number of layers, the thickness of each layer, and the type of material are shortlisted from the available code of practice. The basic material properties required for the structural models are the moduli of BL, base layer and subgrade; those for the distress models involve the various failure criteria. In structural models/finite element models, the use of load intensities, contact radius and contact pressure from the traffic inputs and the number of repetitions is used by distress models. Pavement response includes stresses, strains, and deflection. Those are evaluated from the structural models and are used as inputs for distress models. The distress models include fatigue and rutting criteria. If reliability for a certain distress is beyond the minimum level required, the assumed pavement configuration is unsuitable; likewise, for all shortlisted trials, the procedure is implemented. It is apparent from details presented in Fig. 1 that the finite element method of solution constitutes a primary component of the design process.

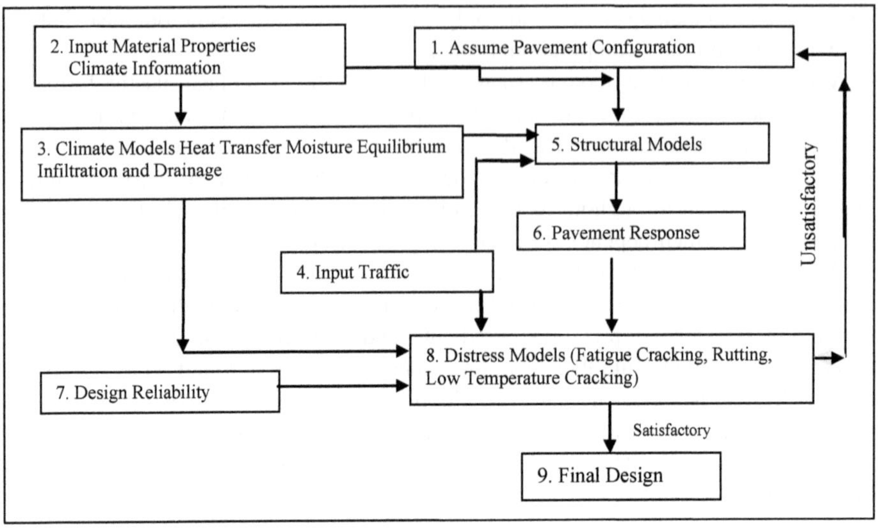

Fig. 1. General methodology for flexible pavement design [14]

It is apparent from details presented in Fig. 1 that the finite element method of solution constitutes a primary component of the design process. The method of analysis is too well known to need any more detailed description, hence only the salient features governing the problem under consideration are discussed herein.

References

1. Sinha, A.K., Chandra, S., Kumar, P.: Finite element analysis of flexible pavement with different subbase materials. Indian Highw. **42**(2), 53–63 (2014) (New Delhi)
2. American Association of State Highway Officials AASHTO: AASHTO Guide for the Design of Pavement Structures. AASHTO, Washington, D.C. (1993)
3. Chiasson, A., Yavuzturk, C., Ksaibati, K.: Linearized approach for predicting thermal stresses in asphalt pavements due to environmental conditions. J. Mater. Civ. Eng. ASCE **20** (2), 118–127 (2008). doi:10.1061/(ASCE)0899-1561
4. Das, A., Pandey, B.B.: The m-e design of bituminous road: and Indian perspective. J. Transp. Eng. ASCE **125**(5), 463–471 (1999). doi:10.1061/(ASCE)0733-947X
5. Saad, B., Mitri, H., Poorooshasb, H.: Three-dimensional dynamic analysis of flexible conventional pavement foundation. J. Transp. Eng. ASCE **131**(6), 460–469 (2005). doi:10.1061/ASCE0733-947X
6. Saad, B., Mitri, H., Poorooshasb, H.: 3D FE analysis of flexible pavement with geosynthetic reinforcement. J. Transp. Eng. ASCE **132**(5), 402–415 (2006). doi:10.1061/ASCE0733-947X2006132:5(402)
7. Das, A.: Analysis of Pavement Structures. CRC Press, Taylor and Francis, Boca Raton (2015)
8. Diefenderfer, B.K., Al-Qadi, I.L., Diefenderfer, S.D.: Model to predict pavement temperature profile: development and validation. J. Transp. Eng. ASCE **132**(2), 162–167 (2006)

9. Desai, C.S., Abel, J.F.: Introduction to the Finite Element Method- A Numerical Method for Engineering Analysis. CBS Publishers and Distributors, Delhi (2005)
10. Gonzalez, A., Saleh, M.F., Ali, A.: Evaluating nonlinear elastic models for unbound granular materials in the accelerated testing facility. Transportation Research Record No. 1990, pp. 141–149 (2007)
11. Wang, G., Roque, R.: Evaluation of truck tyre types on near-surface pavement response based on finite element analysis. Int. J. Pavement Res. Technol. **4**(4), 203–211 (2011)
12. Ziari, H., Khabiri, M.M.: Preventive maintenance of flexible pavement and mechanical properties of steel slag asphalt. J. Environ. Eng. Lands. Manag. **15**(3), 188–192 (2010) (Taylor and Francis). doi:10.1080/16486897.2007.9636928
13. Helwany, S., Dyer, J., Joeleidy, J.: Finite element analysis of flexible pavement. J. Transp. Eng. ASCE **124**(5), 491–499 (1998). doi:10.1061/(ASCE)0733-947X
14. Yang, H.H.: Pavement Analysis and Design. Pearson Education, Inc., and Dorling Kindersley Publishing, Inc., Upper Saddle River, New York (2008)
15. IRC: 37-2012: Guidelines for the design of flexible pavements. Indian Roads Congress, New Delhi
16. Immanuel, S., Timm, D.H.: Measured and theoretical pressures in base and subgrade layers under dynamic truck loading. In: Proceedings of 2006 Airfield and Highway Pavement Specialty Conference, pp. 155–166. American Society of Civil Engineers (2006)
17. Kranthi Kumar, K., Rajasekhar, R.: Reclaimed asphalt pavements in bituminous mixes. Indian Highw. **42**, 12–18 (2014). (New Delhi)
18. Su, K., Sun, L., Hachiya, Y., Maekawa, R.: Analysis of shear stress in asphalt pavement contact pressure. In: 6th ICPT Sapporo, Japan, July 2008
19. Kumar, S.S., Sridhar, R., Reddy, K.S., Bose, S.: Analytical investigation on the influence of loading and temperature on top-down cracking in bituminous layers. J. Indian Road Cong. **69**(1), 71–77 (2008)
20. NCHRP: Evaluation of mechanistic-empirical design procedure. National Cooperative Highway Research Program, NCHRP Project 1–37A. National Research Council, Washington, D.C (2007)
21. Chakroborty, P., Das, A.: Principles of Transportation Engineering. Prentice Hall of India Private Limited, Delhi (2003)
22. Mulungye, R.M., Owende, P.M.O., Mellon, K.: Finite element modeling of flexible pavement on soft soil subgrades. Mater. Des. **28**, 739–756 (2007) (Elsevier)
23. Tarefder, R.A., Saha, N., Stormont, J.: Evaluation of subgrade strength and pavement designs for reliability. J. Transp. Eng. ASCE **136**(4), 379–391 (2010). doi:10.1061/(ASCE) TE.1943-5436.0000103
24. Siddharthan, R.V., Sebaaly, P., El-Desouky, M., Strand, D., Huft, D.: Heavy off-road vehicle tyre-pavement interaction and response. J. Transp. Eng. ASCE **131**(3), 239–247 (2005). doi:10.1061/(ASCE)0733-947X
25. Ranadive, M.S., Katkar, A.B.: Finite element analysis of flexible pavements. Indian Highways **38**(6) (2010)
26. Ranadive, M.S., Tapase, A.B.: Investigation of behavioral aspects of flexible pavement under various conditions by finite element method. In: Yang, Q., Zhang, J.-M., Zheng, H., Yao, Y. (eds.) Constitutive modeling of geomaterials. Springer, Berlin, pp. 765–770 (2013). doi:10.1007/978-3-642-32814-5_100
27. Ranadive, M.S., Tapase, A.: Pavement performance evaluation for different combinations of temperature conditions and bituminous mixes. Innov. Infrastruct. Solut. **1**(40), 1–5 (2016). doi:0.1007/s41062-016-0040-9 (Springer)

28. Ranadive, M.S., Tapase, A.B.: Parameter sensitive analysis of flexible pavement. Int. J. Pavement Res. Technol. (IJPRT) (2016). doi:10.1016/j.ijprt.2016.12.001 (Elsevier, Special Issue on Sustainability on Pavement Engineering)
29. Rahman, M.T., Mahmud, K., Ahsan, S.: Stress-Strain characteristics of flexible pavement using finite element analysis. Int. J. Civ. Struct. Eng. **2**(1), 233–240 (2011)
30. Gogoi, R., Das, A., Chakraborty, P.: Are fatigue and rutting distress modes related? Int. J. Pavement Res. Technol. **6**(4), 269–273 (2013)
31. Reddy, K.S., Pandey, B.B.: Lateral placement of commercial vehicle on national highways. HRB Bulletin No. 7, Indian Road Congress, New Delhi (2006)
32. Masad, S., Little, D., Masad, E.: Analysis of flexible pavement response and performance using isotropic and anisotropic material properties. J. Transp. Eng. ASCE **132**(4), 342–349 (2006)
33. Park, S.-W., Lyttol, R.L.: Effect of stress-dependent modulus and Poisson's ratio on structural responses in thin asphalt pavement. J. Transp. Eng. ASCE **130**(3), 387–394 (2004). doi:10.1061/(ASCE)0733-947X
34. Tapase, A., Ranadive, M.: Performance evaluation of flexible pavement using finite element method. In: ASCE GSP 266, Geo-China 2016: Material, Design, Construction, Maintenance and Testing of Pavement, pp. 9–17 (2016). doi:10.1061/9780784480090.002
35. Sahoo, U.C., Reddy, K.S.: Effect of nonlinearity in the granular layer on critical pavement responses of low volume roads. Int. J. Pavement Res. Technol. **3**(6), 320–325 (2010)
36. Wu, Z., Zhang, Z., King, B., Raghavendra, A., Martinez, M.: Instrumentation and accelerated testing on Louisiana flexible pavements. In: Proceedings of 2006 Airfield and Highway Pavement Specialty Conference, pp. 119–130. American Society of Civil Engineers (2006)
37. Yoder, E.J., Witczak, M.W.: Principles of Pavement Design. Wiley-Interscience Publication, Hoboken (1975)
38. Zienkiewicz, O.C., Taylor, R.L.: The Finite Element Method, vol. 2. McGraw-Hill, New York (1991)
39. Zuo, G., Drumm, E.C., Meier, R.W.: Environmental effects on the predicted service life of flexible pavements. J. Transp. Eng. **133**(1), 47–56 (2007)

Performance Testing of Paving Mixes for Libya's Hot and Arid Conditions, Using Marshall Stability and SUPERPAVE Gyratory Compactor Methods

Fathi S. Almadwi(✉) and Gabriel J. Assaf

Department of Civil Engineering and Construction,
École de Technologie Supérieure (ÉTS), University of Québec,
1100 Rue Notre-Dame Ouest, Montreal, QC H3C, Canada
fathi.madwi.1@ens.etsmtl.ca, Gabriel.Assaf@etsmtl.ca

Abstract. Asphalt Concrete Pavements (ACP) in Libya's southern desert regions suffer two major challenges: the hot and arid climate, with road surface temperatures reaching 65–70 °C, and air humidity below 50%. As such, ACP in Libya develops excessive deformation. The lack of modern testing methods and in-situ monitoring during the use-phase make predicting the performance of new mix designs difficult. This paper aims to provide comparative performance data on paving mixes of different Libya-sourced bitumen grades under simulated climate conditions, and to compare the usefulness of empirical testing methods (Marshall) to Performance Graded methods. Two mixes, one using Bitumen 60/70 (B 60/70) and the other using PG70-10, are assessed with Marshall Stability and Super Gyratory Compactor tests, using modern equipment at the laboratory of the ETS (École de technologie supérieure) faculty of engineering of the University of Québec, Canada. The performance of PG70-10 mixes is found to be superior to that of the performance of the mixes using B 60/70. The PG70-10 mix performed within the requirements of lower-volume roads (≤ 300 vehicles per day). Also, the results of the Super Gyratory Compactor tests are found to be a better indicator of in-service performance than those given by Marshall Stability tests. These results provide a foundation for performance-testing of different paving mixes that varied in sand and filler content; these are available for applications in similar arid climates, and may provide significant savings by allowing engineers to substitute local materials, such as the abundant rounded sand in southern Libya, for more scarce and costly materials, such as manufactured aggregate.

1 Introduction

Libya's roads suffer from excessive deformation (primarily rutting, as well as cracking) under the hot and arid conditions found across the country, especially in the southern desert regions. The asphalt cement binder currently used in roads in the hottest regions is Bitumen 60/70 (B 60/70). The bitumen is produced in local refineries but it underperforms in extreme heat and results in rutting of ACP. In the southern deserts, daytime air temperatures can reach 55 °C, resulting in road surface temperatures

L. Mohammad (ed.), *Advancement in the Design and Performance of Sustainable Asphalt Pavements*, Sustainable Civil Infrastructures, DOI 10.1007/978-3-319-61908-8_23

exceeding 70 °C. The problem of selecting proper materials for roads becomes more complicated when older empirical grading tests such as the Marshall Stability Test are used. Such tests assess material qualities under standard conditions but they do not give results in engineering units that would allow the prediction of in-service performance (cracking and rutting) across a range of conditions. This paper discusses two points. First, it selects a grade of binder that can perform under conditions simulating those found in hot arid regions like southern Libya; second, it compares the results of PG and Marshall tests to facilitate a transition to PG-based selection of materials.

1.1 Empirical vs Performance Graded Classification

The Strategic Highway Research Program (SHRP) conducted research from 1987 to 1993 to improve the design and testing of asphalt materials, resulting in the SUPERPAVE program. Part of this program is the development of Performance Graded (PG) standards for materials. These standards are based on performance across a range of in-service conditions. Since 1999, PG standards for asphalt have been accepted by the American Association of State Highway and Transportation Officials (AASHTO) and the American Society for Testing and Materials (ASTM). This is superior to empirical testing which only measures properties under a standard temperature or other fixed conditions. For example, the Bowen Penetration Machine invented in 1888 measured penetrability at 25 °C; ASTM and ASHTO grading procedures were developed in the 1960s and included tests to measure the viscosity at 60 °C, which better approximates the maximum temperature of a pavement surface (Salem et al. 2014).

Performance Grading specifies a range of temperatures within which a material can be expected to perform to the needs of the project. Specifically, they give average, high, and absolute minimum temperatures. For example, the PG70-10 test indicates when a material undergoes the 7-day average road surface temperature, where it does not exceed 70 °C or the absolute minimum temperature does not go below −10 °C. Therefore, engineers can use PG classification to easily identify which materials are suitable for given service conditions. With these tests, we can select materials that are proven to perform under in-service conditions. Engineers can make these choices before construction and during in-place monitoring of roads. By using the PG grading as a guide, engineers can prevent costly and disruptive failures and reduce the need for premature repairs. When needed, the engineer can select higher grades to account for extra strain on roads, such as temperature spikes outside the normal range, or especially high traffic loads. Standard practice is to go up one PG grade (6 °C increase, e.g. PG64 to PG70) from fast to slow moving traffic, and up two grades from fast to stationary traffic, such as intersections and bus stops (Arora et al. 1997).

Where no data exists, the transition to SUPERPAVE methods is difficult to compare; this is especially true with regards to past methods and future construction using PG material specifications. It is also a problem where extreme conditions are outside the parameters to which existing PG graded construction has been specified. This is also complicated by the lack of access to modern testing equipment for determining PG grades. To improve road-building in countries like Libya, it is useful to generate data on materials in extreme heat and aridity. Likewise, it is important to find low-cost, accessible testing that allows results to be compared with existing SUPERPAVE data.

1.2 Transition from Empirical to SUPERPAVE Methods

Studies have already been done to establish temperature ranges and to determine the required PG grades in countries with conditions similar to those found in Libya.

Al-Abdul Wahhab et al. conducted a study of asphalt pavement rutting in Saudi Arabia where they identified key factors in surface deformation and made recommendations for traffic control and material selection. This study did not use SUPERPAVE grading in its material recommendations (Wahhab et al. 1995). Arora and Kennedy studied conditions and road wear in Saudi Arabia and recommended the use of binder with PG70 or PG76 instead of the PEN60/70 binder in use at the time (Arora et al. 1997).

Asi conducted a study to evaluate mix design methods for use in Jordan. This study created a temperature map for Jordan and conducted a series of tests on samples prepared according to Marshall and SUPERPAVE designs. In particular, they used the following tests: Marshall Stability, Loss of Stability, Indirect Tensile Strength, Loss of Indirect Tensile Strength, Resilient Modulus, Fatigue Life, Rutting Behaviour, and Creep Performance. The map showed that most of Jordan was suitable for PG64-10 and that the southwest, northwest and northeast corners of the country were suitable for PG70-10. This study found that SUPERPAVE mix designs, which prescribed a lower bitumen content than Marshall designs, performed better than the Marshall mixes (Asi 2007).

Alani, Albayati and Abbas addressed the transition to PG grading for road construction in Iraq. They created a temperature zoning map for that country and recommended grades from PG64-16 (equivalent to PEN 60/70 already in use) to PG76-4 for use in the hottest region (Basra) (Alani 2010).

Salem, Uzelac and Matic have done work to map the road surface temperatures in Libya. They used air temperature data from eight locations across the country to estimate the range of road surface temperatures from 2000–2009 and 2012–2013. Then they applied Long Term Pavement Performances (LTPP) equations for high temperature modelling of the data; these were applied at 50% and 98% reliability. With this work, Salem et al. established 7-day average maximum and absolute minimum temperatures for each location. They used the results of the SHRP model, at a 50% reliability level, to select recommended PG grading. The predictions of the SHRP equations were more severe and so provided a greater buffer against stresses of extreme heat and uncontrolled axle loading. Such conditions are often found on roadways in countries like Libya. Uncontrolled axle loading is a critical factor also cited by studies in other desert countries (Ahmed and Roffa 2002; Wahhab et al. 1995]. As shown in Fig. 1, for the central, eastern and most of the western region of the country (88% of the total area), the recommended grade of asphalt binder is PG76-10; for the northwest corner, the recommended grade is PG70-10; for the southwest corner, the recommendation is PG82-10 (Salem et al. 2014).

As this short review demonstrates, the transition to PG grading in road construction is already underway in hot and arid regions such as in north Africa and the Middle East. It is justified because older mix designs have underperformed. There is a need for data to determine which materials may be suitable to the performance parameters identified in these regions. This is important because lower-quality material might pass the Marshall design requirements but might not pass the SUPERPAVE mix design requirements.

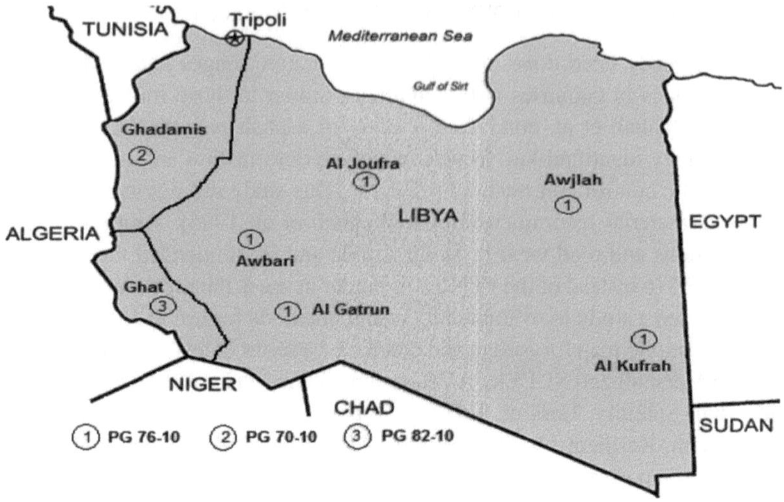

Fig. 1. PG zone map of Libya from Salem et al.

2 Testing Methodology

For this paper, samples were prepared using a mix of 63.333% crushed sand, 33.167% natural sand and 3.50% filler. Half of the samples used B 60/70 binder and the other half used PG70-10 binder. Three tests were performed on the samples: Marshall Stability, SUPERPAVE Gyratory Compactor, and Rutting Analyzer. All these tests provide useful information about the deformation behaviour of the asphalt in-service. They did this by examining volumetric properties in the case of Marshall and Gyratory Compactor tests, and by directly simulating loaded tire passage in the case of the Rutting Analyzer.

2.1 Materials Used for Hot Mix Design

2.1.1 Aggregate

The study used aggregates imported from the southern desert region of Libya and standardized manufactured aggregates for the tests and analyses at ETS in Montreal, Canada. The natural sand was 0–5 mm and the manufactured sand was the same size. The specific gravity, the bulk specific gravity and the water absorption of the two kinds of aggregate are shown in Table 1. Figure 2 shows the grading of natural Libyan sand and manufactured sand.

2.1.2 Asphalt Binder

The two asphalt binders used in the study were PG 70-10 and B 60/70. The PG 70-10 mixing temperature was recommended to be 162 °C and the compaction temperature was recommended to be 154 °C. The B 60/70 was recommended to be mixed at 156 °C and compacted at 143 °C. All recommendations were followed exactly. Figure 3 shows the Brookfield viscosity as compared to temperature equivalents and how they relate to the binders used here.

Table 1. Aggregate properties

Aggregate	Apparent specific gravity	Bulk specific gravity	Absorption (%)
Natural sand 0–5 mm	2.63	2.42	0.33
Manufactured sand 0–5 mm	2.639	2.44	0.58

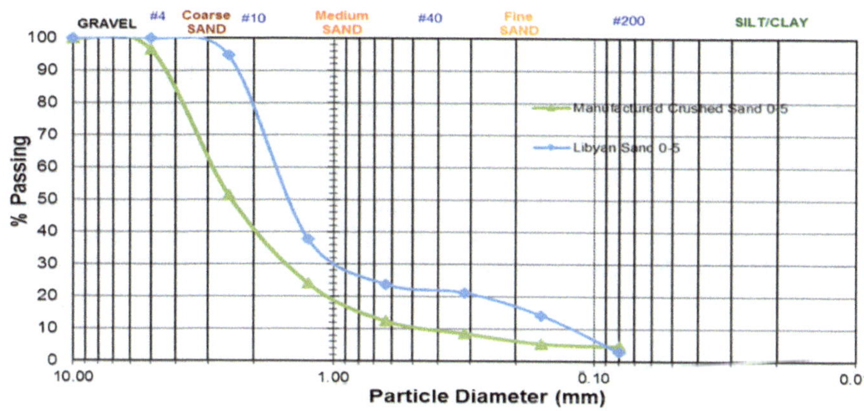

Fig. 2. Grading of Libyan natural sand and manufactured crushed sand.

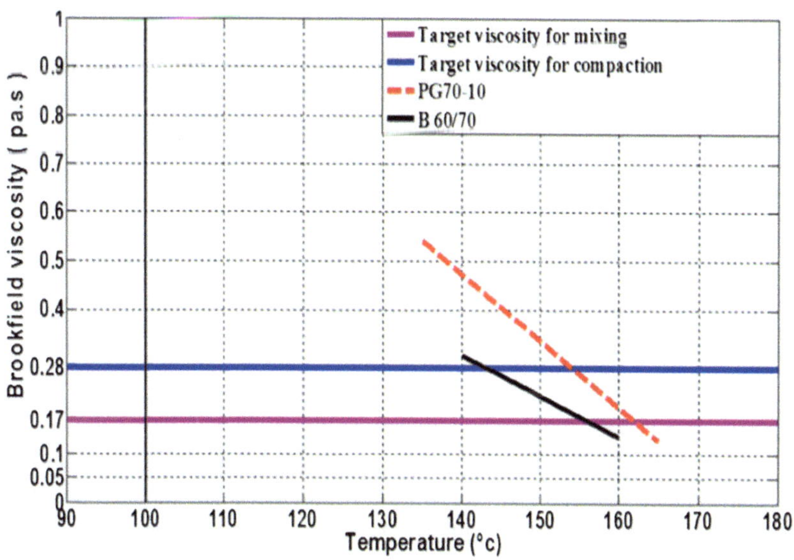

Fig. 3. Brookfield temperature vs viscosity curve for asphalt binders.

2.2 Hot Mix Asphalt Design Tests

2.2.1 Marshall Stability Test

This test is used to determine the following volumetric properties of the sample: the Optimum Bitumen Content (OBC) of the mix; the stability and plastic flow; the percentage of air voids (Va); the percentage of voids in mineral aggregate (VMA); the percentage of voids filled with asphalt (VFA). In order to determine the OBC, tests were performed with a 5% content; these were insufficient to bind the mix. Further tests were performed with 6% content; these resulted in bleeding of bitumen. Therefore, a bitumen content of 5.5% by weight was chosen for the samples in these tests. All tests followed procedures for determining OBC in Marshall mixes according to the "Basic Asphalt Materials Mixture Design and Testing Technician Training Manual", October (2012).

2.2.2 SUPERPAVE Gyratory Compaction Test

This test is used to produce compacted specimens; these provide data on the air void (Va) content at each stage of the compaction process. The samples were measured at 10, 80, 100 and 200 gyrations. The samples that contained PG70-10 binder were tested at 154 °C; the samples that contained B 60/70 were tested at 143 °C. The standards were from the ESG-5 category of hot mix paving materials from the Quebec Ministry of Transport (Ministère des Transports, Quebec: MTQ). These standards specify targets of **Va ≥ 11%** for N-ini, **Va = 4–7%** for N-des, and **Va ≥ 2%** for N-max (Internationale et al. 2015).

2.2.3 Rutting Analyzer Test

This test subjects the samples to a loaded wheel in order to simulate the passage of traffic over time. The machine temperature can be adjusted for binders with different softening points; the weighting and tire pressure of the wheel can be adjusted to simulate different vehicle conditions. The first test consisted of 1000 cycles at 600 kPa tire pressure at room temperature; following that, tests were made of 1000, 3000, 10000 and 30000 cycles (cumulative) at 65 °C. The rutting depth was measured after each stage. The MTQ standard specifies a maximum deformation of 10% at 1000 heated cycles and 15% at 3000 heated cycles (Internationale et al. 2015).

3 Results

All three tests showed that, overall, PG70-10 performed better than B 60/70 under the specified conditions, as indicated in the next sections.

3.1 Marshall Stability Results

As shown in Table 2, samples containing PG70-10 were more stable and more consistent in measures of air void content than samples containing B 60/70. Table 2 also shows that samples containing PG70-10 were within the target range specified by the MTQ standard. All samples containing B 60/70 produced results exceeding the 2–4 mm target for flow, according to the standards referenced above. Each curve in Fig. 4(a)–(c) represents one of the two binders, PG70-10 or B 60/70.

Table 2. Marshall Stability test results

% of mixture design 63.333 crushed sand 33.167 natural sand 3.50 filler		Marshall volumetric properties				
	Sample no	VMA	VFA	Va	Flow	Stability
PG 70-10	S1	14.14	78.3	3.08	4.00	23.654
	S2	14.85	73.90	3.88	3.90	20.831
	S3	14.81	74.12	3.84	3.70	20.800
	S4	14.26	77.50	3.21	3.50	21.812
BITUMEN 60/70	S1	15.58	67.77	5.02	5.30	22.623
	S2	16.52	63.19	6.08	6.40	20.285
	S3	16.63	62.69	6.21	4.40	12.820
	S4	15.59	67.71	5.03	4.60	14.442
MTQ standard		Min. 9–max. 22	Min. 60–max. 80	3–7%	2–4 mm	Min. 8 KN

From Fig. 4(a), it can be seen that the Marshall Stability of PG70-10 is higher than B 60/70 and therefore it is preferred.

In Fig. 4(b), it can be seen that the percentage of Air Voids (Va) in PG70-10 is within the acceptable limit of 4%, whereas the Va% of B 60/70 exceeds this limit.

In Fig. 4(c), it can be seen that the Marshall Flow of PG70-10 is between in the ideal target range of 2–4 mm, whereas the Marshall Flow of B 60/70 exceeds this limit.

From these three tests, it is clear that the 5.5% bitumen in the two binders we tested is the best compromise between the desired properties of the hot mix asphalt.

3.2 SUPERPAVE Gyratory Compactor Results

As shown in Table 3, samples that contained PG70-10 had a smaller percentage of air voids than the samples that contained B 60/70. This was true at all gyration counts. Most importantly, the samples containing PG70-10 had air void content within the 4-7% target range for N-des at 80 and 100 gyrations. Both samples containing B 60/70 exceeded that range at 80 gyrations. Figure 5 clearly illustrates this.

3.3 Rutting Analyzer Results

As shown in Table 4, samples containing PG70-10 showed less rutting than samples containing B 60/70 at all stages of the rutting test (1000–30000 cycles cumulative). At 3000 cycles, the PG70-10 samples lost an average of 2.25% of their original height; in contrast, after 3000 cycles, the B 60/70 samples lost an average of 4.34% of their original height. Figure 6 clearly shows this.

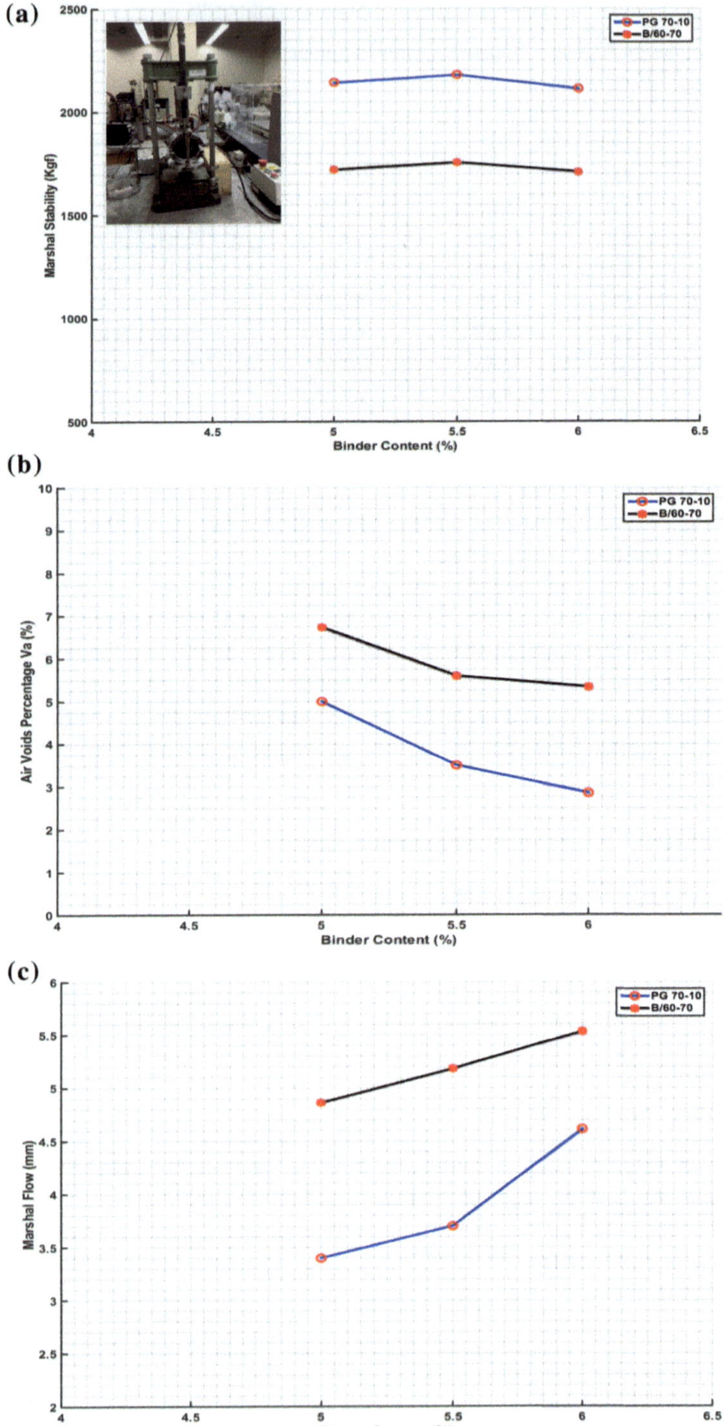

Fig. 4. (a) Marshal stability. (b) Air voids percentage Va (%). (c) Marshal flow

Table 3. SUPERPAVE Gyratory Compactor test results

SUPERPAVE Gyratory Compacter mix design			% Va/gyration			
Binder	Sample no	Compaction temperature	10	80	100	200
PG 70-10	S1	154 (°C)	15.25	6.58	5.73	3.5
	S2	154 (°C)	14.83	5.78	4.38	3.81
BITUMEN 60/70	S1	143 (°C)	15.99	7.40	6.65	4.40
	S2	143 (°C)	16.77	8.23	7.12	5.34
Specification tolerance			≥11	4–7	4–7	≥2

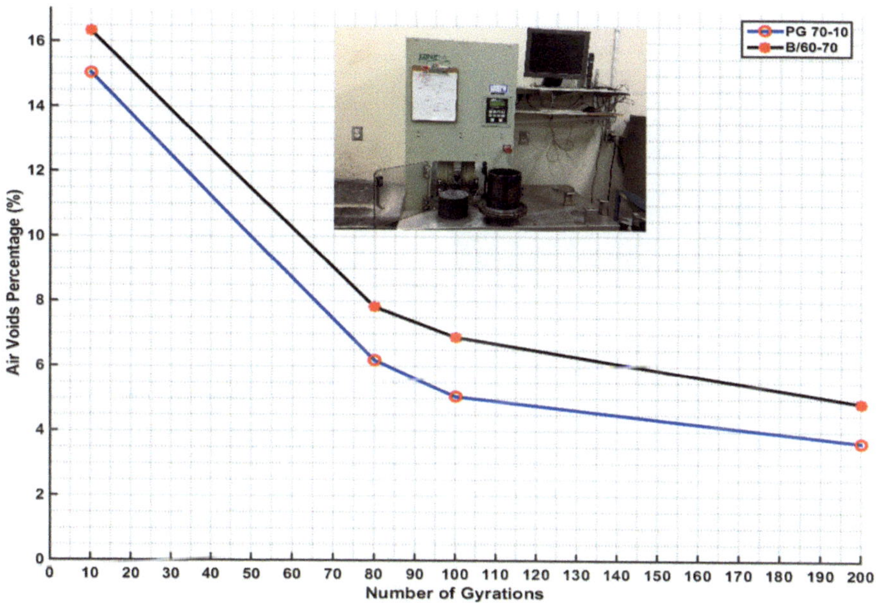

Fig. 5. SUPERPAVE Gyratory Compactor (SGC) test.

Looking at the data from Table 4, as displayed in Fig. 6, it can be seen that the slope of PG70-10 changes noticably after 10 000 cyles. From 1000 to 10000 cyles, the slope of the PG70-10 is comparable to the slope of B 60/70. But from 10000 to 30000 the results from the PG70-10 tests have a much lower slope. This is where the slope of PG70-10 contrasts the most with the slope of the B 60/70 results. This has important implications for a road surface in actual use. Where the B 60/70 undergoes considerable ongoing compaction, the PG70-10 shows much more general stability over time.

Table 4. Rutting analyzer test results

% of mixture design 63.333 crushed sand 33.167 natural sand 3.50 filler			Cold Cycle	Number of cycles (cumulative) at 65 °C				Rutting at 3,000 cycles
Binder	Sample no	Sample height	1000	1000	3000	10000	30000	
PG 70-10	S1	50.31	1.29 (%)	3.17 (%)	4.54 (%)	10.20 (%)	11.30 (%)	2.28 (mm)
	S2	51.21	1.01 (%)	3.14 (%)	4.31 (%)	8.50 (%)	9.90 (%)	2.21 (mm)
BITUMEN 60/70	S1	51.17	2.91 (%)	7.14 (%)	8.27 (%)	13.83 (%)	18.97 (%)	4.23 (mm)
	S2	52.16	3.21 (%)	7.52 (%)	8.51 (%)	13.10 (%)	18.70 (%)	4.44 (mm)

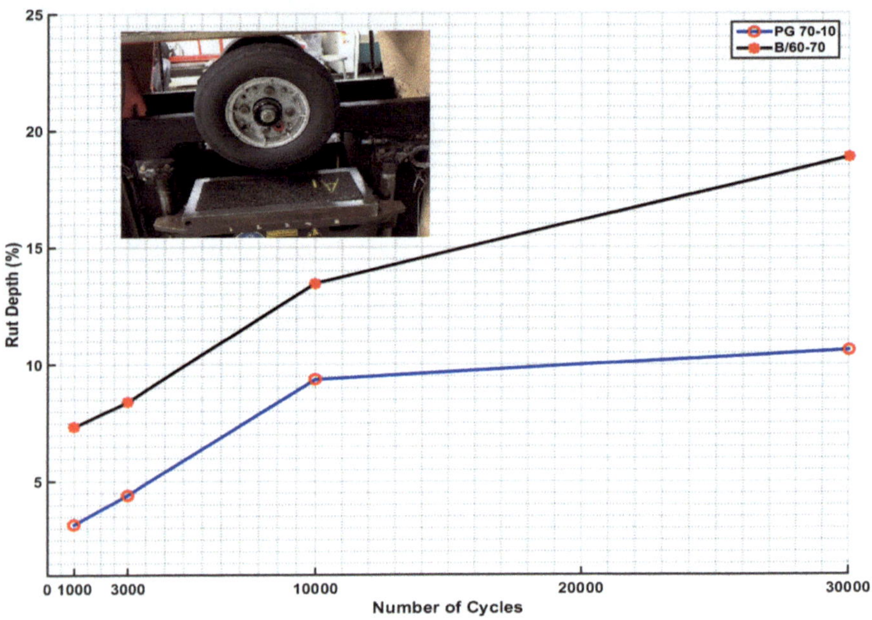

Fig. 6. Rutting analyzer test.

4 Conclusions and Discussion

The lab tests demonstrated the superior performance of PG70-10 binder in asphalt mixes subjected to conditions simulating those found in Libya's southern desert regions. This supports the recommendations found in previous literature that examines deformation in rutting of asphalt pavements in countries like Libya. The test results also show that the SUPERPAVE Gyratory Compactor predicts in-service performance more accurately than the Marshall Stability test.

5 Future Work

Future research will investigate the performance of different additives in the mixes. These additives include polymers or fibres. Other research will include varying the ratio of manufactured to natural sand. The information gained by these studies will provide data resources to engineers planning a new generation of road construction in countries like Libya. This will facilitate the transition to SUPERPAVE-compliant methods. It will also help to construct better-performing roads with lower maintenance costs.

Acknowledgements. The first author wishes to thank ETS for providing the laboratory equipment for this research.

References

1. Ahmed, E., Roffa, A.O.: Faculty of Civil Engineering Department of Roads the Evaluation of Pavement in Desert Region in Libya Hodnocení Vozovek (2002)
2. Alani, P.H.M.H.: The transition To a Pg grading system for asphalt cement in Iraq. J. Eng. **16**(4) (2010)
3. Arora, M.G., Kennedy, T.W., Session, M., Arabia, S.: Shrp Asphalt Binder Specifications for Saudi Environment (1997)
4. Asi, I.M.: Performance evaluation of SUPERPAVE and Marshall asphalt mix designs to suite Jordan climatic and traffic conditions. Constr. Build. Mater. **21**(8), 1732–1740 (2007). doi:10.1016/j.conbuildmat.2006.05.036
5. Basic Asphalt Materials Mixture Design and Testing Technician Training Manual October 2012 Table of Contents (2012)
6. Internationale, O., Normalisation, D.E., Documents, A., Society, A., Testing, F.O.R.: Norme 1 (2015)
7. Salem, H.A., Uzelac, D., Matic, B.: Temperature zoning of Libya desert for asphalt mix design. Appl. Mech. Mater. **638–640**, 1414–1426 (2014). doi:10.4028/www.scientific.net/AMM.638-640.1414
8. Wahhab, H.I.A., Fatani, M.N., Noureldin, A.S., Bubshait, A.: National study of asphalt pavement rutting in Saudi Arabia. Transp. Res. Rec. (1473) (1995)

Author Index

L. Mohammad (ed.), *Advancement in the Design and Performance of Sustainable Asphalt Pavements*, Sustainable Civil Infrastructures, DOI 10.1007/978-3-319-61908-8

MIX
Papier aus verantwortungsvollen Quellen
Paper from responsible sources
FSC® C105338
FSC
www.fsc.org

If you have any concerns about our products,
you can contact us on
ProductSafety@springernature.com

In case Publisher is established outside the EU,
the EU authorized representative is:
Springer Nature Customer Service Center GmbH
Europaplatz 3, 69115 Heidelberg, Germany

Printed by Libri Plureos GmbH
in Hamburg, Germany